Handbook of Dam Engineering

Handbook of Dam Engineering

Edited by Arian Barker

Larsen & Keller
www.larsen-keller.com

Handbook of Dam Engineering
Edited by Arian Barker
ISBN: 979-8-88836-083-5 (Hardback)

Larsen & Keller
Published by Larsen and Keller Education,
5 Penn Plaza,
19th Floor,
New York, NY 10001, USA

Cataloging-in-Publication Data

Handbook of dam engineering / edited by Arian Barker.
p. cm.
Includes bibliographical references and index.
ISBN 979-8-88836-083-5
1. Dams--Design and construction. 2. Dams. 3. Engineering I. Barker, Arian.
TC540 .H36 2023
627.8--dc23

For more information regarding Larsen and Keller Education and its products, please visit the publisher's website www.larsen-keller.com

Table of Contents

Preface

A dam refers to a part of infrastructure system that comprises different structures designed to store, collect and manage water. There are different types of dams, such as buttress dams, gravity dams, embankment dams and arch dams. The reservoirs formed by dams not only prevent floods, but they also provide water for multiple purposes, including industrial use, irrigation, human consumption and aquaculture. Dam engineering is a field of engineering that deals with the problems associated with the building, planning and maintenance of dams. The development and building of dams has benefited the human civilization, by providing a supply of power and fresh water, as well as shielding it from calamities such as floods and storms. This book is a valuable compilation of topics, ranging from the basic to the most complex advancements in the field of dam engineering. It strives to provide a fair idea about this discipline and to help develop a better understanding of the latest advances within this field. Those in search of information to further their knowledge will be greatly assisted by this book.

After months of intensive research and writing, this book is the end result of all who devoted their time and efforts in the initiation and progress of this book. It will surely be a source of reference in enhancing the required knowledge of the new developments in the area. During the course of developing this book, certain measures such as accuracy, authenticity and research focused analytical studies were given preference in order to produce a comprehensive book in the area of study.

This book would not have been possible without the efforts of the authors and the publisher. I extend my sincere thanks to them. Secondly, I express my gratitude to my family and well-wishers. And most importantly, I thank my students for constantly expressing their willingness and curiosity in enhancing their knowledge in the field, which encourages me to take up further research projects for the advancement of the area.

Editor

1

Autonomous Dam Surveillance Robot System Based on Multi-Sensor Fusion †

Chao Zhang [1], Quanzhong Zhan [1], Qi Wang [2,*], Haichao Wu [2], Ting He [1] and Yi An [3]

[1] Information Center of the Ministry of water resources of the P.R.C, Beijing 100053, China; zhangchao@mwr.gov.cn (C.Z.); zcqcz@mwr.gov.cn (Q.Z.); heting@mwr.gov.cn (T.H.)
[2] Beijing Tritalent Intelligence Technology Co., Ltd., Beijing 100078, China; wuhaichao@tritalent.cn
[3] School of Control Science and Engineering, Dalian University of Technology, Dalian 116024, China; anyi@dlut.edu.cn
* Correspondence: wangqi@tritalent.cn
† This paper is an extended version of our paper published in Zhang, C.; Wang, Q.; Zhan, Q.; He, T.; An, Y. Autonomous System Design for Dam Surveillance Robots. In Proceedings of the IEEE Smart World Congress, Leicester, UK, 19–23 August 2019.

Abstract: Dams are important engineering facilities in the water conservancy industry. They have many functions, such as flood control, electric power generation, irrigation, water supply, shipping, etc. Therefore, their long-term safety is crucial to operational stability. Because of the complexity of the dam environment, robots with various kinds of sensors are a good choice to replace humans to perform a surveillance job. In this paper, an autonomous system design is proposed for dam ground surveillance robots, which includes general solution, electromechanical layout, sensors scheme, and navigation method. A strong and agile skid-steered mobile robot body platform is designed and created, which can be controlled accurately based on an MCU and an onboard IMU. A novel low-cost LiDAR is adopted for odometry estimation. To realize more robust localization results, two Kalman filter loops are used with the robot kinematic model to fuse wheel encoder, IMU, LiDAR odometry, and a low-cost GNSS receiver data. Besides, a recognition network based on YOLO v3 is deployed to realize real-time recognition of cracks and people during surveillance. As a system, by connecting the robot, the cloud server and the users with IOT technology, the proposed solution could be more robust and practical.

Keywords: water conversancy robot; dam surveillance; autonomous navigation; fusion localization

---

1. Introduction

In the water conservancy industry, there are many fundamental engineering facilities, such as dams, water and soil conservation, water transfer project, shipping project, water supply project, hydraulic power plants, irrigation facilities, etc. Big water dams are the most comprehensive facilities. They always have many functions, such as flood control, electric power generation, irrigation, water supply, shipping, etc. [1]. Therefore, the safety of big dams is crucial to cities and people around.

In the past, staff must check the environment, structure, and electromechanical facilities of dams regularly every day. This is a necessary way of keeping the dam running safely. However, it cannot realize all-weather all-day monitor and sometimes staffs are at risk since most dams are constructed in remote rural areas. Nowadays, with the great improvement of robot technology, robots have been used in public safety, security check, disaster rescue, and high-voltage lines inspection [2–5]. Many researchers and engineers are also paying attention to utilize underwater robots, unmanned aerial vehicles, unmanned surface vehicles for dams' surveillance and inspection.

However, applications of ground mobile robots are very rare. It is not easy for a ground robot to do the dam surveillance job autonomously. Firstly, the working places for most dams are off-road terrains, which needs a small, powerful, and agile robot mechanics design. Secondly, the GNSS satellite signal cannot fully cover all working area. The robot is difficult to continue path planning without accurate position information. Besides, to make a comprehensive check for a dam, the robot needs carry kinds of sensors. The robot must be capable of real-time data analyzing. In addition to the above, the robot system has to be low cost and robust.

In the autonomous robot area, simultaneous localization and mapping (SLAM) technology has been applied in many practical robot applications, especially laser SLAM. Researchers have proposed many laser-SLAM algorithms to build 2D/3D maps and realize localization [6,7]. Visual odometer (VO) is another hot topic in this area [8,9]. Cameras are relatively cheap and visual algorithms could show better performance in some environments. Many path planning methods are also being proposed to solve the obstacle avoiding problem [10,11]. The navigation technology now is ready for mobile robot design. Developing an autonomous dam surveillance robot system presents a wide variety of challenges. The robot should move on all kinds of ground surfaces and sometimes should move on narrow spaces. GNSS signal cannot cover all the work areas and it is difficult to acquire an accurate position. Online real-time inspection should be applied to the robot and the inspection algorithms should work with small computing resources.

This paper presents a general system for the mobile robot, which includes the robot, the cloud server, and terminals. The system can watch surveillance jobs status and can control the robot remotely when an emergency. As the robot body itself, we design a small powerful wheeled skid-steering mobile platform. Compared to alternative wheel configurations such as Ackerman or axle-articulated, the skid-steering platform shows two major advantages. First, it is simple and robust in mechanical terms. Second, it provides better maneuverability, including zero-radius turning. After analyzing the kinematic model of the skid-steered robot, we can estimate wheel odometry well with an onboard IMU (inertial measurement unit).

In this paper, a low-cost LiDAR sensor is used for laser odometry. This Livox MID40 LiDAR is a high-performance low-cost LiDAR sensor easily available. However, the LiDAR only has a 38.4° FOV and has a novel non-repeat scan mode, which differs a lot with common laser rangefinders. The method LiDAR odometry and mapping (LOAM) is adopted here for odometry estimation from its point cloud data. The method here avoids high computing burden since we only use the data for odometry calculation, not for real-time mapping. Besides, the robot was equipped with a GNSS receiver and an IMU. Therefore, we have position data from different sources: the LiDAR odometry, the longitude and latitude data, the wheel odometry and IMU measured data. To build a robust and fast localization system, we designed two extended Kalman filter (EKF) nodes. The first EKF node fuses IMU data, wheel odometry, LiDAR odometry to get robot transform in the inertial frame. By running it on the robot STM32 MCU board, we can ensure the fusion has real-time performance with high frequency. The second fuses results of first node, IMU data, longitude and latitude data to get robot transform in the world frame. The second node is running on the upper PC. The longitude and latitude data would be the observer position when the GNSS signal is good. After the position is acquired, we use a dynamic window approach (DWA) algorithm to do motion planning.

The paper is organized as follows. Some related work on robot control, fusion localization, object detection methods is summarized in Section 2. The general robot design is shown in Section 3. The mechanical design and control of the mobile robot can be seen in Section 4. All mounted sensors would be introduced in Section 5. Localization solution and some practical results could be found in Section 6. The online detection method is described in Section 7. Finally, a short conclusion is given in Section 8.

2. Related Work

To solve the dam surveillance and inspecting the problem, many robotics and sensor technologies have been deployed. A GPS-based surveillance robot is proposed in [12] to monitor the exterior deformation of the dam. A mechanically scanned imaging sonar (MSIS) is used for locating the underwater robot relative position for dam wall surveillance [13]. Many researchers utilized UAVs for dam image collecting and modeling, which is low-cost and convenient [14,15].

The wheeled skid-steering mobile robot is agile and strong, but its kinematics is complex since the drift always occurs and difficult to measure [16]. In [16], an IMU-based EKF fusion method is also proposed to improve the wheel odometry. To improve real-time motion control and dead-reckoning, a method to experimentally obtain an optimized kinematic model of the robot is presented in [17]. A detail model analysis and a back-stepping controller design are proposed in [18], simulation results for trajectory tracking are good. To find the correct model parameters [19], designed analysis and experimental kinematic scheme for skid-steering wheeled robot based-on a laser scanner sensor. The practical experiments show improvement in dead-reckoning performance.

A real-time kinematic (RTK) GPS sensor can help the robot to get an accurate position. It is commonly used in wide-open space [20]. Laser SLAM is now a relatively mature way to do mapping and localization (LiDAR odometry and mapping). LOAM proposed in [21] is one of the best LiDAR SLAM methods. The LOAM was running on the Velodyne Puck LiDAR which has 16 laser lines. Recently, a novel version of LOAM is proposed in [22], which is running on a low-cost novel Livox LiDAR and shows good performance in experiments. For land robots, dead reckoning is another reliable source of odometry. A robust AHRS could improve the wheel dead reckoning well [23].

Localization is one of the most important elements in the robot system. When the GPS/GNSS signal is too weak to use, a framework combines keyframe-based visual-inertial odometry with novel geometric image-based localization is built to provide a real-time estimate of a UAV's pose. Weak GPS feeds are used as a weak prior for suggesting loop closures [24]. A Monte Carlo Localization system that fuses wheel and visual odometry for the prediction are designed for sewer inspection robots. The update step takes into account the network topology and other active methods. The system works well in practical experiments [25]. To address existing positioning systems do not perform well in urban canyons, tunnels, and indoor parking lots for driverless cars, an EKF based multi-sensor positioning system that combines a GPS sensor, a camera, and in-vehicle sensors assisted by kinematic and dynamic vehicle models [26]. Among all the localization system, EKF is widely used and shows good performances [27,28].

For dam surveillance, people and dam crack detection are important to ensure people and dam safety. Vision-based approaches to detect, track and identify people on a mobile robot in real-time has been used for many years. Thermal and grey images are successfully used to do the people recognition job on a mobile security robot [29]. Deep Learning shows great potential for crack damage detection. A deep architecture of convolutional neural networks (CNNs) method is proposed in [30] to find concrete cracks for civil infrastructures.

Therefore, a wheeled skid-steering mobile robot is built for the dam surveillance, which carries an onboard IMU, wheel encoders, a Livox LiDAR sensor, a GNSS antenna, a monocular industrial camera, and a customized MCU controller. We proposed the robot platform controller design, odometry methods, fusion localization, crack detection network to make the whole system low-cost and robust.

3. General Design

In dam surveillance applications, there are three main modules, which are the robot itself, the cloud server and the front-end terminals. The robot is responsible for moving in the workplace, collecting monitor data and handling some online detection applications. The cloud server takes charge of data processing and transferring. The users can watch the robot status, do manual analysis, and control the robot remotely based on real-time video streaming when necessary.

The robot and the server are connected by mobile internet and WIFI. The server and the terminal users are connected by the Internet and LAN. The general structure diagram is shown in Figure 1. The robot itself and the cloud server are connected via message queuing telemetry transport protocol

(MQTT). Robot navigation data can transfer to the server and the server can send job command and global navigation plan to the robot. This paper mainly focuses on the robot module.

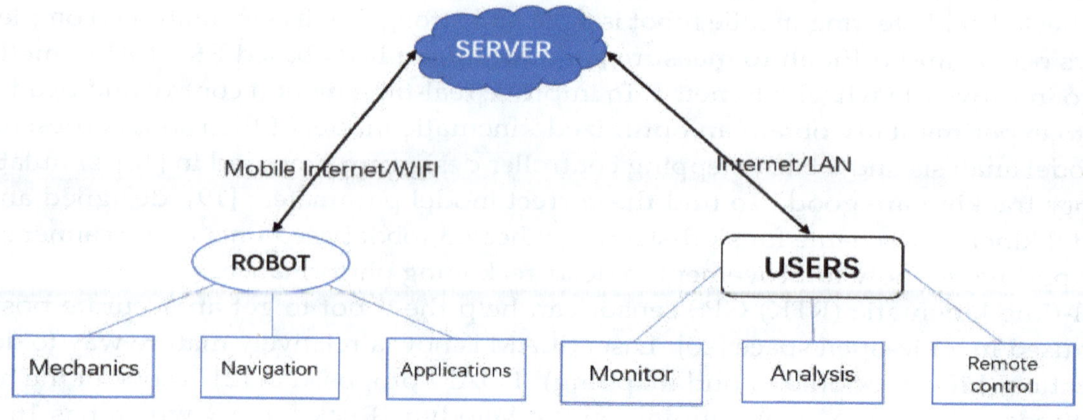

Figure 1. General design scheme for a dam surveillance robot.

A state machine is designed here for robot control, which includes self-check, autonomous navigation, remote control, idle, and emergency states. The robot would get into self-check state once the power is turned on. All sensors, motors, localization data, the high-level computer will then be checked. The robot would get into an idle state and wait for task command from the cloud server if the self-check is passed. Autonomous navigation would be triggered once the task and waypoints are received. The emergency state would be triggered when the robot is stuck or some sensors are running wrong. The logic of the state machine is shown in Figure 2.

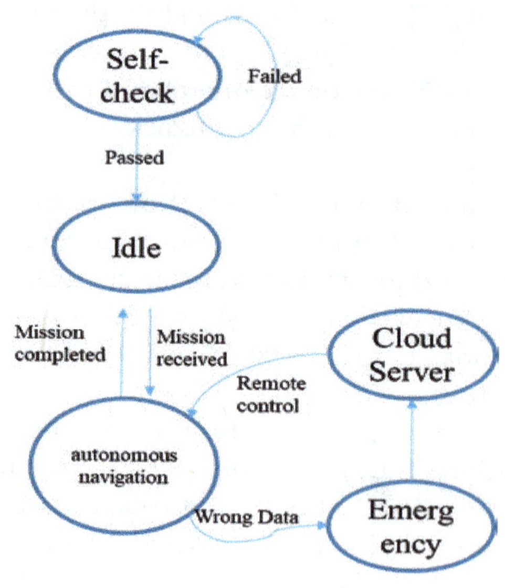

Figure 2. State machine logic structure.

4. Robot Platform

To realize high robustness and agile motion performance, the robot body is built with an all-aluminum chassis which about a total of 10 Kg weight. Two 200W DC brushless high torque motors are deployed to realize four-wheel drive. A 24 V lithium battery with 40 AH capacity is used as the power supply.

All motors are controlled by a customized MCU board, which connects the motors and drivers by CAN bus. The maximal speed is 2 m/s; the maximum payload is 40 Kg; the maximum runtime is more than 4 hours. It is suitable for rugged all-terrain operation with four off-road tires. Figure 3 shows the body design.

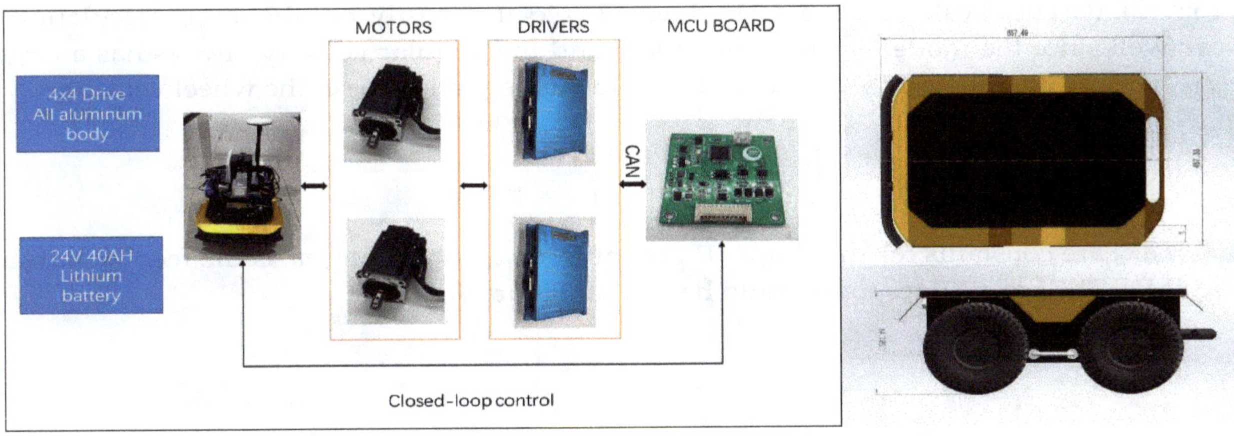

Figure 3. Mechanical layout and onboard control diagram.

As shown in Figure 4, for the robot body, we will use the velocity and the angular velocity in the robot axis as state variables, i.e., $[v_x \ w]^T$. After the kinematic analysis of the robot, we can find the relationship between robot velocities and wheel speed.

$$\begin{bmatrix} v_x \\ w \end{bmatrix} = r\frac{\pi d}{60n}\begin{bmatrix} \frac{1}{2} & \frac{1}{2} \\ -\frac{1}{2c} & -\frac{1}{2c} \end{bmatrix}\begin{bmatrix} W_l \\ W_r \end{bmatrix} \tag{1}$$

where r is so called effective radius of wheels, n is the reduction ratio, $2c$ is a spacing wheel track.

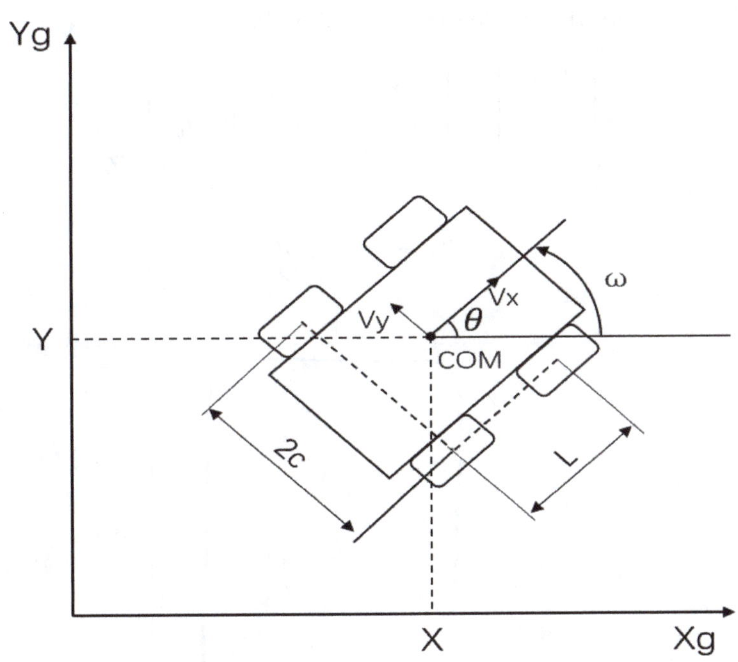

Figure 4. Kinematic model of the robot.

For the robot body, it will get velocity and angular velocity command $[v_x^d \ w^d]^T$ from upper PC. We should control the velocities to track the command. From (1), we have

$$\begin{cases} W_l + W_r = \frac{2v_x^d}{r\frac{\pi d}{60n}} \\ -W_l + W_r = \frac{2w^d c}{r\frac{\pi d}{60n}} \end{cases} \tag{2}$$

From (2), we could calculate the desired wheel velocities easily, but the direct calculation does not work well since the model is a realistic model and the angular velocity always has a large lag. Therefore, we add an onboard IMU and use the following equation to set the wheel velocities.

$$\begin{cases} v_x = v_x^d \\ w = k_p e_w + k_i \int e_w + k_d \dot{e}_w \end{cases} \tag{3}$$

where k_p, k_i, k_d are constants for the simple PID controller, $e_w = w^d - w^i$, w^i is the feedback from the onboard IMU. The practical response could be seen in Figure 5.

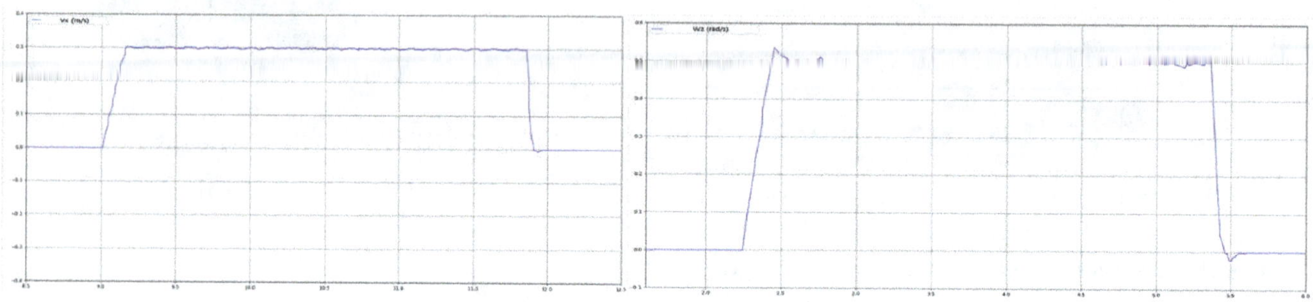

Figure 5. Practical response of the forward velocity and angular velocity.

It is simple to calculate the wheel odometry in the inertial frame. We drive the robot moving a square and back to the origin, the dead reckoning result is shown in Figure 6 and final position value is [0.1229–0.2654 0.03] (should be $[0\ 0\ 0]^T$). The whole distance is about 70 m and we did experiments 10 times, the origin RMSE (root mean square errors) is 0.3135.

$$\begin{cases} \begin{bmatrix} \dot{X} \\ \dot{Y} \end{bmatrix} = \begin{bmatrix} \cos\theta & \sin\theta \\ \sin\theta & \cos\theta \end{bmatrix} \begin{bmatrix} v_x \\ v_y \end{bmatrix} \quad Z \\ \dot{\theta} = w^i \\ v_y = -X_{ICR} \cdot w^i \end{cases} \tag{4}$$

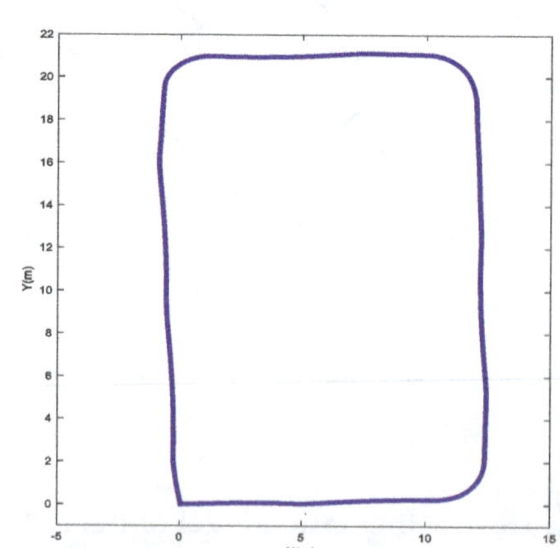

Figure 6. Dead reckoning results.

5. Sensors

The experimental robot with sensors is shown in Figure 7. From the front to the back, there are 1 GNSS receiver, a stereo camera, a Livox LiDAR, a monocular camera, a 4G LTE wireless router, an industrial computer, a GNSS antenna. An onboard MCU controller with IMU is located inside the body box. Motion control is handled by the onboard MCU which also sends odometry data to the high-level industrial computer. The navigation is handled by the computer which is connected to the cloud server.

Figure 7. Experiment robot with sensors.

A navigation map is built mainly by the LiDAR and odometry data using the LOAM SLAM method. The map would be regarded as prior knowledge. The monocular camera in the middle is used to transfer video streaming and do the people and crack detection jobs. The GNSS receiver with antenna can acquire accurate position data in the wide-open area. On the other hand, the LiDAR's point cloud and cameras' depth data are designed to be used in obstacles perception as well in the future.

6. Localization and Navigation

The navigation includes perception, localization, path planning, and a remote controller. The whole diagram is depicted in Figure 8. The mature cost grid map is used for general environment representation. The filtered point cloud and depth data would be obstacles shown on the map.

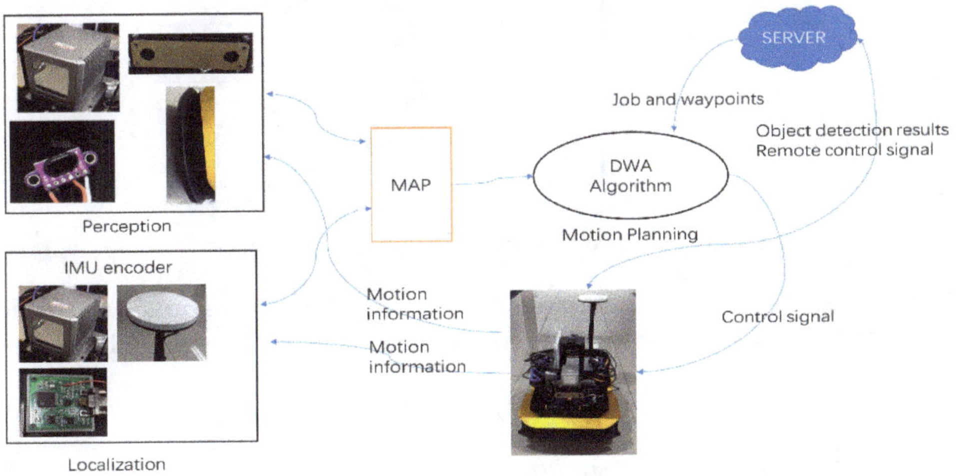

Figure 8. Navigation diagram.

The path planning method is divided into two parts here. The global waypoints are sent to the robot by the cloud server, which is pre-defined by different jobs. The local motion planning is achieved by the DWA algorithm on the grid map.

Localization is one of the most difficult problems for dam robots since the environment is changing and the robot has to move indoor and outdoor. Using one single sensor to acquire a positon is an impossible task. Many sensors would introduce fusion problem. Extended Kalman filter (EKF) is a good way to solve this problem and has been deployed in many robot localization applications.

A two EKF node structure is proposed here to build a robust localization system. The GNSS data is transformed based on the local origin and then input to EKF global node update process. The yaw angle from the LiDAR odometry is input to the EKF update process. The result of the EKF local node is input to the EKF global predict process. The LiDAR odometry data are input to the EKF local update process. The wheel odometry data and the IMU measurement is input to the EKF local predict process. It is noted here that the local EKF node is running on the embedded MCU at 200Hz and it handles in 3D space with 16 states.

It is important to get the 3D pose information for the rubber ground surface. However, the global EKF node is running at 30Hz on the computer and it only produces in 2D space. The robot position and orientation in the map are acquired with the EKF global updating. The fusion localization structure diagram is shown in Figure 9. Where W_l^d, W_l^d are the wheel velocity command, W_l, W_l are the feedback wheel velocity from encoders, V_x^d, ω^d are the velocity command generated from motion planning, $V_x, \omega z$ are the wheel odometry outputs, x, y, yaw are the LiDAR odometry, X, Y, Yaw are the final global pose estimation result.

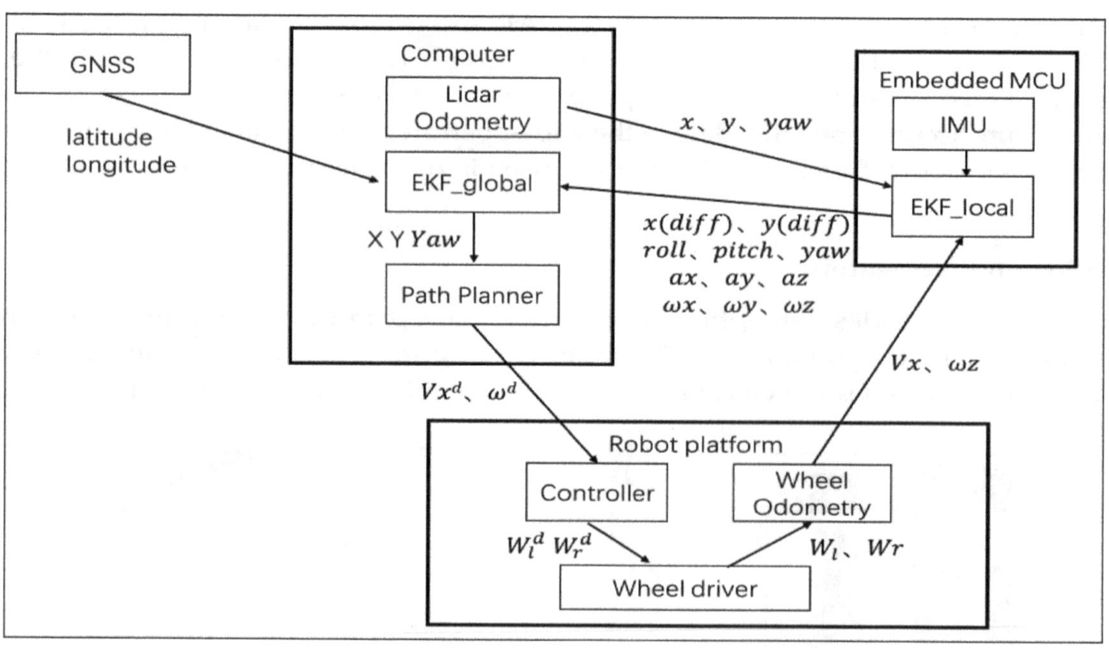

Figure 9. Fusion localization structure diagram.

With the kinematic models and zero-velocity constraints, we are now ready to design a kinematic-model-based robot positioning scheme for the local EKF node. The general state update equation is

$$x_t = Fx_{t-1} + Gu_t \tag{5}$$

where $x_t = [q_0\ q_1\ q_2\ q_3\ x\ y\ z\ V_x\ V_y\ V_z\ \delta\varphi_{bias}\ \delta\theta_{bias}\ \delta\psi_{bias}\ \delta V_{xbias}\ \delta V_{ybias}\ \delta V_{zbias}]^T$ is the state variables, $q_0\ q_1\ q_2\ q_3$ are the quaternion, $x\ y\ z$ are the position in the inertial frame, $V_x\ V_y\ V_z$ are the corresponding velocities, $\delta\varphi_{bias}$ $\delta\theta_{bias}$ $\delta\psi_{bias}$ are the rotation zero-offset, δV_{xbias} δV_{ybias} δV_{zbias} are the acceleration zero-offset. We have the following detail state update equations.

$$\begin{bmatrix} q_0(k+1) \\ q_1(k+1) \\ q_2(k+1) \\ q_3(k+1) \end{bmatrix} = q(h) \bullet \begin{bmatrix} q_0(k) \\ q_1(k) \\ q_2(k) \\ q_3(k) \end{bmatrix} \tag{6}$$

$$q(h) = \begin{bmatrix} 1 \\ \frac{\delta\phi}{2} - \frac{\delta\phi\ _{bias}}{2} \\ \frac{\delta\theta}{2} - \frac{\delta\theta_{bias}}{2} \\ \frac{\delta\psi}{2} - \frac{\delta\psi_{bias}}{2} \end{bmatrix} \tag{7}$$

where $\delta\phi$, $\delta\theta$, $\delta\psi$ are the radian variations during dt.

$$\begin{bmatrix} V_x(k+1) \\ V_y(k+1) \\ V_z(k+1) \end{bmatrix} = \begin{bmatrix} V_x(k) \\ V_y(k) \\ V_z(k) \end{bmatrix} + \begin{bmatrix} g_x(k) \\ g_y(k) \\ g_z(k) \end{bmatrix} \bullet dt + T_b^n \delta V \tag{8}$$

where $g_x(k) g_y(k) g_z(k)$ are the components of gravity.

$$\delta V = \begin{bmatrix} \delta V_x \\ \delta V_y \\ \delta V_z \end{bmatrix} - \begin{bmatrix} \delta V_{xbias} \\ \delta V_{ybias} \\ \delta V_{zbias} \end{bmatrix} \tag{9}$$

where δV_{xbias}, δV_{ybias}, δV_{zbias} are the velocity variations during dt.

$$\begin{bmatrix} x(k+1) \\ y(k+1) \\ z(k+1) \end{bmatrix} = \begin{bmatrix} x(k) \\ y(k) \\ z(k) \end{bmatrix} + \begin{bmatrix} V_x(k) \\ V_y(k) \\ V_z(k) \end{bmatrix} \bullet dt \tag{10}$$

$$\begin{bmatrix} \delta\phi_{bias}(k+1) \\ \delta\theta_{bias}(k+1) \\ \delta\psi_{bias}(k+1) \end{bmatrix} = \begin{bmatrix} \delta\phi_{bias}(k) \\ \delta\theta_{bias}(k) \\ \delta\psi_{bias}(k) \end{bmatrix} \tag{11}$$

$$\begin{bmatrix} \delta V_{xbias}(k+1) \\ \delta V_{ybias}(k+1) \\ \delta V_{zbias}(k+1) \end{bmatrix} = \begin{bmatrix} \delta V_{xbias}(k) \\ \delta V_{ybias}(k) \\ \delta V_{zbias}(k) \end{bmatrix} \tag{12}$$

The general time update equation is

$$P_k = F_{k-1} P_{k-1} F^T{}_{k-1} + G_{k-1} Q_{k-1} G^T{}_{k-1} + Q_s \tag{13}$$

where, P_k is the covariance matrix, $F_k = \left(\frac{\partial f}{\partial x}\right)_k$ is state Jacobian, $G_k = \left(\frac{\partial f}{\partial u}\right)_k$ is control input Jacobian, Q_{k-1} is sensor process noise, Q_s is an additional process noise. The state transform matrix have

$$F_k = \begin{bmatrix} 1 & F_{0,1} & F_{0,2} & F_{0,3} & 0 & 0 & 0 & 0 & 0 & 0 & F_{0,10} & F_{0,11} & F_{0,12} & 0 & 0 & 0 \\ F_{1,0} & 1 & F_{1,2} & F_{1,3} & 0 & 0 & 0 & 0 & 0 & 0 & F_{1,10} & F_{1,11} & F_{1,12} & 0 & 0 & 0 \\ F_{2,0} & F_{2,1} & 1 & F_{2,3} & 0 & 0 & 0 & 0 & 0 & 0 & F_{2,10} & F_{2,11} & F_{2,12} & 0 & 0 & 0 \\ F_{3,0} & F_{3,1} & F_{3,2} & 1 & 0 & 0 & 0 & 0 & 0 & 0 & F_{3,10} & F_{3,11} & F_{3,12} & 0 & 0 & 0 \\ F_{4,0} & F_{4,1} & F_{4,2} & F_{4,3} & 1 & 0 & 0 & 0 & 0 & 0 & 0 & 0 & 0 & F_{4,13} & F_{4,14} & F_{4,15} \\ F_{5,0} & F_{5,1} & F_{5,2} & F_{5,3} & 0 & 1 & 0 & 0 & 0 & 0 & 0 & 0 & 0 & F_{5,13} & F_{5,14} & F_{5,15} \\ F_{6,0} & F_{6,1} & F_{6,2} & F_{6,3} & 0 & 0 & 1 & 0 & 0 & 0 & 0 & 0 & 0 & F_{6,13} & F_{6,14} & F_{6,15} \\ 0 & 0 & 0 & 0 & dt & 0 & 0 & 1 & 0 & 0 & 0 & 0 & 0 & 0 & 0 & 0 \\ 0 & 0 & 0 & 0 & 0 & dt & 0 & 0 & 1 & 0 & 0 & 0 & 0 & 0 & 0 & 0 \\ 0 & 0 & 0 & 0 & 0 & 0 & dt & 0 & 0 & 1 & 0 & 0 & 0 & 0 & 0 & 0 \\ 0 & 0 & 0 & 0 & 0 & 0 & 0 & 0 & 0 & 0 & 1 & 0 & 0 & 0 & 0 & 0 \\ 0 & 0 & 0 & 0 & 0 & 0 & 0 & 0 & 0 & 0 & 0 & 1 & 0 & 0 & 0 & 0 \\ 0 & 0 & 0 & 0 & 0 & 0 & 0 & 0 & 0 & 0 & 0 & 0 & 1 & 0 & 0 & 0 \\ 0 & 0 & 0 & 0 & 0 & 0 & 0 & 0 & 0 & 0 & 0 & 0 & 0 & 1 & 0 & 0 \\ 0 & 0 & 0 & 0 & 0 & 0 & 0 & 0 & 0 & 0 & 0 & 0 & 0 & 0 & 1 & 0 \\ 0 & 0 & 0 & 0 & 0 & 0 & 0 & 0 & 0 & 0 & 0 & 0 & 0 & 0 & 0 & 1 \end{bmatrix} \tag{14}$$

Therefore, we can acquire

$$F_{0,1} = \frac{\partial q_0(t+\Delta t)}{\partial q_1} = \delta\phi_{bias}/2 - \delta\phi/2; \tag{15}$$

$$F_{0,2} = \frac{\partial q_0(t+\Delta t)}{\partial q_2} = \delta\theta_{bias}/2 - \delta\theta/2; \tag{16}$$

$$F_{0,3} = \frac{\partial q_0(t+\Delta t)}{\partial q_3} = \delta\psi_{bias}/2 - \delta\psi/2; \tag{17}$$

$$F_{0,10} = \frac{\partial q_0(t+\Delta t)}{\partial \delta\phi_{bias}} = \frac{q_1}{2}; \tag{18}$$

$$F_{0,11} = \frac{\partial q_0(t+\Delta t)}{\partial \delta\theta_{bias}} = \frac{q_2}{2}; \tag{19}$$

$$F_{0,12} = \frac{\partial q_0(t+\Delta t)}{\partial \delta\psi_{bias}} = \frac{q_3}{2}; \tag{20}$$

$$F_{1,0} = \frac{\partial q_1(t+\Delta t)}{\partial q_0} = \delta\phi/2 - \delta\phi_{bias}/2; \tag{21}$$

$$F_{1,2} = \frac{\partial q_1(t+\Delta t)}{\partial q_2} = \delta\psi/2 - \delta\psi_{bias}/2; \tag{22}$$

$$F_{1,3} = \frac{\partial q_1(t+\Delta t)}{\partial q_3} = \delta\theta_{bias}/2 - \delta\theta/2; \tag{23}$$

$$F_{1,10} = \frac{\partial q_1(t+\Delta t)}{\partial \delta\phi_{bias}} = -\frac{q_0}{2} \tag{24}$$

$$F_{1,11} = \frac{\partial q_1(t+\Delta t)}{\partial \delta\theta_{bias}} = \frac{q_3}{2}; \tag{25}$$

$$F_{1,12} = \frac{\partial q_1(t+\Delta t)}{\partial \delta\psi_{bias}} = -\frac{q_2}{2}; \tag{26}$$

$$F_{2,0} = \frac{\partial q_2(t+\Delta t)}{\partial q_0} = \delta\theta/2 - \delta\theta_{\ bias}/2; \tag{27}$$

$$F_{2,1} = \frac{\partial q_2(t+\Delta t)}{\partial q_1} = \delta\psi_{bias}/2 - \delta\psi/2; \tag{28}$$

$$F_{2,3} = \frac{\partial q_2(t+\Delta t)}{\partial q_3} = \delta\phi/2 - \delta\phi_{bias}/2; \tag{29}$$

$$F_{2,10} = \frac{\partial q_2(t+\Delta t)}{\partial \delta\phi_{\ bias}} = -\frac{q_3}{2}; \tag{30}$$

$$F_{2,11} = \frac{\partial q_2(t+\Delta t)}{\partial \delta\theta_{bias}} = -\frac{q_0}{2}; \tag{31}$$

$$F_{2,12} = \frac{\partial q_2(t+\Delta t)}{\partial \delta\psi_{bias}} = \frac{q_1}{2}; \tag{32}$$

$$F_{3,0} = \frac{\partial q_3(t+\Delta t)}{\partial q_0} = \delta\psi/2 - \delta\psi_{bias}/2; \tag{33}$$

$$F_{3,1} = \frac{\partial q_3(t+\Delta t)}{\partial q_1} = \delta\theta/2 - \delta\theta_{bias}/2; \tag{34}$$

$$F_{3,2} = \frac{\partial q_3(t+\Delta t)}{\partial q_2} = \delta\phi_{\ bias}/2 - \delta\phi/2; \tag{35}$$

$$F_{3,10} = \frac{\partial q_3(t+\Delta t)}{\partial \delta\phi_{\ bias}} = \frac{q_2}{2}; \tag{36}$$

$$F_{3,11} = \frac{\partial q_3(t+\Delta t)}{\partial \delta\theta_{bias}} = -\frac{q_1}{2}; \tag{37}$$

$$F_{3,12} = \frac{\partial q_3(t+\Delta t)}{\partial \delta\psi_{bias}} = -\frac{q_0}{2}; \tag{38}$$

$$F_{4,0} = \frac{\partial v_x(t+\Delta t)}{\partial q_0} = 2*\mathrm{q0}*(\delta V_x - \delta V_{xbias}) - 2*\mathrm{q3}*(\delta V_y - \delta V_{ybias}) + 2*\mathrm{q2}*(\delta V_z - \delta V_{zbias}\Big); \tag{39}$$

$$F_{4,1} = \frac{\partial v_x(t+\Delta t)}{\partial q_1} = 2*\mathrm{q1}*(\delta V_x - \delta V_{xbias}) + 2*\mathrm{q2}*(\delta V_y - \delta V_{ybias}) + 2*\mathrm{q3}*(\delta V_z - \delta V_{zbias}\Big); \tag{40}$$

$$F_{4,2} = \frac{\partial v_x(t+\Delta t)}{\partial q_2} = 2*\mathrm{q1}*(\delta V_y - \delta V_{ybias}) - 2*\mathrm{q2}*(\delta V_x - \delta V_{xbias}) + 2*\mathrm{q0}*(\delta V_z - \delta V_{zbias}); \tag{41}$$

$$F_{4,3} = \frac{\partial v_x(t+\Delta t)}{\partial q_3} = 2*\mathrm{q1}*(\delta V_z - \delta V_{zbias}) - 2*\mathrm{q0}*(\delta V_y - \delta V_{ybias}) - 2*\mathrm{q3}*(\delta V_x - \delta V_{xbias}\Big); \tag{42}$$

$$F_{4,13} = \frac{\partial v_x(t+\Delta t)}{\partial \delta V_{xbias}} = -q0^2 - q1^2 + q2^2 + q3^2 \tag{43}$$

$$F_{4,14} = \frac{\partial v_x(t+\Delta t)}{\partial \delta V_{ybias}} = 2*\mathrm{q0}*\mathrm{q3} - 2*\mathrm{q1}*\mathrm{q2} \tag{44}$$

$$F_{4,15} = \frac{\partial v_x(t+\Delta t)}{\partial \delta V_{zbias}} = -2(q_0*q_2 + q_1*q_3); \tag{45}$$

$$F_{5,0} = \frac{\partial v_y(t+\Delta t)}{\partial q_0} = 2*\mathrm{q3}*(\delta V_x - \delta V_{xbias}) + 2*\mathrm{q0}*(\delta V_y - \delta V_{ybias}) - 2*\mathrm{q1}*(\delta V_z - \delta V_{zbias}\Big); \tag{46}$$

$$F_{5,1} = \frac{\partial v_y(t+\Delta t)}{\partial q_1} = 2*\mathrm{q2}*(\delta V_x - \delta V_{xbias}) - 2*\mathrm{q1}*(\delta V_y - \delta V_{ybias}) - 2*\mathrm{q0}*(\delta V_z - \delta V_{zbias}); \tag{47}$$

$$F_{5,2} = \frac{\partial v_y(t+\Delta t)}{\partial q_2} = 2*\mathrm{q1}*(\delta V_x - \delta V_{xbias}) + 2*\mathrm{q2}*(\delta V_y - \delta V_{ybias}) + 2*\mathrm{q3}*(\delta V_z - \delta V_{zbias}); \tag{48}$$

$$F_{5,3} = \frac{\partial v_y(t+\Delta t)}{\partial q_3} = 2*\mathrm{q0}*(\delta V_x - \delta V_{xbias}) - 2*\mathrm{q3}*(\delta V_y - \delta V_{ybias}) + 2*\mathrm{q2}*(\delta V_z - \delta V_{zbias}); \tag{49}$$

$$F_{5,13} = \frac{\partial v_y(t+\Delta t)}{\partial \delta V_{xbias}} = -2*\mathrm{q0}*\mathrm{q3} - 2*\mathrm{q1}*\mathrm{q2} \tag{50}$$

$$F_{5,14} = \frac{\partial v_y(t+\Delta t)}{\partial \delta V_{ybias}} = -q0^2 + q1^2 - q2^2 + q3^2 \tag{51}$$

$$F_{5,15} = \frac{\partial v_y(t+\Delta t)}{\partial \delta V_{zbias}} = 2(q_0 * q_1 - q_2 * q_3); \tag{52}$$

$$F_{6,0} = \frac{\partial v_z(t+\Delta t)}{\partial q_0} = 2*\mathrm{q1}*(\delta V_y - \delta V_{ybias}) - 2*\mathrm{q2}*(\delta V_x - \delta V_{xbias}) + 2*\mathrm{q0}*(\delta V_z - \delta V_{zbias}); \tag{53}$$

$$F_{6,1} = \frac{\partial v_z(t+\Delta t)}{\partial q_1} = 2*\mathrm{q3}*(\delta V_x - \delta V_{xbias}) + 2*\mathrm{q0}*(\delta V_y - \delta V_{ybias}) - 2*\mathrm{q1}*(\delta V_z - \delta V_{zbias})); \tag{54}$$

$$F_{6,2} = \frac{\partial v_z(t+\Delta t)}{\partial q_2} = 2*\mathrm{q3}*(\delta V_y - \delta V_{ybias}) - 2*\mathrm{q0}*(\delta V_x - \delta V_{xbias}) - 2*\mathrm{q2}*(\delta V_z - \delta V_{zbias})); \tag{55}$$

$$F_{6,3} = \frac{\partial v_z(t+\Delta t)}{\partial q_3} = 2*\mathrm{q1}*(\delta V_x - \delta V_{xbias}) + 2*\mathrm{q2}*(\delta V_y - \delta V_{ybias}) + 2*\mathrm{q3}*(\delta V_z - \delta V_{zbias})); \tag{56}$$

$$F_{6,13} = \frac{\partial v_z(t+\Delta t)}{\partial \delta V_{xbias}} = 2*\mathrm{q0}*\mathrm{q2} - 2*\mathrm{q1}*\mathrm{q3} \tag{57}$$

$$F_{6,14} = \frac{\partial v_z(t+\Delta t)}{\partial \delta V_{ybias}} = -2*\mathrm{q0}*\mathrm{q1} - 2*\mathrm{q2}*\mathrm{q3} \tag{58}$$

$$F_{6,15} = \frac{\partial v_z(t+\Delta t)}{\partial \delta V_{zbias}} = -q0^2 + q1^2 + q2^2 - q3^2 \tag{59}$$

The control input Jacobian matrix is

$$G_k = \begin{bmatrix}
-\frac{q1}{2} & -\frac{q2}{2} & -\frac{q3}{2} & 0 & 0 & 0 \\
\mathrm{SG}(1) & -\frac{q3}{2} & \frac{q2}{2} & 0 & 0 & 0 \\
\frac{q3}{2} & \mathrm{SG}(1) & -\frac{q1}{2} & 0 & 0 & 0 \\
-\frac{q2}{2} & \frac{q1}{2} & \mathrm{SG}(1) & 0 & 0 & 0 \\
0 & 0 & 0 & \mathrm{SG}(4) - \mathrm{SG}(3) - \mathrm{SG}(2) + \mathrm{SG}(5) & \mathrm{SG}(8) - 2*\mathrm{q0}*\mathrm{q3} & \mathrm{SG}(7) + 2*\mathrm{q0}*\mathrm{q2} \\
0 & 0 & 0 & \mathrm{SG}(8) + 2*\mathrm{q0}*\mathrm{q3} & \mathrm{SG}(3) - \mathrm{SG}(2) - \mathrm{SG}(4) + \mathrm{SG}(5) & \mathrm{SG}(6) - 2*\mathrm{q0}*\mathrm{q1} \\
0 & 0 & 0 & \mathrm{SG}(7) - 2*\mathrm{q0}*\mathrm{q2} & \mathrm{SG}(6) + 2*\mathrm{q0}*\mathrm{q1} & \mathrm{SG}(2) - \mathrm{SG}(3) - \mathrm{SG}(4) + \mathrm{SG}(5) \\
0 & 0 & 0 & 0 & 0 & 0 \\
0 & 0 & 0 & 0 & 0 & 0 \\
0 & 0 & 0 & 0 & 0 & 0 \\
0 & 0 & 0 & 0 & 0 & 0 \\
0 & 0 & 0 & 0 & 0 & 0 \\
0 & 0 & 0 & 0 & 0 & 0 \\
0 & 0 & 0 & 0 & 0 & 0 \\
0 & 0 & 0 & 0 & 0 & 0 \\
0 & 0 & 0 & 0 & 0 & 0 \\
0 & 0 & 0 & 0 & 0 & 0 \\
0 & 0 & 0 & 0 & 0 & 0 \\
0 & 0 & 0 & 0 & 0 & 0 \\
0 & 0 & 0 & 0 & 0 & 0 \\
0 & 0 & 0 & 0 & 0 & 0 \\
0 & 0 & 0 & 0 & 0 & 0 \\
0 & 0 & 0 & 0 & 0 & 0 \\
0 & 0 & 0 & 0 & 0 & 0
\end{bmatrix} \tag{60}$$

where

$$SG = \begin{bmatrix}
\frac{q_0}{2} \\
{q_3}^2 \\
{q_2}^2 \\
{q_1}^2 \\
{q_0}^2 \\
2*q_2*q_3 \\
2*q_1*q_3 \\
2*q_1*q_2
\end{bmatrix} \tag{61}$$

The process noise matrix is

$$Q_k = \begin{bmatrix} q_{0,0} & SQ(9) & SQ(8) & SQ(7) & 0 & 0 & 0 & 0 & 0 & 0 & 0 & 0 & 0 & 0 & 0 & 0 \\ SQ(9) & q_{1,1} & SQ(6) & SQ(5) & 0 & 0 & 0 & 0 & 0 & 0 & 0 & 0 & 0 & 0 & 0 & 0 \\ SQ(8) & SQ(6) & q_{2,2} & SQ(4) & 0 & 0 & 0 & 0 & 0 & 0 & 0 & 0 & 0 & 0 & 0 & 0 \\ SQ(7) & SQ(5) & SQ(4) & q_{3,3} & 0 & 0 & 0 & 0 & 0 & 0 & 0 & 0 & 0 & 0 & 0 & 0 \\ 0 & 0 & 0 & 0 & q_{4,4} & SQ(3) & SQ(2) & 0 & 0 & 0 & 0 & 0 & 0 & 0 & 0 & 0 \\ 0 & 0 & 0 & 0 & SQ(3) & q_{5,5} & SQ(1) & 0 & 0 & 0 & 0 & 0 & 0 & 0 & 0 & 0 \\ 0 & 0 & 0 & 0 & SQ(2) & SQ(1) & q_{6,6} & 0 & 0 & 0 & 0 & 0 & 0 & 0 & 0 & 0 \\ 0 & 0 & 0 & 0 & 0 & 0 & 0 & 0 & 0 & 0 & 0 & 0 & 0 & 0 & 0 & 0 \\ 0 & 0 & 0 & 0 & 0 & 0 & 0 & 0 & 0 & 0 & 0 & 0 & 0 & 0 & 0 & 0 \\ 0 & 0 & 0 & 0 & 0 & 0 & 0 & 0 & 0 & 0 & 0 & 0 & 0 & 0 & 0 & 0 \\ 0 & 0 & 0 & 0 & 0 & 0 & 0 & 0 & 0 & 0 & 0 & 0 & 0 & 0 & 0 & 0 \\ 0 & 0 & 0 & 0 & 0 & 0 & 0 & 0 & 0 & 0 & 0 & 0 & 0 & 0 & 0 & 0 \\ 0 & 0 & 0 & 0 & 0 & 0 & 0 & 0 & 0 & 0 & 0 & 0 & 0 & 0 & 0 & 0 \\ 0 & 0 & 0 & 0 & 0 & 0 & 0 & 0 & 0 & 0 & 0 & 0 & 0 & 0 & 0 & 0 \\ 0 & 0 & 0 & 0 & 0 & 0 & 0 & 0 & 0 & 0 & 0 & 0 & 0 & 0 & 0 & 0 \\ 0 & 0 & 0 & 0 & 0 & 0 & 0 & 0 & 0 & 0 & 0 & 0 & 0 & 0 & 0 & 0 \end{bmatrix} \tag{62}$$

where,

$$SQ = \begin{bmatrix} \text{dvzCov} * (\text{SG}(6) - 2 * \text{q0} * \text{q1}) * (\text{SG}(2) - \text{SG}(3) - \text{SG}(4) + \text{SG}(5)) - \text{dvyCov} * (\text{SG}(6) + 2 * \text{q0} * \text{q1}) * (\text{SG}(2) - \text{SG}(3) + \text{SG}(4) - \text{SG}(5)) + \text{dvxCov} * (\text{SG}(7) - 2 * \text{q0} * \text{q2}) * (\text{SG}(8) + 2 * \text{q0} * \text{q3}) \\ \text{dvzCov} * (\text{SG}(7) + 2 * \text{q0} * \text{q2}) * (\text{SG}(2) - \text{SG}(3) - \text{SG}(4) + \text{SG}(5)) - \text{dvxCov} * (\text{SG}(7) - 2 * \text{q0} * \text{q2}) * (\text{SG}(2) + \text{SG}(3) - \text{SG}(4) - \text{SG}(5)) + \text{dvyCov} * (\text{SG}(6) + 2 * \text{q0} * \text{q1}) * (\text{SG}(8) - 2 * \text{q0} * \text{q3}) \\ \text{dvzCov} * (\text{SG}(6) - 2 * \text{q0} * \text{q1}) * (\text{SG}(7) + 2 * \text{q0} * \text{q2} - \text{dvyCov} * (\text{SG}(8) - 2 * \text{q0} * \text{q3}) * (\text{SG}(2) - \text{SG}(3) + \text{SG}(4) - \text{SG}(5)) - \text{dvxCov} * (\text{SG}(8) + 2 * \text{q0} * \text{q3}) * (\text{SG}(2) + \text{SG}(3) - \text{SG}(4) - \text{SG}(5)) \\ (\text{dayCov} * \text{q1} * \text{SG}(1))/2 - (\text{dazCov} * \text{q1} * \text{SG}(1))/2 - (\text{daxCov} * \text{q2} * \text{q3})/4 \\ (\text{dazCov} * \text{q2} * \text{SG}(1))/2 - (\text{daxCov} * \text{q2} * \text{SG}(1))/2 - (\text{dayCov} * \text{q1} * \text{q3})/4 \\ (\text{daxCov} * \text{q3} * \text{SG}(1))/2 - (\text{dayCov} * \text{q3} * \text{SG}(1))/2 - (\text{dazCov} * \text{q1} * \text{q2})/4 \\ (\text{daxCov} * \text{q1} * \text{q2})/4 - (\text{dazCov} * \text{q3} * \text{SG}(1))/2 - (\text{dayCov} * \text{q1} * \text{q2})/4 \\ (\text{dazCov} * \text{q1} * \text{q3})/4 - (\text{daxCov} * \text{q1} * \text{q3})/4 - (\text{dayCov} * \text{q2} * \text{SG}(1))/2 \\ (\text{dayCov} * \text{q2} * \text{q3})/4 - (\text{daxCov} * \text{q1} * \text{SG}(1))/2 - (\text{dazCov} * \text{q2} * \text{q3})/4 \\ SG(1)^2 \\ q1^2 \end{bmatrix} \tag{63}$$

$$\begin{aligned} q_{0,0} &= (dayCov * q2^2)/4 + (dazCov * q3^2)/4 + (daxCov * SQ(11))/4 \\ q_{1,1} &= (dazCov * q2^2)/4 + (dayCov * q3^2)/4 + daxCov * SQ(10) \\ q_{2,2} &= (daxCov * q3^2)/4 + dayCov * SQ(10) + (dazCov * SQ(11))/4 \\ q_{3,3} &= (daxCov * q2^2)/4 + (dayCov * SQ(11))/4 + dazCov * SQ(10) \\ q_{4,4} &= dvyCov * (SG(8) - 2 * q0 * q3)^2 + dvzCov * (SG(7) + 2 * q0 * q2)^2 \\ &\quad + dvxCov * (SG(2) + SG(3) - SG(4) - SG(5))^2 \\ q_{5,5} &= dvxCov * (SG(8) + 2 * q0 * q3)^2 + dvzCov * (SG(6) - 2 * q0 * q1)^2 \\ &\quad + dvyCov * (SG(2) - SG(3) + SG(4) - SG(5))^2 \\ q_{6,6} &= dvxCov * (SG(7) - 2 * q0 * q2)^2 + dvyCov * (SG(6) + 2 * q0 * q1)^2 \\ &\quad + dvzCov * (SG(2) - SG(3) - SG(4) + SG(5))^2 \end{aligned} \tag{64}$$

where, $daxCov, dayCov, dazCov$ are the radius variance, and $dvxCov$, $dvyCov$, $dvzCov$ are the velocities variance.

The Kalman feedback matrix is

$$K_k = P_k^- H_k \left[H_k P_k^- H_k^T + R_k \right]^{-1} \tag{65}$$

where

$$H_k = \left(\frac{\partial z_p}{\partial x} \right)_k \tag{66}$$

The covariance update equation is

$$P_k^+ = [I - K_k H_k] P_k^- \tag{67}$$

Finally, we could get the state update equation

$$x_k^+ = x_k^- + K_k v \tag{68}$$

The global EKF node is designed in the same method, the details are omitted here. The localization result in the practical environment can be found in Figure 10. The body coordinate represents the filtered pose of the robot now. The robot was running a circle in the practical environment. We can see that the localization performs well even sometimes the GNSS receiver is out of the lock and the result is ready for navigation. The localization and path planning could fail sometimes, even if so many sensors have been deployed. In this application, we designed a remote controller module. Users can control the robot remotely with real-time video streaming.

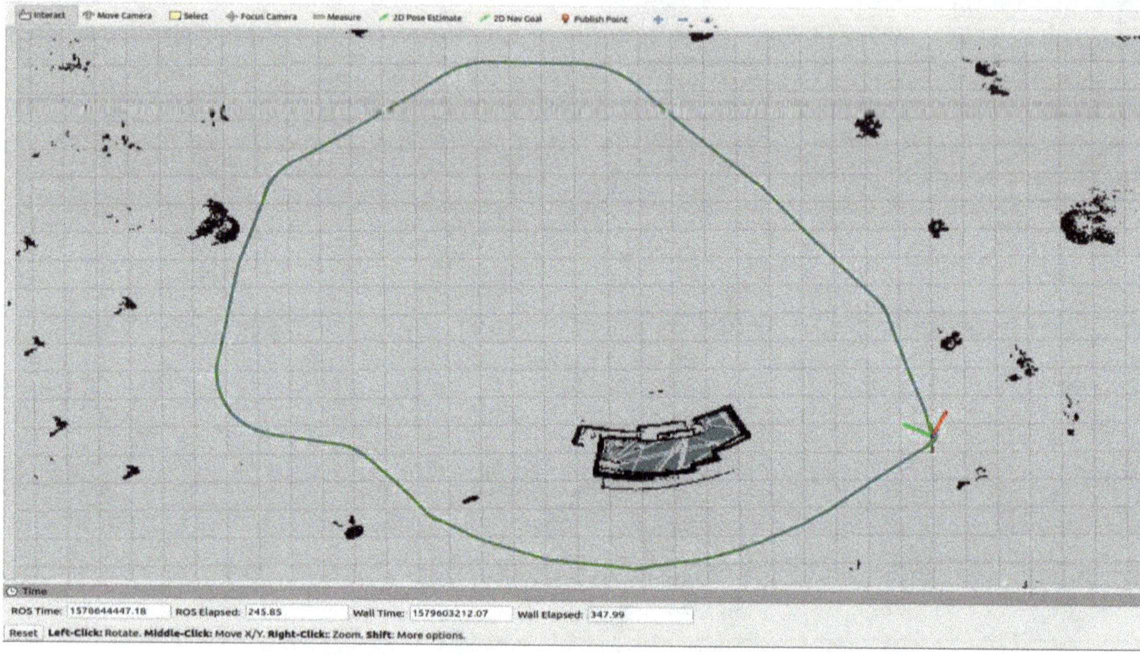

Figure 10. Practical localization performance on map.

The Figure 11 shows a picture of an experiment environment for a surveillance job, this picture is captured from the camera on the robot. In this environment, the road is narrow and the GNSS signal is weak because of many trees. However, the proposed robot system works well. The practical results could be seen in Figure 12. The goal and the blue line are the planned global path. The green line is the local path planning result. We can see the robot locate itself well and find where to go correctly.

Figure 11. Experiment environment.

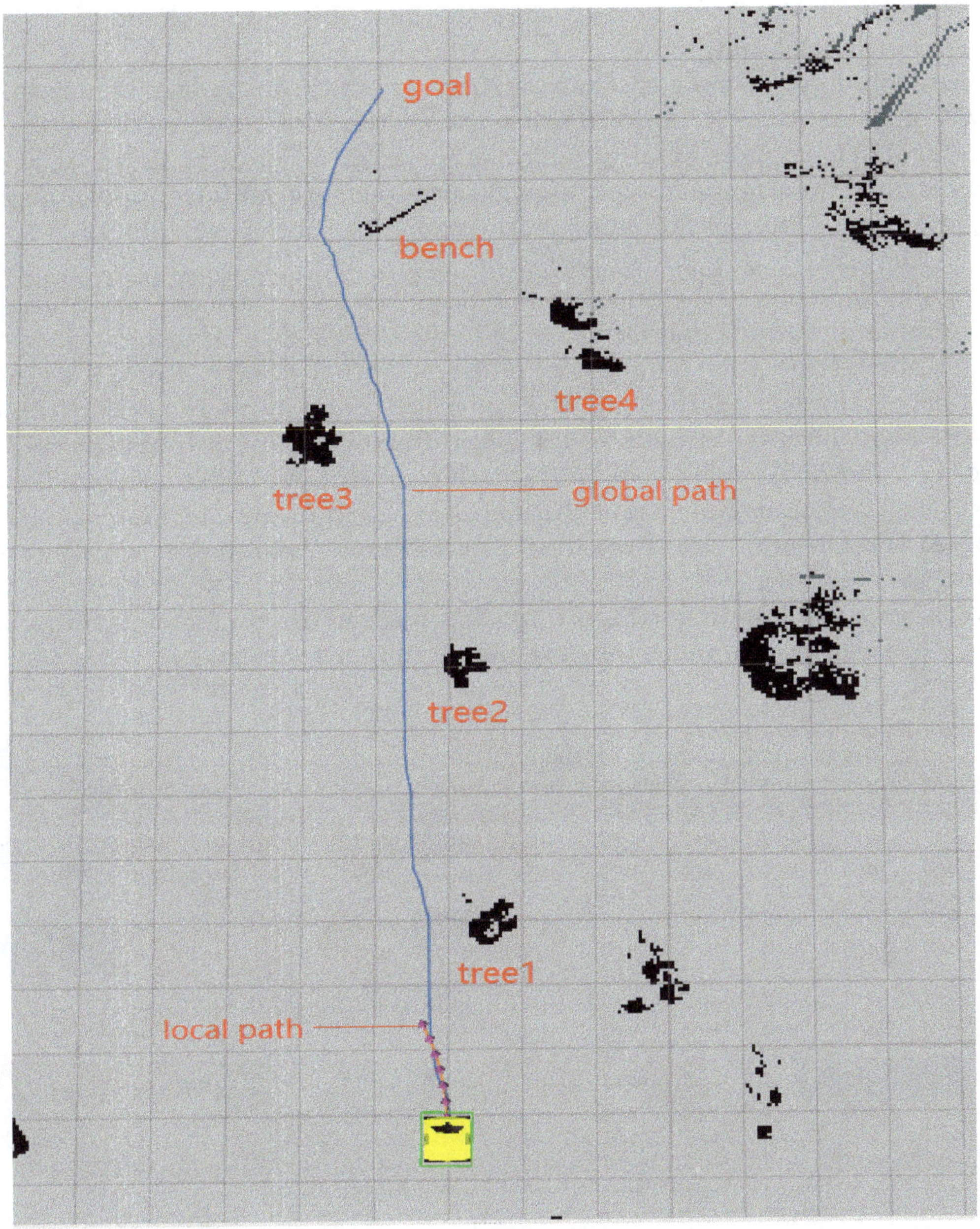

Figure 12. Navigation results.

7. Environment Inspection

The dam crack identification and pedestrian detection are the key technologies related to the safety of the dam. We use the trained model of YOLO V3 [31] to recognize them. We use the dataset online to train the YOLO V3 algorithm and then recognize the dam crack. The YOLO V3 algorithm includes a darknet-53 module, eight DBL components, three convolutional layers, two upsampling layers, and two tensor concat layers. The darknet-53 includes a DBL component and five residual learning units res1, res2, res8, res8, and res4.

The DBL contains a convolutional layer, a BN layer and a leaky ReLU. The convolutional parameters of the convolutional layer are the kernel size 3×3, stride 1, same padding and output channels 32. The res1, res2, res8, and res4 contain 1, 2, 8, and 4 basic units of the residual learning, respectively, and each basic unit of the residual learning contains two DBL and an identity map. The convolutional parameters of the first DBL are the kernel size 1×1, stride 1, and the output channels are equal to the number of basic units of the residual learning. The convolutional parameters of the second DBL are the kernel size 3 × 3, stride 2, same padding, and output channels are equal to 2 times basic units of the residual learning.

The three convolutional layers are CConv1, CConv2, and CConv3. The filter sizes of the CConv1, CConv2, and CConv3 are 512, 256, and 128 respectively. The convolutional layers CConv1, CConv2, and CConv3 respectively contain six convolutional operations. The convolutional parameters of the first convolutional operation are the convolutional kernel size 1 × 1, and the number of output channels is equal to the filter size. The convolutional parameters of the second convolutional operation are the convolutional kernel size 3 × 3, and the number of output channels is equal to 2 times the filter size. The convolutional parameters of the third convolutional operation are the convolutional kernel size 1 × 1, and the number of output channels is equal to the filter size. The convolutional parameters of the fourth convolutional operation are the convolutional kernel size 3 × 3, and the number of output channels is equal to 2 times the filter size. The convolutional parameters of the fifth convolutional operation are the convolutional kernel size 1 × 1, and the number of output channels is equal to the filter size. The convolutional parameters of the sixth convolutional operation are the convolutional kernel size 3 × 3, and the number of output channels is equal to 2 times the filter size. The upsampling layers Upsample1 and Upsample2 sample the input feature maps and the input feature maps of the concat layer into the same size.

The network structure is shown in Figure 13. This method can do the detection job at 10 Hz with less than 20% CPU occupying. Some results are shown in Figures 14 and 15.

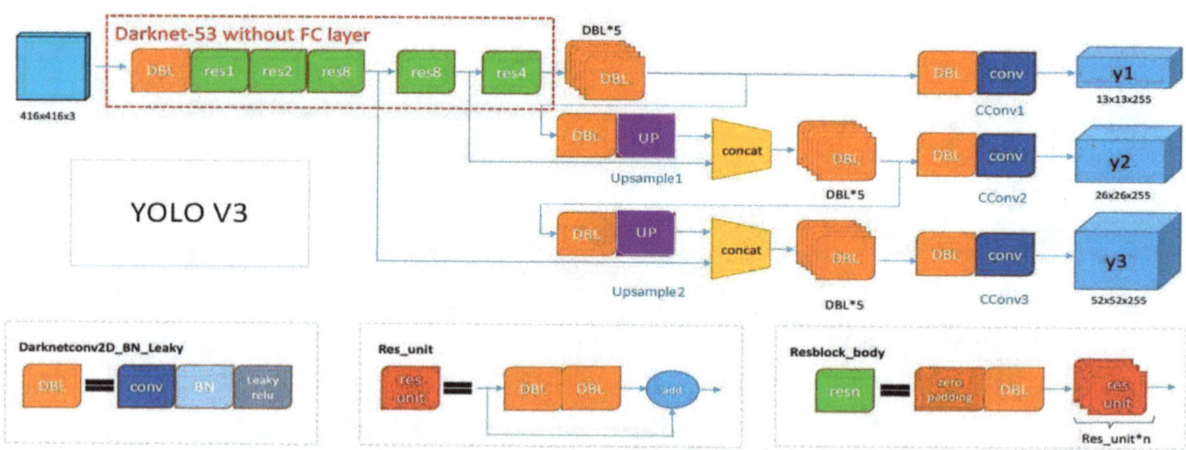

Figure 13. YOLO V3 network structure.

Figure 14. People detection results.

Figure 15. Dam crack detection results.

8. Conclusions

In this paper, a general robot system is proposed for dam surveillance. The robot is connected to cloud servers and terminal users by mobile internet and IoT network. Like the robot itself, this paper introduces mechanics layout, sensor selection, and the navigation method. A simple controller and a wheel odometry calculation are proposed and achieve good performance. A two-node EKF structure localization framework is proposed to solve localization problem, which fuses LiDAR SLAM, wheel odometry, IMU, and GNSS signals. For unexpected circumstances, a remote controller based on real-time video streaming is deployed as an emergency supplement. To make the whole system able to work all-time, a control state machine is also introduced. A YOLO v3 network is trained and deployed to detect dam crack and people around. From the practical experiments, this system can work well and is capable of the surveillance job for dams. Afterward, we will pay more attention to specific dams' surveillance jobs, such as intrusion detection and dam deformation detection. We believe robots will greatly improve the work efficiency for water conservancy in the future.

Author Contributions: Data curation, C.Z. and Q.W.; Formal analysis, T.H. and H.W.; Funding acquisition, Q.Z.; Methodology, C.Z., Y.A., and Q.W.; Software, Y.A. and H.W.; Supervision, Q.Z.; Writing—original draft, C.Z. All authors have read and agreed to the published version of the manuscript.

References

1. Su, H.; Hu, J.; Wu, Z. A study of safety evaluation and early-warning method for dam global behavior. *Struct. Health Monit.* **2012**, *11*, 269–279.
2. Räty, T.D. Survey on contemporary remote surveillance systems for public safety. *IEEE Trans. Syst. Man Cybern. Part C (Appl. and Rev.)* **2010**, *40*, 493–515. [CrossRef]
3. Theodoridis, T.; Hu, H. Toward intelligent security robots: A survey. *IEEE Trans. Syst. Man Cybern. Part C (Appl. and Rev.)* **2012**, *42*, 1219–1230. [CrossRef]
4. Yoshida, T.; Nagatani, K.; Tadokoro, S.; Nishimura, T.; Koyanagi, E. Improvements to the rescue robot quince toward future indoor surveillance missions in the Fukushima Daiichi nuclear power plant. In *Field and Service Robotics*; Kazuya, Y., Satoshi, T., Eds.; Springer: Berlin/Heidelberg, Germany, 2014; pp. 19–32.
5. Debenest, P.; Guarnieri, M. Expliner—From prototype towards a practical robot for inspection of high-voltage lines. In Proceedings of the 2010 IEEE 1st International Conference on Applied Robotics for the Power Industry, Montreal, QC, Canada, 5–7 October 2010; pp. 1–6.

6. Hess, W.; Kohler, D.; Rapp, H.; Andor, D. Real-time loop closure in 2D LIDAR SLAM. In Proceedings of the 2016 IEEE International Conference on Robotics and Automation (ICRA), Stockholm, Sweden, 16–21 May 2016; pp. 1271–1278.
7. Grisetti, G.; Stachniss, C.; Burgard, W. Improved techniques for grid mapping with rao-blackwellized particle filters. *IEEE Trans. Rob.* **2007**, *23*, 34–36. [CrossRef]
8. Qin, T.; Li, P.; Shen, S. Vins-mono: A robust and versatile monocular visual-inertial state estimator. *IEEE Trans. Rob.* **2018**, *34*, 1004–1020. [CrossRef]
9. Leutenegger, S.; Lynen, S.; Bosse, M.; Siegwart, R.; Furgale, P. Keyframe-based visual-inertial odometry using nonlinear optimization. *Int. J. Rob. Res.* **2015**, *34*, 314–334. [CrossRef]
10. Kunchev, V.; Jain, L.; Ivancevic, V.; Finn, A. Path planning and obstacle avoidance for autonomous mobile robots: A review. In *Knowledge-Based Intelligent Information and Engineering Systems*; Gabrys, B., Howlett, R.J., Jain, L.C., Eds.; Springer: Berlin/Heidelberg, Germany, 2006; pp. 537–544.
11. Duchoň, F.; Babinec, A.; Kajan, M.; Beňo, P.; Florek, M.; Fico, T.; Jurišica, L. Path planning with modified a star algorithm for a mobile robot. *Procedia Eng.* **2014**, *96*, 59–69. [CrossRef]
12. Jiang, C.G.; Peng, J.G. Research, manufacture and application of GPS-based surveying robot automatic monitoring system for dam safety. In Proceedings of the 2009 IEEE International Conference on Intelligent Computing and Intelligent Systems, Shanghai, China, 20–22 November 2009; pp. 151–155.
13. Kazmi, W.; Ridao, P.; Ribas, D.; Hernández, E. Dam wall detection and tracking using a mechanically scanned imaging sonar. In Proceedings of the 2009 IEEE International Conference on Robotics and Automation, Kobe, Japan, 12–17 May 2009; pp. 3595–3600.
14. Buffi, G.; Manciola, P.; Grassi, S.; Barberini, M.; Gambi, A. Survey of the Ridracoli Dam: UAV–based photogrammetry and traditional topographic techniques in the inspection of vertical structures. *Geomatics Nat. Hazard. Risk* **2017**, *8*, 1562–1579. [CrossRef]
15. Ridolfi, E.; Buffi, G.; Venturi, S.; Manciola, P. Accuracy analysis of a dam model from drone surveys. *Sensors* **2017**, *17*, 1777. [CrossRef]
16. Yi, J.; Wang, H.; Zhang, J.; Song, D.; Jayasuriya, S.; Liu, J. Kinematic modeling and analysis of skid-steered mobile robots with applications to low-cost inertial-measurement-unit-based motion estimation. *IEEE Trans. Rob.* **2009**, *25*, 1087–1097.
17. Mandow, A.; Martinez, J.L.; Morales, J.; Blanco, J.L.; Garcia-Cerezo, A.; Gonzalez, J. Experimental kinematics for wheeled skid-steer mobile robots. In Proceedings of the 2007 IEEE/RSJ International Conference on Intelligent Robots and Systems, San Diego, CA, USA, 29 October–2 November 2007; pp. 1222–1227.
18. Kozłowski, K.; Pazderski, D. Modeling and control of a 4-wheel skid-steering mobile robot. *Int. J. Appl. Math. Comput. Sci.* **2004**, *14*, 477–496.
19. Wang, T.; Wu, Y.; Liang, J.; Han, C.; Chen, J.; Zhao, Q. Analysis and experimental kinematics of a skid-steering wheeled robot based on a laser scanner sensor. *Sensors* **2015**, *15*, 9681–9702. [CrossRef] [PubMed]
20. Lenain, R.; Thuilot, B.; Cariou, C.; Martinet, P. Rejection of sliding effects in car like robot control: Application to farm vehicle guidance using a single RTK GPS sensor. In Proceedings of the 2003 IEEE/RSJ International Conference on Intelligent Robots and Systems (IROS 2003), Las Vegas, NV, USA, 27–31 October 2003; pp. 3811–3816.
21. Zhang, J.; Singh, S. LOAM: Lidar Odometry and Mapping in Real-time. *Rob. Sci. Syst.* **2014**, *2*, 9.
22. Lin, J.; Zhang, F. A fast, complete, point cloud based loop closure for LiDAR odometry and mapping. *arXiv* **2019**, arXiv:1909.11811.
23. Sabet, M.T.; Daniali, H.M.; Fathi, A.R.; Alizadeh, E. Experimental analysis of a low-cost dead reckoning navigation system for a land vehicle using a robust AHRS. *Rob. Autom. Syst.* **2017**, *95*, 37–51. [CrossRef]
24. Surber, J.; Teixeira, L.; Chli, M. Robust visual-inertial localization with weak GPS priors for repetitive UAV flights. In Proceedings of the 2017 IEEE International Conference on Robotics and Automation (ICRA), Singapore, 29 May–3 June 2017; pp. 6300–6306.
25. Alejo, D.; Caballero, F.; Merino, L. A Robust Localization System for Inspection Robots in Sewer Networks. *Sensors* **2019**, *19*, 4946. [CrossRef] [PubMed]
26. Min, H.; Wu, X.; Cheng, C.; Zhao, X. Kinematic and Dynamic Vehicle Model-Assisted Global Positioning Method for Autonomous Vehicles with Low-Cost GPS/Camera/In-Vehicle Sensors. *Sensors* **2019**, *19*, 5430. [CrossRef]

27. Rezaei, S.; Sengupta, R. Kalman filter-based integration of DGPS and vehicle sensors for localization. *IEEE Trans. Control Syst. Technol.* **2007**, *15*, 1080–1088. [CrossRef]
28. Suhr, J.K.; Jang, J.; Min, D.; Jung, H.G. Sensor fusion-based low-cost vehicle localization system for complex urban environments. *IEEE Trans. Intell. Transp. Syst.* **2016**, *18*, 1078–1086. [CrossRef]
29. Treptow, A.; Cielniak, G.; Duckett, T. Active people recognition using thermal and grey images on a mobile security robot. In Proceedings of the 2005 IEEE/RSJ International Conference on Intelligent Robots and Systems, Edmonton, AB, Canada, 2–6 August 2005; pp. 2103–2108.
30. Cha, Y.J.; Choi, W.; Büyüköztürk, O. Deep learning-based crack damage detection using convolutional neural networks. *Comput.-Aided Civ. Infrastruct. Eng.* **2017**, *32*, 361–378. [CrossRef]
31. Redmon, J.; Farhadi, A. Yolov3: An incremental improvement. *arXiv* **2018**, arXiv:1804.02767.

2

Mining and Tailings Dam Detection in Satellite Imagery using Deep Learning

Remis Balaniuk [1,*], Olga Isupova [2] and Steven Reece [3]

[1] Graduate Program in Governance, Technology and Innovation, Universidade Católica de Brasília, Brasília 71966-700, Brazil

[2] Department Computer Science, University of Bath, Bath BA2 7PB, UK; oi260@bath.ac.uk

[3] Department Engineering Science, Oxford University, Oxford OX1 3PJ, UK; reece@robots.ox.ac.uk

* Correspondence: remisb@tcu.gov.br

Abstract: This work explores the combination of free cloud computing, free open-source software, and deep learning methods to analyze a real, large-scale problem: the automatic country-wide identification and classification of surface mines and mining tailings dams in Brazil. Locations of officially registered mines and dams were obtained from the Brazilian government open data resource. Multispectral Sentinel-2 satellite imagery, obtained and processed at the Google Earth Engine platform, was used to train and test deep neural networks using the TensorFlow 2 application programming interface (API) and Google Colaboratory (Colab) platform. Fully convolutional neural networks were used in an innovative way to search for unregistered ore mines and tailing dams in large areas of the Brazilian territory. The efficacy of the approach is demonstrated by the discovery of 263 mines that do not have an official mining concession. This exploratory work highlights the potential of a set of new technologies, freely available, for the construction of low cost data science tools that have high social impact. At the same time, it discusses and seeks to suggest practical solutions for the complex and serious problem of illegal mining and the proliferation of tailings dams, which pose high risks to the population and the environment, especially in developing countries.

Keywords: tailings dam detection; surface mines detection; environmental impact of mining; remote sensing; machine learning; deep learning; cloud computing

1. Introduction

About 3.5 billion people live in countries rich in oil, gas, or minerals (https://www.worldbank.org/en/topic/extractiveindustries). Natural resources have the potential to drive growth, progress, and poverty reduction in developing countries. Despite its economic importance, mining is associated with major risks, both related to accidents and environmental damage. Tailings dams are a by-product of mining, storing toxic waste liquid and solids. The failure of tailings dams can have a catastrophic effect, with huge amounts of water and solid material being released suddenly, which can cause great loss of human life and huge damage to the environment, buildings, and plantations. The risks posed by long-term containment and the high number of recent tailings dam failures have led to a growing awareness of the need for improved safety measures [1].

We are not aware of any extensive inventory of active tailings dams worldwide. The lack of comprehensive tailings dam inventories prevents a significant analysis of technical failures, which could serve as a learning tool and help prevent future incidents. Existing records are very incomplete in crucial data elements: design height of dam; design footprint; construction type (e.g., upstream, downstream, and center line); age; design life; construction status; ownership status; capacity; release volume; runout; etc.

Most developed countries adopt frameworks for mining regulation that attempt to mitigate mining disasters whilst respecting community needs, the environment and keeping hazard risks under control [2]. Unfortunately, many developing countries still face a myriad of challenges, such as weak governance, lack of transparency, and lack of accountability in extractive industries, preventing inclusive growth, while threatening both people and the environment. Federal governments' lack of governance opens the door to informal and illegal mining and the fraudulent or incomplete reporting of activities by mining companies, causing major tax collection losses [3] but also increasing risks to the environment and the population.

Many illegal or informal mining operations take place in and around forests, sometimes overlapping with areas of high conservation value and high watershed stress that are home to vast amounts of biodiversity [4,5]. These mining activities are often difficult to detect from the ground. However, satellite or airborne sensing coupled with geographic information systems (GIS) can be used to assess where hazardous natural or man-made phenomena occur or are likely to occur over a wide area. This information can feed into risk assessments or illegal activities detection, along with information on natural resources, population, and infrastructure. It can be used by governments and communities in order to prevent or stop dangerous or illegal activities but also to design mitigation strategies to reduce vulnerabilities to acceptable levels.

Unfortunately, the use of GIS technologies in complex and country-wide scale analysis is not currently available to all. It requires access to huge remote sensing imagery datasets, robust computation infrastructure, expensive software tools, and teams of dedicated experts to interpret the vast amounts of imagery. Governments of developing countries are not always prepared, or do not have the financial resources, to deploy such a sophisticated infrastructure.

The main objective of this work was to contribute to the control and supervision of mining projects, as well as to the prevention of illegal or dangerous mining. We seek to meet this objective by proposing a scalable and low-cost method for detecting tailings dams and mines in wide areas combined with a risk assessment strategy for the identified mines. The scientific objective was to explore the combination of recent advances in machine learning and freely available multispectral satellite imagery, seeking to propose solutions adapted to a very specific and atypical context, where one wants to identify "objects", specifically dams and mines, of very different shapes, dimensions, and hue in low resolution imagery covering huge geo-spatial areas.

2. Background

New technological trends may help to reduce both the need for significant in-country technical expertise and computation infrastructure. Free cloud computing resources, free open-source software tools and deep learning methods can be used to build inexpensive and powerful data analysis tools. Easy access to these powerful technologies, coupled with the support that can easily be obtained from numerous online technical forums, tutorials, and online training opens new frontiers for their unrestricted use by governments, entities, and communities committed to the social good. Research and training carried out by universities and research centres, often in cooperative projects between different countries, has also contributed to a rapid expansion in the use of these new technologies (This work was made possible through a collaboration between the University of Oxford in the UK and Brazilian researchers). Anyone can access the latest free Earth observation imagery, from optical and radar satellite imagery to weather data and digital elevation maps. Forty years worth of free satellite images can be obtained from USGS-NASA Landsat (https://earthexplorer.usgs.gov/) or European Space Agency (ESA)'s Copernicus Open Access Hub (https://scihub.copernicus.eu/). Google Earth Engine (https://earthengine.google.com/) can be used at no cost for the searching, previewing, and preprocessing of these large free GIS image sets.

The analysis of remote sensing images in combination with machine learning methods has seen a massive increase in popularity in recent years. Machine learning was applied to several tasks, including scene and object classification, object detection, land cover, use and classification (LULC), object-based image segmentation, and analysis (OBIA) [6]. To explore the combination of free cloud computing, free software, and deep learning in analysing a real large-scale problem, in this paper, we demonstrate an application in automatic country-wide identification of surface mines and mining tailings dams in Brazil, followed by the classification of their potential environmental impact. The approach uses open-source and freely available software including Python [6] and TensorFlow [7] that can be run for free on Google cloud infrastructure Colaboratory (or Colab) (https://colab.research.google.com/notebooks/welcome.ipynb). Our paper is accompanied by open-source Python code and data that can be adapted to other Earth observation (EO) applications by government departments and their supervisory bodies and also by local communities and non-governmental organisations. The code supporting this publication can be accessed at: https://github.com/remis/mining-discovery-with-deep-learning. Although this study focuses on mining activity in Brazil, the proposed approach is applicable to any region where satellite imagery is available.

Section 2.1 describes the environmental and social impact of illegal mining and motivates a Brazil case-study. After presenting related work in Section 2.2, we describe the data processing methodology, specifically, our convolutional neural network approach, and its application to mine detection and tailings dam environmental risk assessment in Section 3. The efficacy of our approach is demonstrated on the Brazil case-study in Section 4. Section 5 summarizes the main results obtained in this research. We conclude in Section 6.

2.1. Mining Processes and the Brazil Case Study

Surface mining accounts for two-thirds of the world's metals used in industrial and technological applications and produces geochemically abundant metals such as aluminium, iron, magnesium, coal, manganese, and titanium, and also geochemically scarce metals, which include copper, lead, zinc, gold, and silver. Large-scale quarrying and open-cast mining is also used for stone, sand, gravel, crushed stone, and phosphates extraction. Surface mining usually uses heavy equipment, such as earthmovers, to remove overburden, followed by large machines, such as dragline or bucket wheel excavators, to extract the mineral. Large mining companies typically use appropriate methods and equipment and assume responsibility for environmental damage or impacts. Illegal mining, by contrast, uses makeshift methods and never takes responsibility for the damage it causes. In some rainforest territories, illegal miners use ferries to search the riverbed for precious metals. In others, as well as the rafts, there are almost industrial scale mining fronts, where backhoes and dredgers dig craters in the woods. This is the case for many of the mines in the Tapajós region of Pará, Brazil, where satellite and aerial images show large open "scars" in the forest, even on protected indigenous land (see, for example, Figure 1 showing an aerial photo taken in 2019 by journalists flying over a huge illegal mining area in the Kayapós indigenous reserve, in the state of Pará (https://www.bbc.com/portuguese/brasil-49053678).

Metal mining uses a huge amount of water in the crushing and ore separation process and leaves behind residual rocks that have been reduced to the consistency of gravel or sand, called tailings. Tailings are a mixture of particles suspended in liquid, usually toxic and potentially radioactive, and can be deposited in large ponds. These ponds, called "tailings dams", are enormous lake-like structures, square-kilometres in size, and are held in place by earth-filled dams. The dams in mountainous valleys can reach as high as 300 m. These dams are designed for long-term containment and are some of the largest man-made structures on Earth.

According to Brazilian legislation, those interested in exploring the subsoil for the extraction of ores must obtain a mineral exploitation concession through the mining concession regime, aimed at the extraction, processing, and commercialization of the mineral asset. Before obtaining this concession, the entrepreneur must follow a series of steps, starting with the request for a research authorization [8].

Figure 1. Illegal mining fronts in Kayapó Indigenous Land, Pará. Photo: Ibama/BBC News Brasil.

The Brazilian National Mining Agency (ANM) is responsible for supervising the mining activity in Brazil, from the authorization of mining to supervision of the enterprise, including the collection of taxes from mineral exploration and the inspection of tailings dams. Unfortunately, ANM suffers from a poor administrative structure with insufficient staff numbers, a lack of training, technological deficiencies, and an annual budget that has been cut by successive governments [9]. In 2019, the agency had only 35 inspectors to inspect the 790 ore dams officially declared across the country (https://brasil.estadao.com.br/noticias/geral,pais-tem-apenas-35-fiscais-de-barragem-de-mineracao,70002699885).

ANM's work on controversial projects, including mining in conservation areas and indigenous lands, is also hampered by legal loopholes. There is no consensus in the normative scope for mineral research in protected areas. The Brazilian federal constitution authorizes preliminary research and mining on indigenous lands, but requires specific regulations from the National Congress, after hearing from the affected communities [10]. As this regulation does not yet exist, instead of rejecting proposals from mining companies, ANM can only attempt to delay them, arguing that they wait for the creation of an appropriate law. Meanwhile, mining companies, some armed with provisional authorizations from the courts, continue their enterprises with impunity.

According to a World Wildlife Fund for Nature (WWF) report [11], in 2018, there were 5675 requests for mining concessions awaiting the approval of the ANM that overlap totally or partially with protected areas within the scope of the Legal Amazon. This demonstrates the magnitude of the threat to these protected areas and the pressure to remove restrictions on mining.

2.2. Related Work

Remote sensing imagery has been used for surface mine monitoring for several decades [12]. Recent increased availability and diversity of satellite image sets has sparked a wider range of studies to monitor environmental impact of mining activities using visual or computer-assisted interpretation with automated pattern recognition techniques [13–16]. This work is, in part, driven by the lack of databases containing geographic coordinates, characteristics, and attributes of mines. These characteristics of mines are important for environmental management and monitoring. The United States Geological Survey (USGS) provides the Major Mineral Deposits of the World (MMDW) database (https://mrdata.usgs.gov/major-deposits/) containing locations of known surface mining areas around the world. Yu et al. [13] used this database to investigate the effectiveness of remote sensing datasets to identify and map land cover changes related to surface mining activities, but reported a lack of precision related to geographic location, geometry and spatial extent of deposits. Soulard et al. [17] employed semi-automated procedures to map the country-wide mining footprint and change detection in US using mine seed points derived from official surveys and datasets. However, to the best of our knowledge, no study has been done on automated discovery and classification of unregistered surface mining sites and tailings dams on wide areas.

Space-borne sensors with a wide range of spatiotemporal, radiometric and spectral resolutions have become valuable data sources for large-scale Earth observation applications. However, data availability from multispectral and hyperspatial sensors introduces new challenges in data mining, processing, backup, and retrieval. The complexity and volume of these datasets require advanced data processing, flexible interfaces, and computational power. Cloud-based platforms such as Google Earth and powerful open source software libraries are creating new opportunities and bringing in new users and applications of remote sensing in environmental sciences and natural resources management. For a survey on processing remote sensing data in cloud computing environments, we refer readers to [18].

Machine learning methods have gained strength with these new resources. Deep learning methods have obtained state-of-the-art results in many Earth observation applications, such as image classification [19], semantic segmentation [20], phenological studies [21], poverty mapping [22], precision agriculture [23], and detection of pivot irrigation systems with very-high spatial and temporal resolution imagery [24]. An extensive survey on deep-learning-driven remote sensing image scene understanding can be found in [25]. In deep learning, convolutional neural networks (CNNs) play a central role for processing visual-related problems [26]. CNNs have been successfully employed, for instance, to classify hyperspectral images directly in the spectral domain [27], pixel-wise classification of satellite imagery [28], and detection of informal settlements in very-high resolution (VHR) satellite images [29].

3. Materials and Methods

3.1. Data Acquisition

Deep learning methods require rich training datasets to fit, evaluate, and test the neural network models. Preparation of these datasets requires a considerable number of good quality images containing the targets of interest. For applications in remote sensing it is imperative to obtain a representative collection of satellite imagery and address several corresponding technical issues such as cloud cover, select appropriate spectral bands, and undertake image registration. For supervised learning it is also paramount to have trusted sources of ground truth labelled imagery. In this work, we used, as a starting point, two official georeferenced databases released by the Brazilian federal government. The first is a catalogue of 612 officially reported tailings dams in Brazil, made available by the Brazilian National Mining Agency on its website (https://app.anm.gov.br/sigbm/publico). This database contains some information about each dam, including its owner and spatial coordinates of the dam's location. There is also some technical information including the main ore mined, height, volume, construction method, risk category, and potential damage that could arise if the dam were to fail. The second database is a catalogue of 268 legally operating ore mines in Brazil obtained from the same agency, but is not publicly available.

Unfortunately, both of these databases have quality issues. A significant number of coordinates point to locations near the mines or tailings dams but not exactly to them, or they are completely off target. A visual inspection of each reference using Google Earth images was required to address this problem and correct the target coordinates. Ferreira et al. [30] used the same databases to create the *BrazilDAM*, a public dataset based on Sentinel-2 and Landsat-8 satellite images covering the tailings dams catalogued by the ANM. Once the main database was created with the central coordinates of the mines, work was done to collect points of interest around these locations. Large mines can occupy tens of square kilometres and contain multiple dams. Thus, seeking to increase the number of examples (data augmentation), several coordinates at strategic points of the mining sites were recorded.

Our machine learning classifiers require samples of all target classes to train them. This includes the background class, which are objects other than dams and mines. Thus, it was also necessary to create a database with coordinates of locations where no mines or tailings dams were present. A careful choice of representative locations was necessary in order to capture the diversity of the background class such as cities, construction sites, and natural lakes. Consequently, two databases were created: one containing 1397 coordinates pointing to mines or tailings dams and another containing 1463 background locations. From this set of points, bounding boxes centred on these (latitude, longitude) coordinates defined the sub-imagery used for training the classifier.

As our goal was to implement an approach using only freely available resources that can be used to identify mines and dams in any region of the Brazilian territory, we chose to use only open access satellite imagery. The main providers are NASA and the European Space Agency (ESA), offering the Landsat Mission datasets (https://landsat.gsfc.nasa.gov/data/) and the Copernicus–Sentinel program datasets (https://sentinel.esa.int/web/sentinel/sentinel-data-access), respectively. Landsat satellites provide the longest continuously acquired set of space-based remote sensing data, with Landsat-8 being the latest member of the constellation. The Sentinels are a constellation of satellites deployed by ESA as part of the Copernicus program, which include high-resolution optical images from Sentinel-2A and 2B. The Sentinel-2s are sun-synchronous and multispectral instruments (MSI). The spatial resolution of a remote sensor image is strongly related to the level of detail that can be retrieved from a scene. Image resolution is usually measured using the distance between adjacent pixel centres measured on the ground. The spatial resolution of most bands on Landsat-8 is 30 m whereas the main visible and near-infrared Sentinel-2A bands have a spatial resolution of 10 m. The revisit time of a satellite system (i.e., the time elapsed between subsequent observations of the same area of interest) is another decisive factor of choice. The revisit period of Landsat-8 is 16 days whereas Sentinel-2A's revisit period is 10 days. Given their better resolution and sampling frequencies we chose to use the Sentinel images. Sentinel band resolutions and wavelengths used in this study are shown in Table 1.

To access and process the huge Sentinel-2 COPERNICUS/S2 image collection, we used the Google Earth Engine platform (GEE). The GEE [31] is a cloud computing platform designed to store and process large spatial and remote sensing data sets for free. The platform offers a friendly and convenient interface for the development of algorithms using JavaScript and an application programming interface (API) with image pre-processing features that allow solving some common problems in geoprocessing, such as cloud cover. In addition to the computing infrastructure, Google also archived all Landsat and Sentinel image data sets and linked them to the cloud computing engine. The GEE Image Collections are stacks of raster images that can be processed as whole on single API calls, making it easy to process complex operations over a selection of images and their bands. This structure allows operations like filtering, mapping, reducing, compositing, and iterating on large image sets. The results can be immediately visualized on screen or exported to a Google Drive folder. The same API can be used from a Python Jupyter Notebook interface on Google Colaboratory.

In order to acquire and prepare the image sets, we followed the workflow presented in Figure 2. In the first step (1), the desired data collection is selected and then date and cloud cover filters are applied. To ensure acquisition of good quality images, a broad time-frame of two years, starting January 2018, was chosen. Using the metadata of the images, it was also possible to select only the images within this time-frame with less than 20 percent cloud cover. In the second step (2), the image collection is processed in order to obtain a single raster image. A pixel based function was applied to the image stack to remove pixels containing cloud or cloud shadow patterns evident in the QA60 bit band. Finally, to define a single raster image a pixel based median operation was applied to the image stack. The median value at a location was the most common valid (i.e., no cloud) pixel value from all images at that location. This operation is based on the assumption that clouds are not in the same place permanently.

Table 1. The used spectral bands of Sentinel-2.

Bands	Resolution m	Wavelength µm
B01-Aerosols	60	0.44
B02-Blue	10	0.50
B03-Green	10	0.56
B04-Red	10	0.66
B05-Red edge 1	20	0.70
B06-Red edge 2	20	0.74
B07-Red edge 3	20	0.78
B08-NIR	10	0.83
B09-Water vapor	60	0.94
B10 Cirrus	60	1.37
B11-SWIR 1	20	1.61
B12-SWIR 2	20	2.20

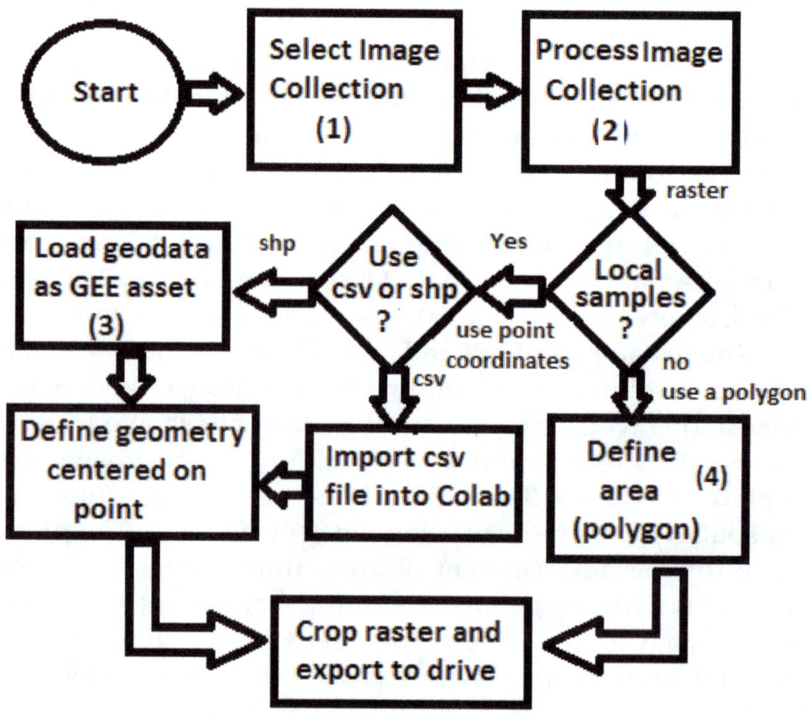

Figure 2. Workflow for accessing and processing Sentinel 2 images.

For the input to our classification algorithm, the coordinates of the points of interest must be in tabular format, and made available as comma-separated values (csv file) or Esri shapefile (shp). To facilitate the processing of shapefiles, we imported them into GEE as feature collections before loading them into Google Colab as arrays using GeoPandas (https://geopandas.org/) (i.e., workflow step 3). A small square image was extracted around each coordinate centred on the coordinate. Each of these sub-images contains enough information to determine the presence, or not, of a mine or dam. Two sets of image crops were prepared. The first one had 2 × 2 km (0.018 × 0.018 degrees) image extracts, each comprising 201 × 201 pixels and 12 spectral bands. These small image extracts were used to train and validate the mine discovery classifier. The 2 × 2 km bounding box size was as large as possible within the memory limitations of the GPUs available in the Google Colab environment. The second set of cropped images, comprised only the point coordinates for which the mining ore was annotated, had 200 × 200 m (0.0018 × 0.0018 degrees) image extracts, each comprising 21 × 21 pixels and 12 spectral bands. In this case, the 200 × 200 m bounding box size was chosen in order to focus on specific locations in the mine where it was possible to identify excavations, water, or ore residues exposed on the surface. This second dataset was used to train the environmental impact classification model. Figure 3 shows an example of a mine and its tailings dam and Figure 4 shows the same spot on a cropped image from the Sentinel-2 collection obtained using the workflow in Figure 2.

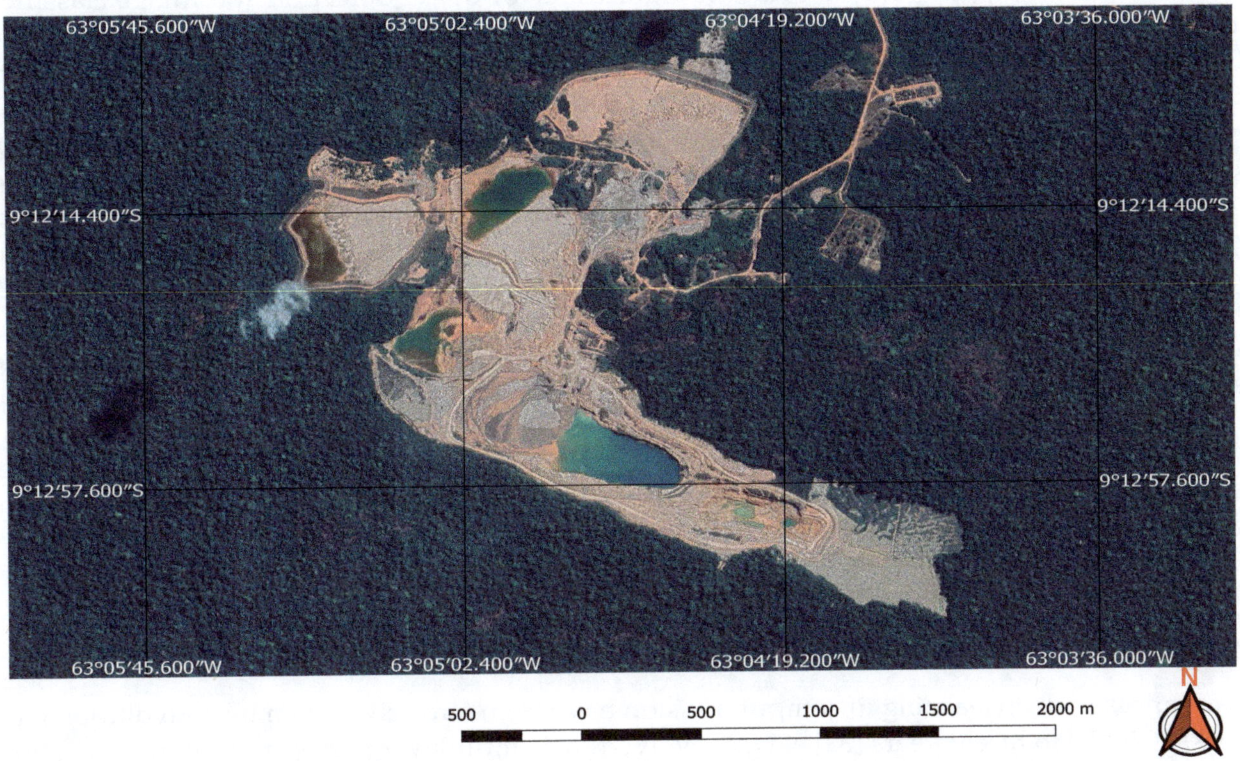

Figure 3. High resolution Google Earth image showing mine and associated dam.

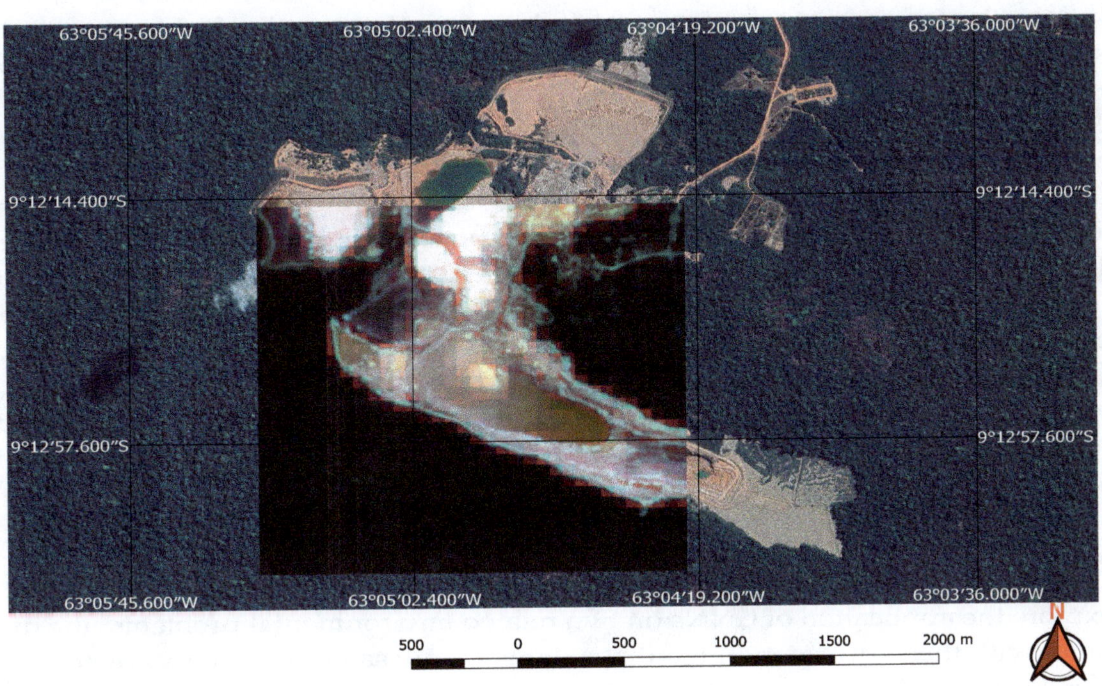

Figure 4. Sentinel 2 image of the mine and dam shown in Figure 3.

3.2. Deep Learning

Deep learning methods have obtained an unprecedented success in computer vision. High accuracy has been achieved on complex tasks such as the ImageNet Large Scale Visual Recognition Challenge (ILSVRC) to evaluate algorithms for object detection and image classification at large scale [32]; however, several aspects need to be taken into consideration when applying these methods in real life cases. State-of-the-art computer vision algorithms use thousands or sometimes millions of images with known annotations for training and evaluation on large scale problems. For real life cases this can be unfeasible due to the lack of examples or dubious interpretations of the object classes present in an image. A level of automation classification accuracy is also task-specific in real life applications. In applications where a classification mistake has a high cost and may cause harm, inaccuracies inherent in machine learning methods may not be tolerated. In applications where automated classification methods are used only to support expert decisions, a small number of false positive or false negative classifications may be tolerated. In this research, which looks for indications of the presence of surface mines or tailings dams as well as of the type of associated environmental impact they have, it would perhaps be impracticable to obtain a highly accurate system given the limited amount of training examples and the numerous variations in shape, hue, areas, and volumes of different mines and ore categories. Thus, we designed a complete analysis strategy to support the analyst in the discovery and classification of mines and dams in the most diverse regions of the planet, prioritizing the reduction of false negatives, so as not to lose relevant indications, and seeking to maintain false positives at acceptable levels.

3.2.1. The Convolutional Neural Network

Advances in deep learning in computer vision have been strongly driven by convolutional neural networks (CNNs) in recent years [33]. CNNs evolved from multilayer perceptrons, which are networks consisting of an input and an output layer, as well as multiple hidden layers, and use a stochastic gradient decent based optimization algorithm to learn their parameters efficiently. A Neural network is defined by its layer architecture and activation functions. A CNN typically comprises a series of convolutional layers, each performing local "filter" operations that detect features in the output of previous layers. The neural net learns the filters that, in traditional computer vision algorithms, are hand-engineered. Other types of layers are also used in CNNs such as pooling, fully connected, and regularization layers. Pooling layers combine the outputs of groups of neurons in the previous layer in order to reduce data dimensionality. A fully-connected layer connects all neurons in one layer to all neurons of the next layer. They usually appear after the convolutional layers so that the fully-connected layers capture complex relationships between high-level features extracted by the convolutional layers. CNNs may be prone to overfitting. To mitigate this undesired effect, they are usually supplied with regularization layers. One method to reduce overfitting is *dropout* where, at each training cycle, the outputs of some nodes are ignored with a given probability to prevent the nodes from co-adapting to the data. The activation function is responsible for transforming the weighted sum of a node's input into that node's output value. The activation function commonly used in CNNs is the Rectified Linear Unit (RELU). For a review of the main concepts related to deep learning please refer to [34].

3.2.2. Visual Pattern Mining

We explore the application of CNNs on two related environmental problems: the discovery of surface mines with their corresponding tailings dams in raw satellite imagery covering large areas (described in Section 3.3) and the classification of the environmental impact of the discovered mines and dams (Section 3.4). These problems required some further adaptation of the CNN methodology, which we describe next.

Computer vision problems are usually categorized as *image classification* when the goal is to classify the whole image, as *object detection* when the goal is to detect and classify one or more objects within an image displaying their position with bounding boxes, and *semantic segmentation* when the goal is to recognize the object class for each pixel of an image. These models traditionally require training data in the form of classified images, objects encased in bounding boxes or highlighted by masks.

However, we use, as input, only point valued data—the spatial coordinates of the dam locations. We do not have more informative masks nor bounding boxes available for image segmentation nor object detection, preventing us from using models such as the Unet [35] or region-based CNNs [36] for example. As Mines and dams frequently do not have clear edges to delimit them in the imagery (see, for example, Figures 5 and 6) and their sizes and shapes vary enormously we cannot easily create bounding boxes nor masks to augment the training data.

An alternative machine learning approach to computer vision problems is visual pattern mining (VPM) [37] which we can adapt for point labelled data. This paper demonstrates the VPM approach has sufficient accuracy for dam localisation using point labelled data.

In essence, VPM classifies parts of an object. VPM is often implemented as a two phase approach. Individual parts of an object are classified in Phase 1 and then Phase 2 classifies the object by using spatial relationships of the parts identified in Phase 1.

In Phase 2 the classification of visual patterns is obtained from the combination of multiple filters from the final convolutional layer of the Phase 1 CNN. Visual Pattern Mining is widely used for mid-level feature representations on image classification tasks and visual summarisation. VPM code is freely available via the Python implementation PatternNet (https://github.com/KnurpsBram/PyTorch-PatternNet).

We adapted the pattern mining method to both mine and dam identification in satellite imagery as well as the classification of different types of construction and the corresponding environmental impact. Our adaptation uses Phase 1 of the VPM approach to identify parts of dams or mines.

We do not distinguish the nature of the individual parts themselves, only that a sub-image of interest either contains or does not contain a part of a dam or mine. This allows us to fix the size of the CNN input image, and thus provides us with an efficient classifier that scales over large areas of interest. The new algorithm was trained on small image samples, some of which contained a part of a dam or mine, annotated with the presence or not of a mine or tailings dam and the ore type if it was known. The data acquisition process was detailed in Section 3.1.

Each image sample covers a 2×2 km area (201×201 pixels). This size was chosen after several modelling cycles where we tried to find a compromise between the prediction performance and the GPU's memory capacity.

We implemented the VPM method as a fully convolutional network (FCN) for computational efficiency and scalability. Since an FCN contains only convolution and pooling layers, with fully-connected layers converted to convolution layers, it can be executed efficiently on variable size images. This is particularly useful for our research problems, where training needs to be done on small image patches, showing parts of a mine or a dam, but classification is performed on big images covering wide areas of Brazil. When applied to wide areas, output of the FCN is a map comprising clusters of class markings, each cluster marking a dam or mine.

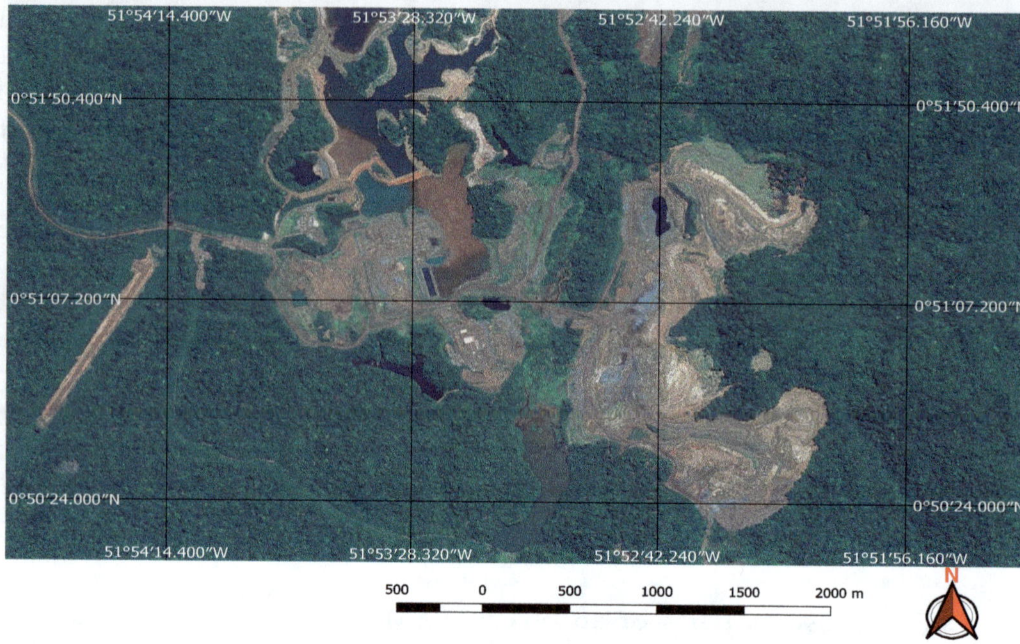

Figure 5. Example of a gold mine.

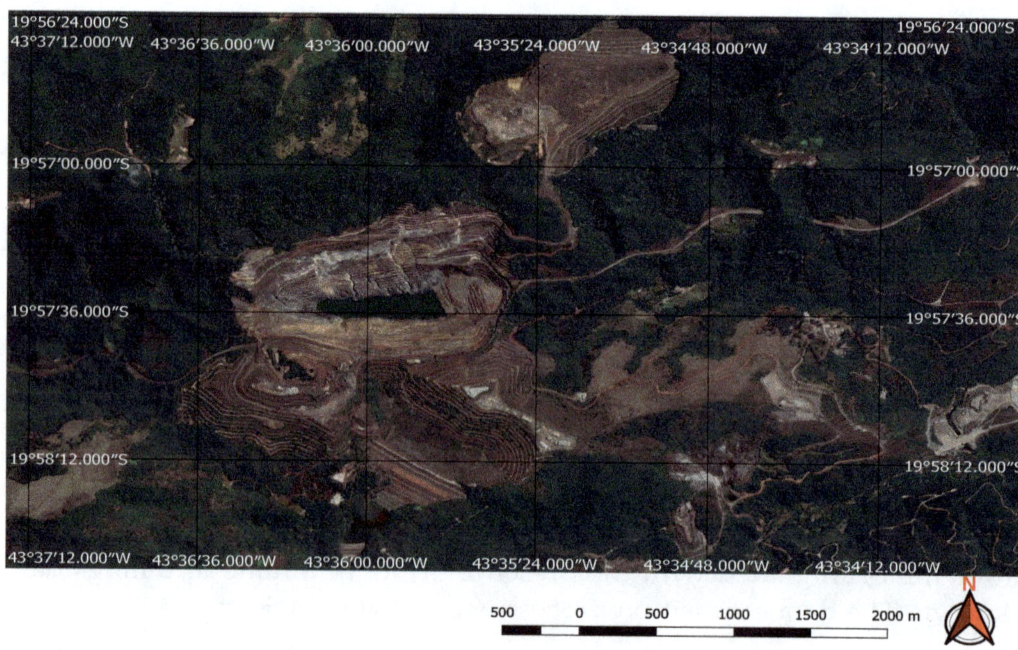

Figure 6. Example of an iron mine.

The FCN algorithm was implemented using open-sourced computational resources, which are freely available and easily accessible online. CNN-based applications can be developed and deployed using open source libraries such as the Keras TensorFlow's high-level application programming interface (API), used for building and training deep learning models [7]. It is useful for fast prototyping, state-of-the-art research, and production. The Keras TensorFlow API is available on Google Colaboratory (Colab), a free cloud service based on Jupyter Notebooks that offers robust graphics processing unit (GPU) based computing at no cost [38]. On Google Colab, a developer has access to a flexible runtime fully configured for deep learning, avoiding all the workload required to prepare the infrastructure and environment that machine learning projects typically require. Large files can be uploaded, accessed and saved using Google Drive, greatly simplifying data input–output operations. The existence of such a platform both simplifies and reduces the cost of developing and running sophisticated machine learning methods, including deep learning, thus allowing its use for social good by governmental and non-governmental organizations in developing countries and not just by conventional academic and business organisations.

3.3. Mine and Tailings Dam Discovery

Our FCN architecture comprises five convolution layers, three pooling layers, and a dropout layer with a 50% dropout rate for regularization. The model contains 259,810 parameters, all trainable and distributed on the network layers, as presented in Table 2. As shown in the table, the 10 layers are arranged in sequence, with the input of images occurring from level 1 and the prediction response at level 10. We used Adam stochastic optimisation [39] to iteratively update the network weights, and minimise the categorical cross-entropy loss function, due to its computational efficiency and little memory requirements. All convolution layers use the rectified linear unit (RELU) activation function and are initialised with a normal weight distribution. The input training data are composed of $201 \times 201 \times 12$ float arrays. They correspond to the 12 spectral bands of 2×2 km raw Sentinel-2 image samples. These images are normalised with each band having zero mean and unit standard deviation across the entire training data. Only the number of bands in the imagery is fixed (i.e., 12) in the FCN input layer, the width and height of the input images are not specified to allow different sized images for training and classification. The output layer is a convolution with a 1×1 kernel and two filters, corresponding to two target classes: *mine* and *not mine*. The *mine* class corresponds to scenes containing surface mines and/or tailings dams.

Table 2. Mine and dam discovery model summary (# means quantity).

Layer	Name	Type	Filters#	Kernel	Param#
1	input	InputLayer			0
2	conv2d_1	Convolution	32	3×3	3488
3	maxpooling1	MaxPooling		3×3	0
4	conv2d_2	Convolution	64	3×3	18,496
5	maxpooling2	MaxPooling		3×3	0
6	conv2d_3	Convolution	64	3×3	36,928
7	maxpooling3	MaxPooling		3×3	0
8	conv2d_4	Convolution	64	7×7	200,768
9	dropout	Dropout 50%			0
10	conv2d_5	Convolution	2	1×1	130

Our FCN architecture is rather conventional and the implementation of its training, validation, and testing procedures was straightforward on the Keras TensorFlow API. The fact that the input data consists of raster images, stored as TIFF (Tagged Image File Format) files, requires proper load and translation steps in order to prepare the float arrays (or tensors) required as input to the CNN. We used the Geospatial Data Abstraction Library (GDAL) (https://gdal.org/) to implement these steps. A total of 2860 $201 \times 201 \times 12$ images were used to train and test the FCN.

3.4. Environmental Impact Classification

Various principles and processes are utilised within mining that determine the impact they have on the environment [40]:

- Low environmental impact: extraction and any processing of sand, gravel, or rock where the environmental impact is usually restricted to dust or noise. It is usually related to extraction for construction purposes or as raw materials for industry. The processing does not go beyond crushing or screening of sand and gravel.
- High environmental impact: extraction of minerals and ores in facilities, which produce large amounts of waste consisting of finely crushed rock (tailings) and discharges of polluted waste

water. This category comprises mining sites where the valuable substances in the ores must undergo processing and where large amounts of tailings must be deposited. Within this category, a sector of great environmental importance is the mining of sulphide ores containing copper, zinc, lead, or nickel, often in combination with pyrite. These ores are the most common ores on a global scale. The environmental impact of this type of mining can last a long time after the end of activities (An exception to this simplified classification of environmental impact concerns gold mines. Although its process is similar to that used in the extraction of gravel, based on washing, crushing, and screening, the environmental impact can be great due to the damage it can cause to water courses and the use of products that are harmful to the environment such as mercury; however, in low resolution satellite images it is not possible to distinguish a gold mine from a simple exploration of sand or gravel.)

In order to provide an indication of the potential environmental impact of a discovered mine or tailings dam we explored again the usefulness of deep learning based classifiers. Using the Brazilian officially declared mines and tailings dams databases, containing the ore type declared by their owners, we chose 1275 representative point coordinates and acquired Sentinel-2 image patches, using the protocol described in Section 3.1. Each image patch has 21 × 21 pixels and 12 spectral bands. The classes considered in this second predictive model, their related materials and the number of image samples used as the training dataset are:

1. high impact waste: lead, iron, niobium, zinc, copper, phosphate, nickel, schist, manganese, and bauxite; 425 samples.
2. low impact waste: sand, gravel, clay, and stones; 425 samples.
3. no ore: soil, rocks, and vegetation; 425 samples.

This second model used a conventional CNN network architecture, consisting of two convolution layers followed by a max pooling layer each and two dense layers. The input is an image patch of a dam or mine and the output is an one-hot encoding of three classes: high, low impact, and no-ore. Due to the smaller 21 × 21 training image patches the convolution layers used a reduced convolutional filter (kernel) of size 2 × 2. The max pooling layers also used a 2 × 2 kernel. The first dense layer has 1024 elements. The second has 3 elements, corresponding to the output classes, and uses a softmax activation. The architecture, comprising 2,373,219 parameters, all trainable, is presented in Table 3, where the 9 layers are arranged in sequence, with the input of images occurring from level 1 and the prediction response at level 9. Adam stochastic optimisation is used to minimise the softmax cross-entropy function.

Table 3. Environmental impact model summary.

Layer	Name	Type	Filters#	Kernel	Param#
1	input	InputLayer			0
2	conv2d_1	Convolution	32	2 × 2	1568
3	maxpooling1	MaxPooling		2 × 2	0
4	conv2d_2	Convolution	64	2 × 2	8256
5	maxpooling2	MaxPooling		2 × 2	0
6	flatten	Flatten			0
7	dense_1	Dense	1024		2,360,320
8	dropout	Dropout 50%			0
9	dense_2	Dense	3		3075

4. Experiments

It is important to note that the content presented here summarize the most significant results obtained after a long process of attempts, with several sets of images, modelling strategies, and CNN architectures implemented and tested. It would not be feasible to show here details of all the studied configurations and comparisons of all the intermediate results obtained.

4.1. Validating the Mine and Dam Localization Model

The FCN described in Section 3.3 takes approximately 100 seconds to train on Colab GPUs. Figure 7 shows the iterative evolution of the model accuracy on ten re-sampling, training, and validation procedures (k-fold cross validation). We obtained an average accuracy on the training data of 99.23 ± 0.7% after 30 epochs. In each re-sampling step, the FCN was validated on a separate dataset containing 286 201 × 201 × 12 images. The mean validation accuracy, F1 score, precision, recall, area under the receiver operating characteristic curve (ROC AUC), and Cohen's kappa [41] are shown in Table 4. The mean confusion matrix is presented in Table 5.

Table 4. Efficacy of the mine and dam localization model for 10-fold cross validation.

Score	Mean
mean validation accuracy	97.44 ± 1.3%
mean F1 score	0.9749
mean recall	0.9747
mean AUC	0.9746
mean Cohen's kappa	0.9488

Table 5. Mean confusion matrix of the mine and dam localization model for 10-fold cross validation.

Classes	Mine (Predicted)	Not Mine (Predicted)
mine (actual)	136.1	3.6
not mine (actual)	3.7	142.6

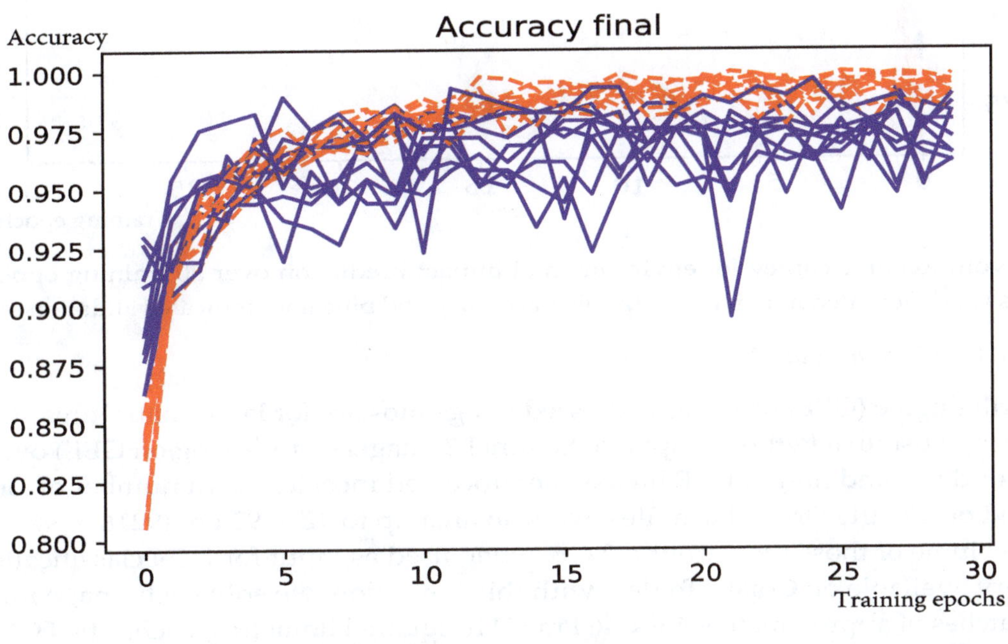

Figure 7. Evolution of accuracy for mines and dams detection over 30 training epochs and 10-fold cross validation. Red lines indicate training accuracy and blue lines validation accuracy.

4.2. Validating the Environmental Impact Classification Model

Training the CNN described in Section 3.4 over 30 epochs takes on average 27 s on Colab GPUs. Figure 8 shows the iterative evolution of the model accuracy. The mean validation accuracy and Cohen's kappa are shown in Table 6. The mean confusion matrix is presented in Table 7.

Table 6. Efficacy of the environmental impact classification model for 10-fold cross validation.

Score	Mean
mean validation accuracy	90.42 ± 3.5%
mean Cohen's kappa	0.8550

Table 7. Mean confusion matrix of the environmental impact classification model for 10-fold cross validation.

Classes	High Impact (Predicted)	Low Impact (Predicted)	No Ore (Predicted)
high impact (actual)	40.7	1.3	0.5
low impact (actual)	2.2	35.9	4.4
no ore (actual)	1.1	2.7	38.7

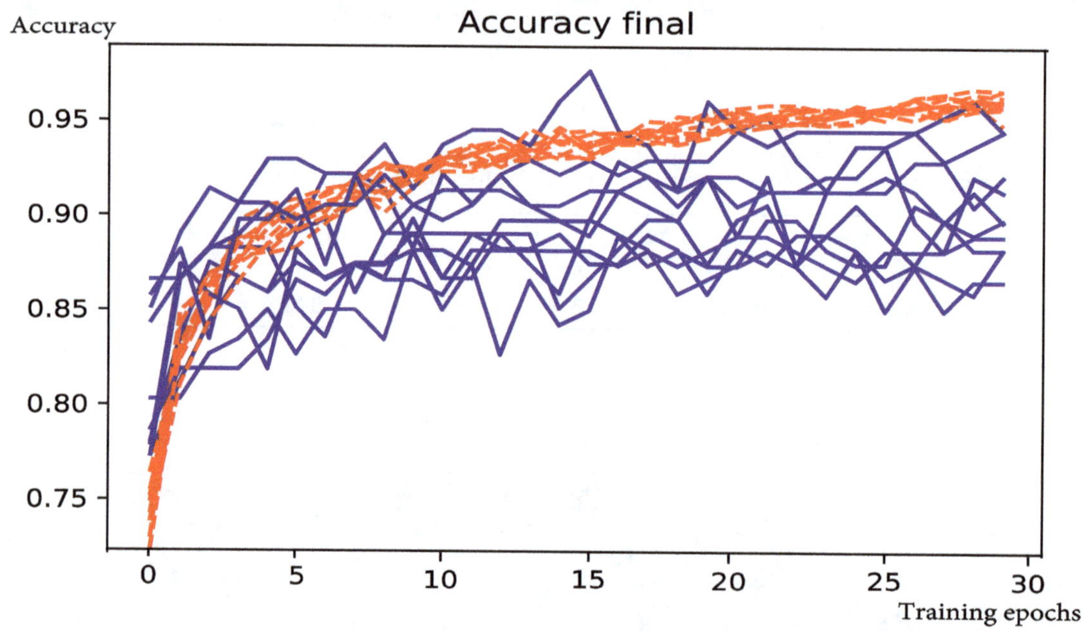

Figure 8. Evolution of accuracy for environmental impact prediction over 30 training epochs and 10-fold cross validation. Red lines indicate training accuracy and blue lines indicate validation accuracy.

4.3. Wide-Area Maps of Dams and Mines

Google Earth Engine (GEE) provides processed image mosaics for large spatial areas (our paper is accompanied by open-source Python scripts for Sentinel 2 imagery processing on GEE) over which we wish to search for dams and mines. GEE breaks the processed mosaics into a number of GeoTIFF files that can be saved on Google Drive. Each file covers an area up to 92 × 92 km (9216 × 9216 pixels and 12 bands). The volume of those files, around 3 GB, when used as input for FCN classification, exceeds the GPU memory available on Colab. To deal with this limitation, we split each image into four near equally sized patches of approximately 46 × 46 km (2116 square kilometres) each. The FCN output for each patch is a 164 × 164 grid of probability distributions, representing the probability of each class on a certain region of the input image patch.

For each grid point predicted as belonging to the *mine* class we also classify its environmental impact (i.e., low, high, or no ore). The environmental impact CNN uses as input a patch with 21 × 21 pixels (corresponding to an area of 210 × 210 m) from the same mosaic taken around the location of the discovered mine. The CNN output is a probability distribution, representing the probabilities of each environmental impact class on that location. If the *no ore* class has the highest probability the predicted mine spot is considered a false positive and discarded. To determine the geographic coordinates of the detected mines from the FCN output map, we used Rasterio (https://rasterio.readthedocs.io/en/latest/), a Python API based on Numpy N-dimensional arrays and GeoJSON, to compute the coordinates of discovered targets.

The locations of the mines and tailings dams are exported as georeferenced delimited text files (csv and kml), which can be scrutinized using GIS software or Google Earth. We suggest an analyst visually inspect the CNN discovered mines and tailings dams against the raw satellite imagery or against a base map, like Google Earth's, to validate the discovered mines or tailings dams. Figure 9 shows the predicted indications of dams and mines pin pointed on a Google Earth base map. It is necessary to understand that each yellow pin points to the centre of an implicit area where a mine or dam pattern was found. In this way, some marks are found in the vicinity of a mine or dam and not exactly in it because the pattern found in that specific prediction could be outside the centre of the considered area. This does not mean that they are false positives. In this way, each mine or dam will be identified by a cloud of nearby marks indicating the region where the mine is located.

Figure 9. Mines and tailings dams indicated by yellow marks on Google Earth.

We applied our approach to two large areas on the Brazilian territory searching for mines and dams and classifying their potential environmental impact. These areas are indicated in Figure 10. The first area covering 67,110 square-kilometres is located in the state of Minas Gerais. It is one of the areas with the highest concentration of mines and tailings dams in Brazil and includes the cities of Brumadinho and Mariana, where the cited tragic dam collapses occurred. The second area covers 170,267 square-kilometres and it is located in the state of Para where protected indigenous land and environment conservation units struggle against illegal mining and deforestation activities. It includes the Kayapo's protected land depicted in Figure 1. Both areas are represented by red squares in Figure 10 and the coordinates of the officially registered tailings dams from the Brazilian Agência Nacional de Águas (ANA) database are shown by orange dots. Raster images from both areas were acquired and processed using the workflow in Figure 2. In the "local samples" decision, we follow the "no" path where the areas are defined directly as geo-polygons on step (4) before the crop and export.

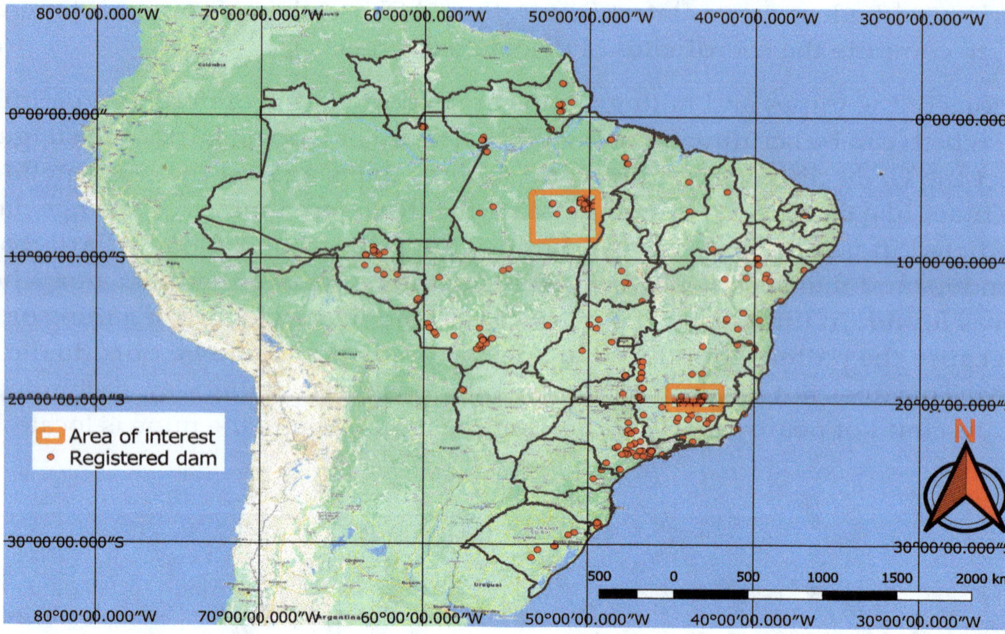

Figure 10. Areas of interest: Minas Gerais (south) and Para (north).

The FCN model was run on the 237,000 square kilometres of satellite imagery of Brazil and was able to identify all visible mines on those areas, even small ones of less than one square kilometre in extent. However, unlike the tests performed on the initial collection of examples, a considerable number of false positives were also present in this large-scale prediction. An example of the typical output obtained using the discovery model is shown in Figure 11. This green rectangle covers a 5624 square kilometre area of intense mining activity in Minas Gerais, Brazil. It includes the cities of Brumadinho and Mariana where the recent dam collapses occurred. Yellow and red dots show predicted mines. Blue ellipses indicate where the predictions were confirmed by visual inspection. All other dots are false positives. Two true positives from this area are shown in Figures 12 and 13. Figure 12 shows the large mine covering 46 square kilometres appearing on the top right corner of Figures 11 and 13 shows a small one of only 0.5 square kilometres.

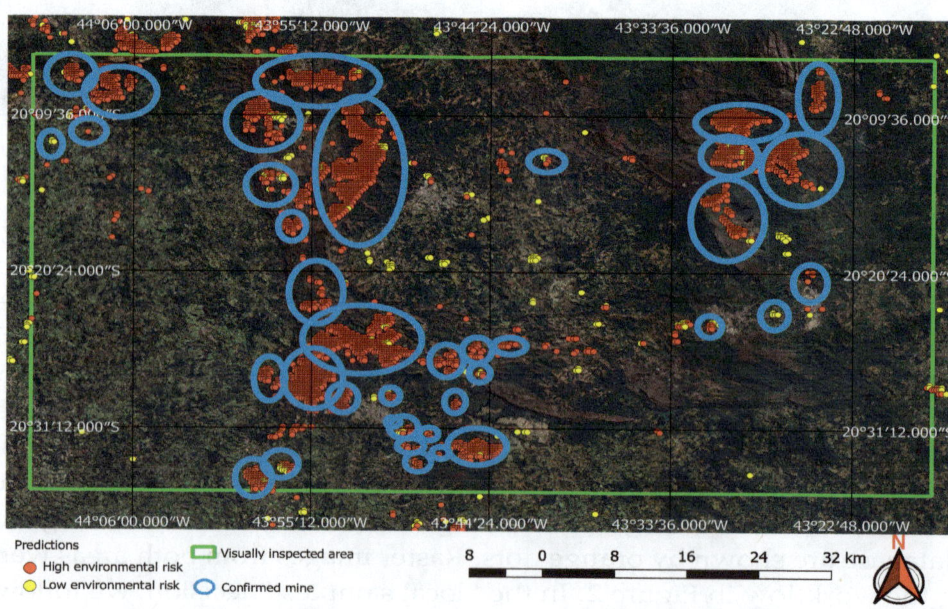

Figure 11. Visual inspection of neural network predictions.

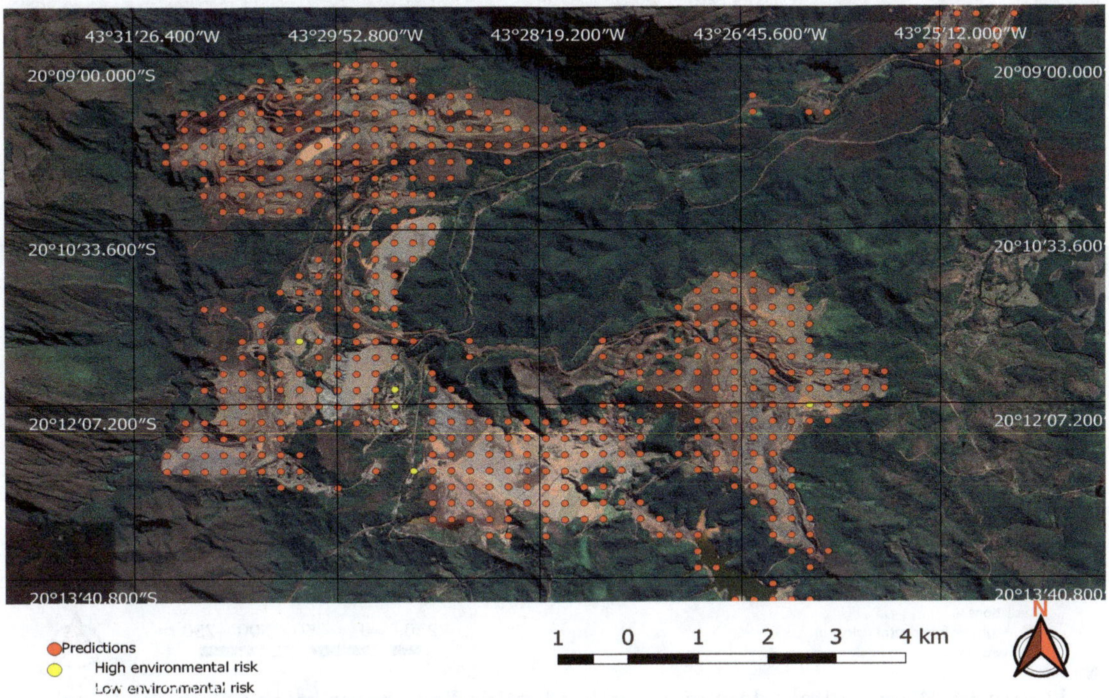

Figure 12. Large mine with potential high environmental impact.

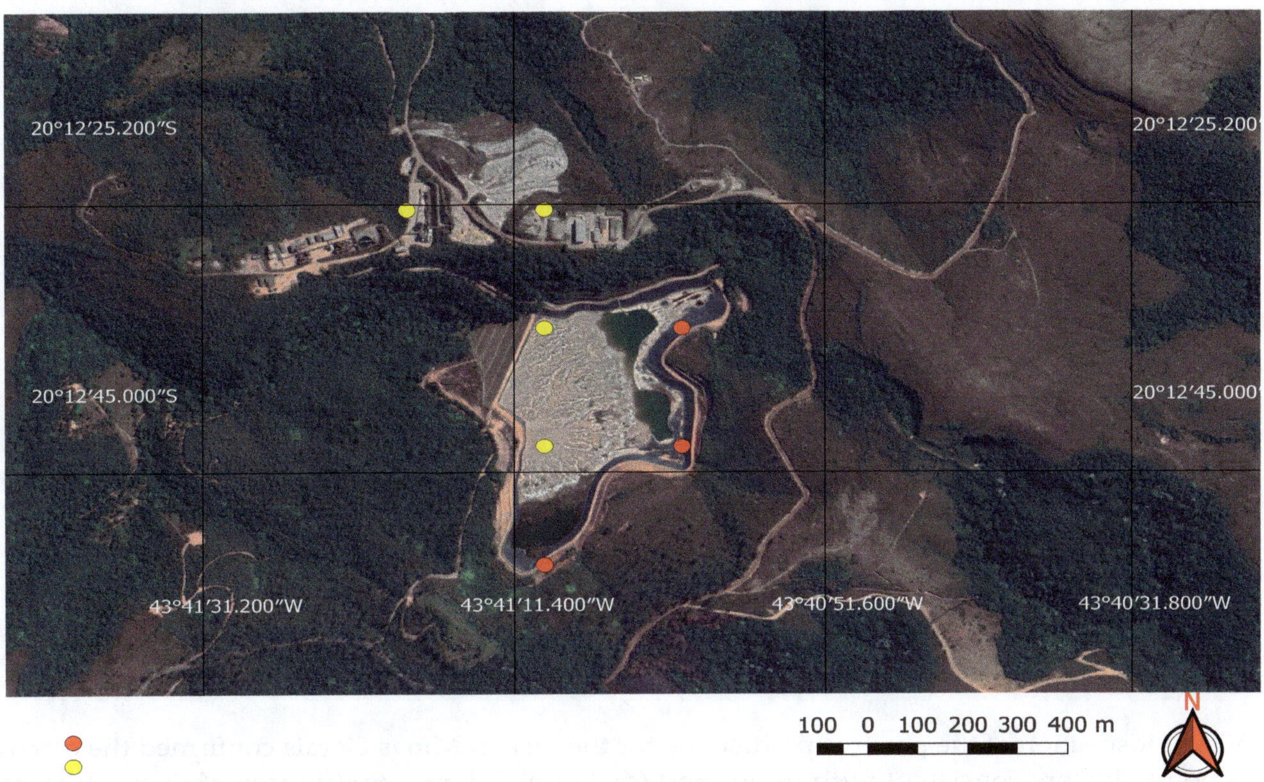

Figure 13. Small mine with potential low environmental impact.

False positives occur in some specific situations. Two common situations are water supply dams, depicted in Figure 14, and river banks shown in Figure 15. Feeding these false positives back into the training data and retraining the network could reduce the false-positive rate. However, as the false-positive rate was acceptable it was our choice to correct these maps manually.

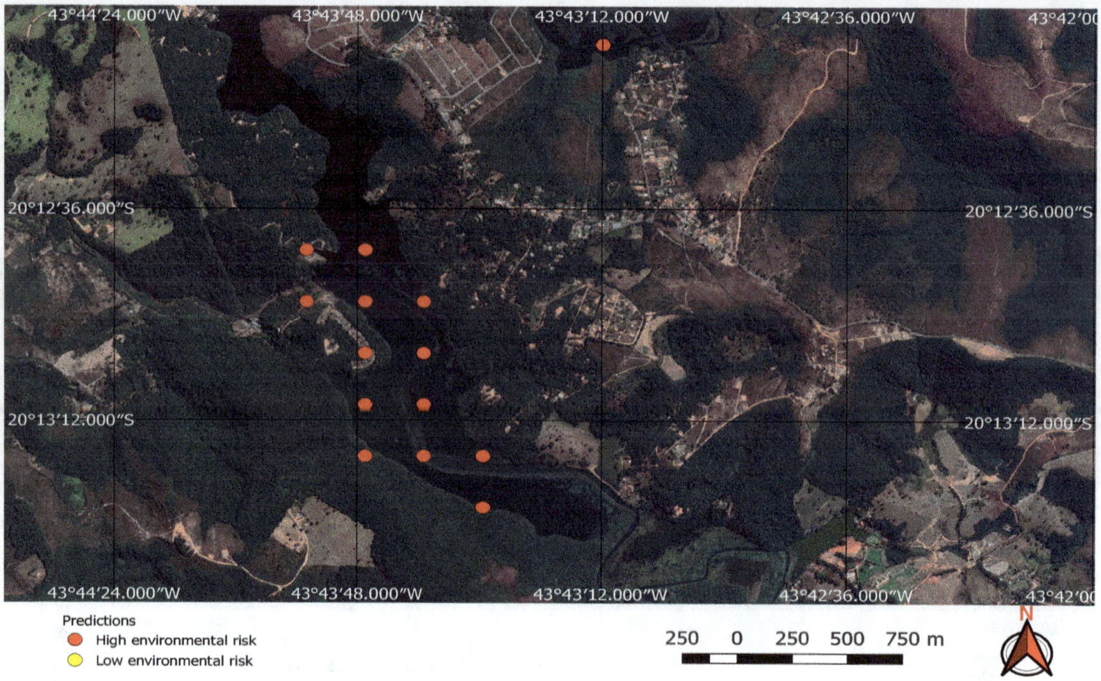

Figure 14. Water supply dam incorrectly identified as a mine by the neural network.

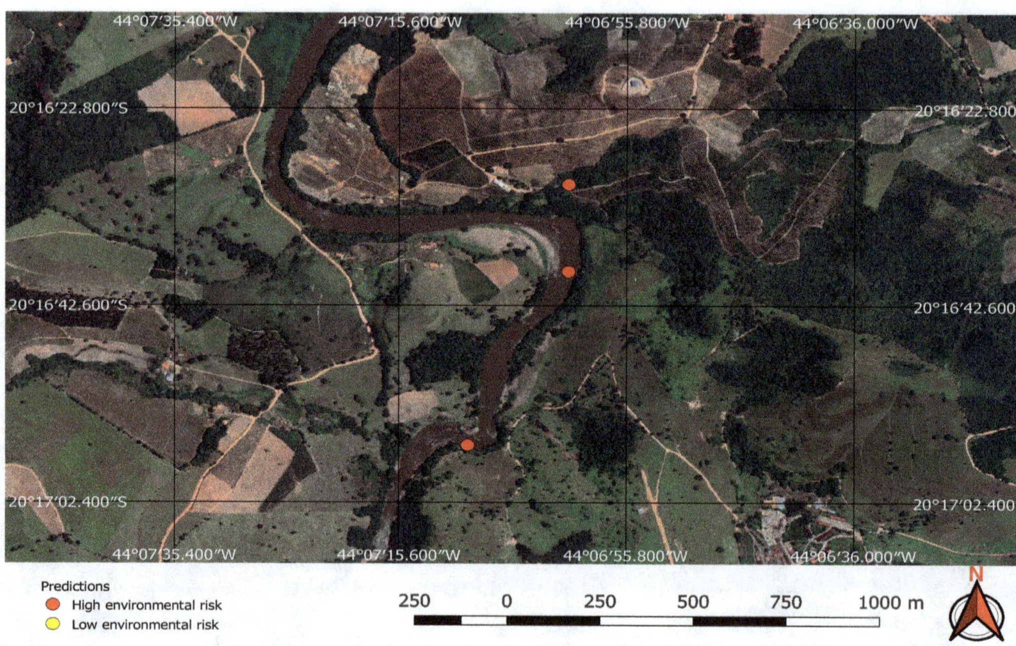

Figure 15. River bank incorrectly identified as a mine by the neural network.

Visual inspection of the positive predictions for the area in Minas Gerais confirmed the discovery of 154 point clusters consistent with mines and/or tailings dams. Sixty three of these clusters are newly discovered mines, not indicated on the official databases of the Brazilian government and not having an official mining concession. Visual inspection for the second area in the state of Para also confirmed a number of relevant positive predictions. Two hundred large unregistered gold mining areas were identified, many of them inside and around protected indigenous land, as can be seen in Figures 16–18 (Part of the discovered mines are in areas where a mining authorization was requested but the concession had not been granted, which makes them illegal mines.) Large iron-ore mines were also detected inside two different environment conservation areas.

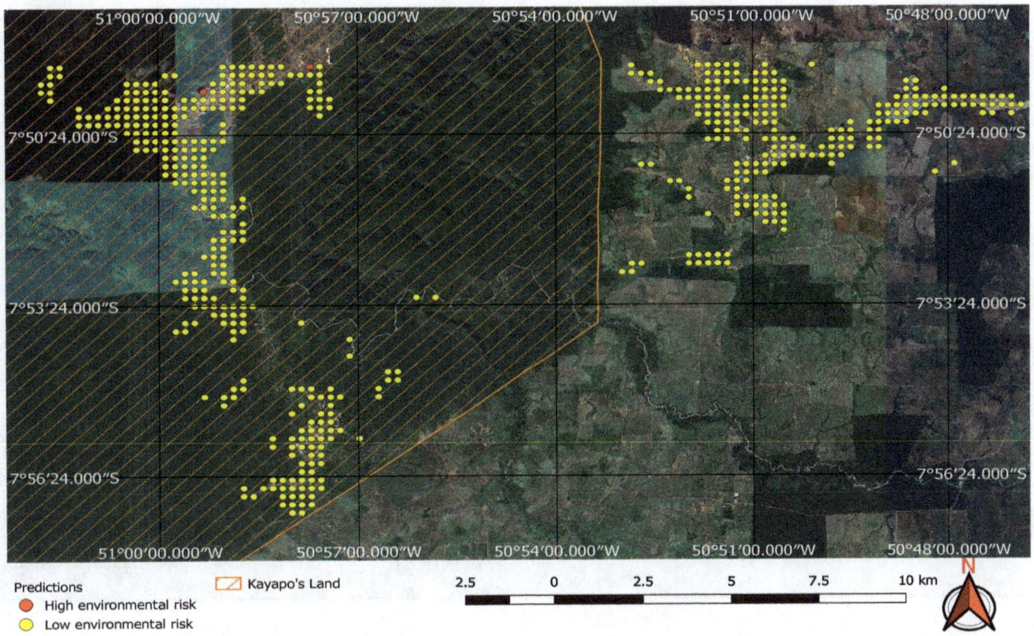

Figure 16. Mines identified on the Kayapo's protected land.

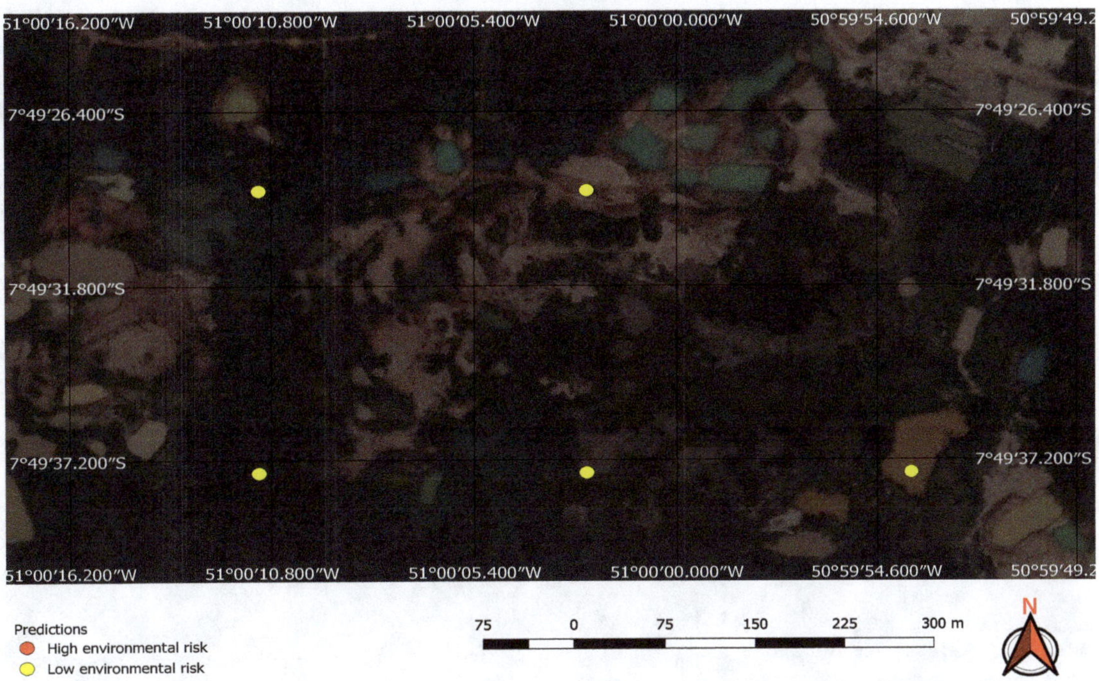

Figure 17. Detail of a discovered mining spot on the Kayapo's protected land.

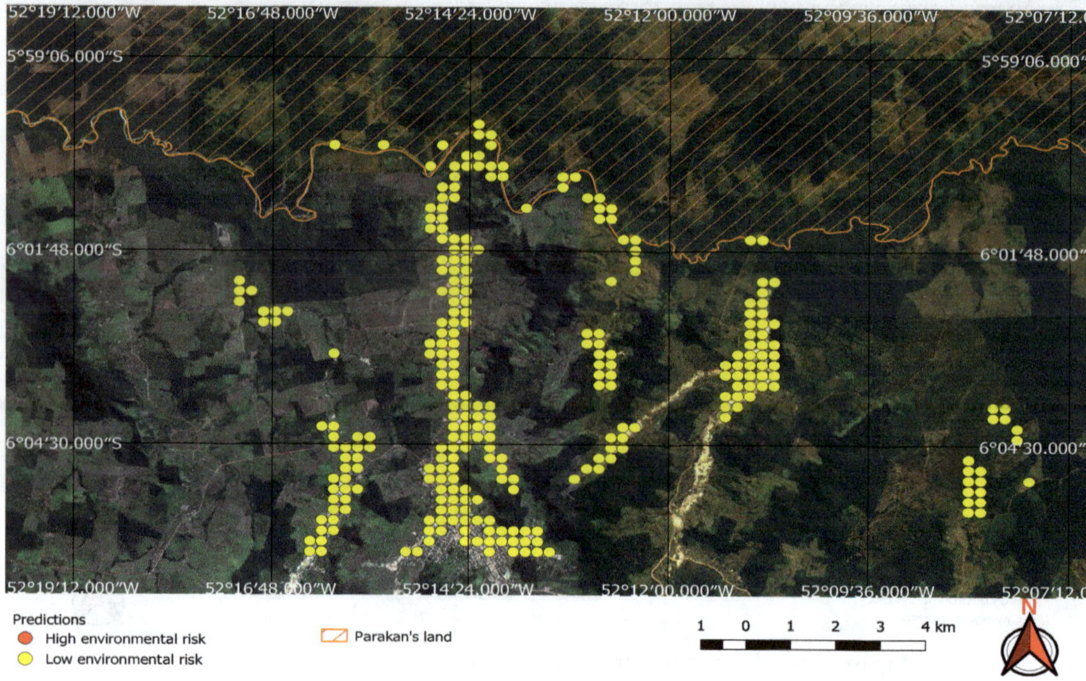

Figure 18. Mining identified by the neural network close to the Parakan's protected land.

5. Discussion

Using the mine and dams detection model to find potential mining sites, and then applying the environment risk model to the mine images enabled the classification of the environment as shown in, for example, Figures 19 and 20 , where two previously known mining sites were identified and correctly classified according to their environment impact. Figure 19 shows a high impact tailings dam in an iron mine and Figure 20 shows a low impact gravel quarry.

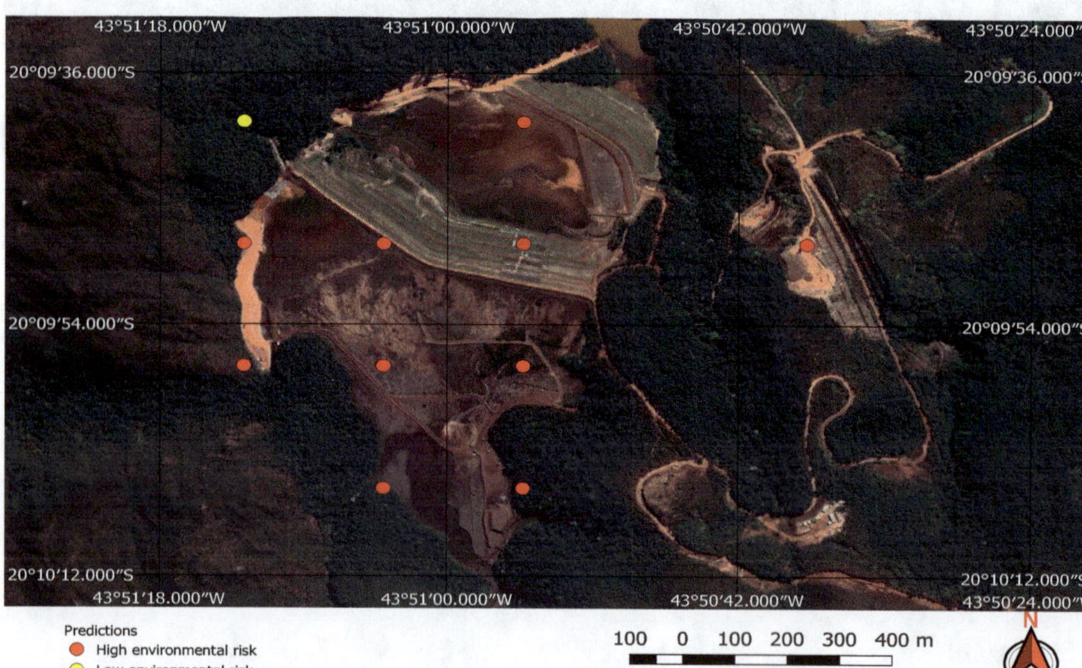

Figure 19. High environmental impact: iron mine tailings dam.

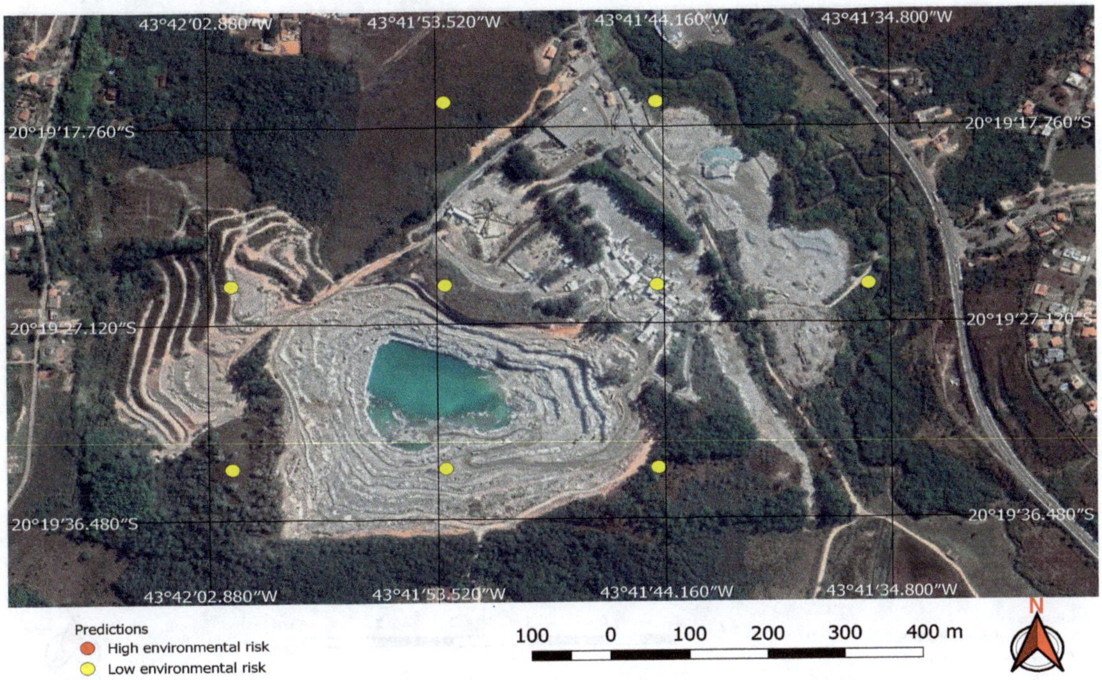

Figure 20. Low environmental risk gravel quarry correctly identified by the neural network.

We have demonstrated that mines can be detected over a wide area using free low resolution satellite imagery and automation. Although the majority of mines and dams are visible in images of good resolution, and can therefore be identified by a specialist without the aid of an algorithm, it is unlikely that this specialist will be able to do it in a cost effective way in very large areas. What is more, they will need high-resolution imagery, which is either paid for or usually out of date when it becomes freely available. Figures 21 and 22 show that some mines discovered at the Kayapo's reserve cannot be seen on Google Earth's outdated imagery.

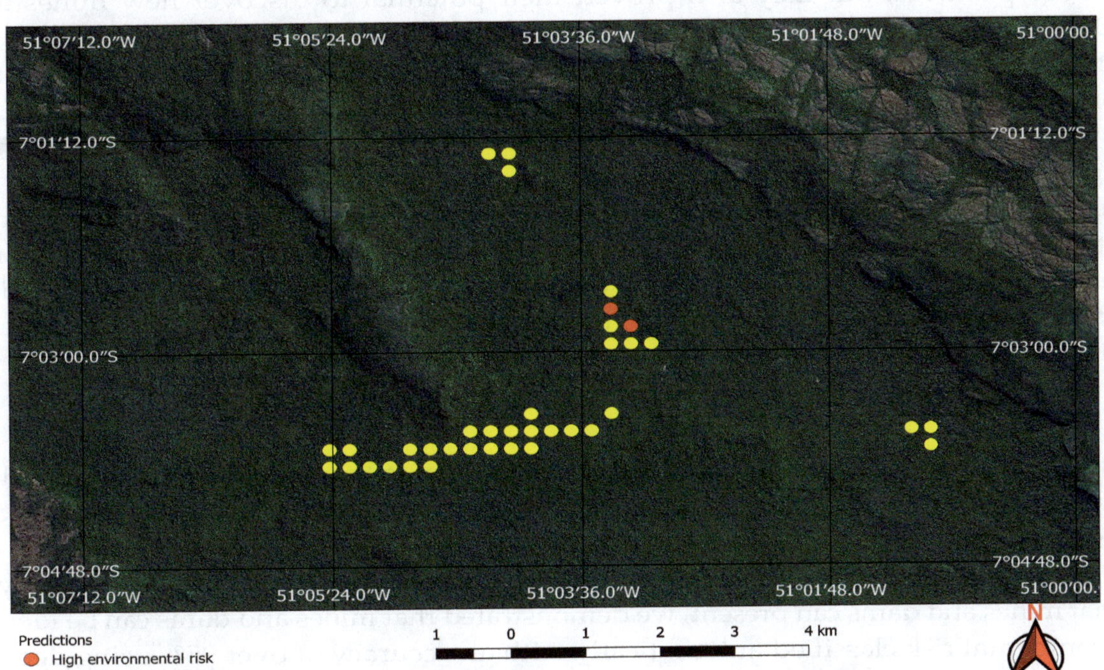

Figure 21. True positives on the Kayapo's protected land not visible on Google Earth.

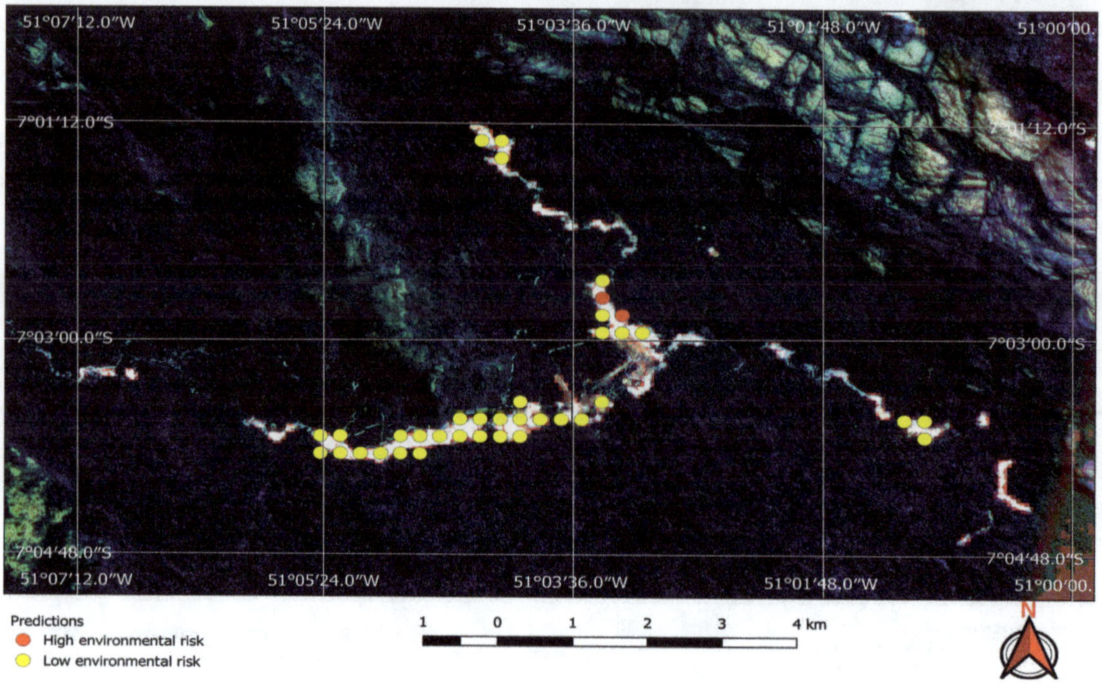

Figure 22. True positives on Kayapo's protected land visible on Sentinel 2 images.

The limited number of confirmed locations of mines and dams that we have also limits the number and diversity of examples we have for training the CNNs, which poses a risk of overfitting. Data augmentation strategies were used, but due to the enormous diversity of format, scale, and shades of colour that a mine or dam can have, the number of images used is still modest. Due to the difficulty in expanding this dataset, we opted for a hybrid strategy of validation of the predictive models. Several CNN architectures were explored and the testing and validation process was carried out both on images from the dataset itself but also in the large areas already described. These experiments in large areas effectively showed results below those obtained in the training and test bases, mainly with regard to false positives, but they also proved their potential to discover new mines and dams, confirmed by visual inspection.

The two-stage prediction strategy also strengthens the method. The first model, which predicts the location of mines and dams, generates inputs for the second model, which predicts the environmental impact of the ore explored there. This second prediction stage serves as a filter for false positives when no ore is identified at the indicated location.

We tested the use of pre-trained CNNs, but with results below those obtained by training them from scratch. This is mainly due to the fact that we used images with 12 spectral bands as input and not conventional RGB images. In addition, as already discussed, the visual patterns associated with mines and dams are very specific, not consistent with those found in the conventional image sets used to train the CNNs available for transfer learning.

6. Conclusions

Our work has found 263 tailings dams hitherto unknown to the Brazilian government, distributed across huge areas of the country; a task that would be very time consuming, prone to error, and costly if performed manually.

Despite the limited number of available training examples and the enormous diversity of visual patterns that mines and dams can present, we demonstrated that mines and dams can be identified and their environmental risk classified automatically with an accuracy of over 95% using deep learning, free moderate resolution satellite images, and free computing infrastructure in the cloud.

We developed two deep learning models: one to identify mines and dams and the other to classify their environmental risk potential. Mines and dams occupy a tiny fraction of the satellite imagery.

Thus, training was performed on small image patches, containing exemplar mines and dams, and also examples of 'background' imagery patches. To speed up the identification of dams in wide areas of interest, the trained CNN was expressed as an FCN in the usual way. The environmental risk classification CNN model operates on the mines and dams identified via the FCN. We demonstrated that a conventional CNN is perfectly adequate for this task.

We have integrated our deep learning models into a pipeline for processing multispectral satellite imagery over large geo-spatial areas and provide open-source Python scripts and image datasets, as well as models already trained and ready to be used on new imagery. Our models were trained with images from mines and tailings dams in Brazil, which certainly creates a bias given the local characteristics of the soil, vegetation, and the mining processes used there. The user can choose between applying the models as they are on their own data, retrain the CNNs from scratch using new imagery of mines and tailings dams, or improve the models using additional training with new images (i.e., transfer learning). Our code is publicly available at: https://github.com/remis/mining-discovery-with-deep-learning.

Author Contributions: Conceptualization, R.B.; Data curation, O.I.; Formal analysis, S.R.; Methodology, O.I. and S.R.; Project administration, S.R.; Software, R.B. and O.I.; Writing—original draft, R.B.; Writing—review and editing, S.R. All authors have read and agreed to the published version of the manuscript.

Abbreviations

The following abbreviations are used in this manuscript:

GEE	Google Earth Engine
FCN	Fully Convolutional Neural Network
CNN	Convolutional Neural Network
VPM	Visual Pattern Mining
API	Application Programming Interface
FAPDF	Fundação de Apoio à Pesquisa do Distrito Federal
EPSRC	Engineering and Physical Sciences Research Council
GPU	Graphics Processing Unit
ANM	Brazilian National Mining Agency
RGB	RGB Color Scheme—Red Green Blue

References

1. ICOLD. *Tailings Dams—Risk of Dangerous Occurrences, Lessons Learnt From Practical Experiences (Bulletin 121)*; Commission Internationale des Grands Barrages: Paris, France, 2001; Volume 155.
2. Cameron, P.D.; Stanley, M.C. *Oil, Gas, and Mining: A Sourcebook for Understanding the Extractive Industries*; The World Bank: Washington, DC, USA, 2017, [CrossRef]
3. Borges, A.; de São Paulo, O.E. *TCU Revela Sonegação em áreas de Mineração*; Estado de São Paulo: São Paulo, Brazil, 2014. Available online: https://economia.estadao.com.br/noticias/geral,tcu-revela-sonegacao-em-areas-de-mineracao,1543471 (accessed on 25 November 2020).
4. Reed, E.; Miranda, M.; WWF Macroeconomics for Sustatinable Development Program Office. Assessment of the Mining Sector and Infrastructure Development in the Congo Basin Region. Available online: http://awsassets.panda.org/downloads/congobasinmining.pdf (accessed on 25 November 2020).
5. Fellet, J.; Costa, C. *Imagens Mostram AvançO do Garimpo Ilegal na Amazônia em 2019*; BBC News Brasil: São Paulo, Brazil, 2019.
6. Guttag, J.V. *Introduction to Computation and Programming Using Python: With Application to Understanding Data*; MIT Press: Cambridge, MA, USA, 2016.

7. Abadi, M.; Barham, P.; Chen, J.; Chen, Z.; Davis, A.; Dean, J.; Devin, M.; Ghemawat, S.; Irving, G.; Isard, M.; et al. TensorFlow: A system for large-scale machine learning. In Proceedings of the 12th USENIX Symposium on Operating Systems Design and Implementation (OSDI 16), Savannah, GA, USA, 2–4 November 2016; pp. 265–283.
8. Atos do Poder Executivo. DECRETO Nº 9.406, DE 12 DE JUNHO DE 2018. DIÁRIO OFICIAL DA UNIÃO Publicado em: 13/06/2018 | Edição: 112 | Seção: 1 | Página: 1. 2018. Available online: http://www.planalto.gov.br/ccivil_03/_ato2015-2018/2018/decreto/D9406.htm (accessed on 25 November 2020).
9. Ministério Público Federal-Procuradoria da República no Estado de Minas Gerais. Ação Civil Pública com Pedido de Tutela Provisória de Urgência em Face da UNIÃO, Pessoa Jurídica de Direito púBlico Interno e da Agência Nacional de Mineração. 2019. Available online: http://www.mpf.mp.br/mg/sala-de-imprensa/docs/acp_anm_uniao-1 (accessed on 25 November 2020).
10. Sales, C.D. Licenciamento Ambiental da Atividade de Mineração em Unidades de Conservação no Amazonas: IncidêNcia, Suporte JurÍDico Administrativo e Aperfeiçoamentos. Master's Thesis, INPA, Manaus, Brazil, 2018.
11. WWF-Brasil. Mineração na Amazônia Legal e Áreas Protegidas-Situação dos Direitos Minerários e Sobreposições. 2018. Available online: https://www.wwf.org.br/?67842%2FMineracao-em-areas-protegidas-titulos-sao-risco-potencial-diz-estudo-do-WWF-Brasil (accessed on 25 November 2020).
12. Legg, C.A. Applications of remote sensing to environmental aspects of surface mining operations in the United Kingdom. In *Remote Sensing: An Operational Technology for the Mining and Petroleum Industries*; Springer: Dordrecht, The Netherlands, 1990; pp. 159–164. [CrossRef]
13. Yu, L.; Xu, Y.; Xue, Y.; Li, X.; Cheng, Y.; Liu, X.; Porwal, A.; Holden, E.J.; Yang, J.; Gong, P. Monitoring surface mining belts using multiple remote sensing datasets: A global perspective. *Ore Geol. Rev.* **2018**, *101*, 675–687. [CrossRef]
14. Redondo-Vega, J.; Gómez-Villar, A.; Santos-González, J.; González-Gutiérrez, R.; Álvarez-Martínez, J. Changes in land use due to mining in the north-western mountains of Spain during the previous 50 years. *CATENA* **2017**, *149*, 844–856. [CrossRef]
15. Petropoulos, G.P.; Partsinevelos, P.; Mitraka, Z. Change detection of surface mining activity and reclamation based on a machine learning approach of multi-temporal Landsat TM imagery. *Geocarto Int.* **2013**, *28*, 323–342. [CrossRef]
16. Connette, K.L.; Connette, G.; Bernd, A.; Phyo, P.; Aung, K.; Tun, Y.; Thein, Z.; Horning, N.; Leimgruber, P.; Songer, M. Assessment of Mining Extent and Expansion in Myanmar Based on Freely-Available Satellite Imagery. *Remote Sens.* **2016**, *8*, 912. [CrossRef]
17. Soulard, C.E.; Acevedo, W.; Stehman, S.V.; Parker, O.P. Mapping Extent and Change in Surface Mines Within the United States for 2001 to 2006. *Land Degrad. Dev.* **2015**, *27*, 248–257. [CrossRef]
18. Prasad, S. *Remotely Sensed Data Characterization, Classification, and Accuracies*; Chapter Processing Remote-Sensing Data in Cloud Computing Environments; CRC Press: Boca Raton, FL, USA, 2015. [CrossRef]
19. Penatti, O.A.; Nogueira, K.; Dos Santos, J.A. Do deep features generalize from everyday objects to remote sensing and aerial scenes domains? In Proceedings of the IEEE Conference on Computer Vision and Pattern Recognition Workshops, Boston, MA, USA, 7–12 June 2015; pp. 44–51.
20. Volpi, M.; Tuia, D. Dense semantic labeling of subdecimeter resolution images with convolutional neural networks. *IEEE Trans. Geosci. Remote. Sens.* **2016**, *55*, 881–893. [CrossRef]
21. Nogueira, K.; dos Santos, J.A.; Menini, N.; Silva, T.S.; Morellato, L.P.C.; Torres, R.D.S. Spatio-Temporal Vegetation Pixel Classification By Using Convolutional Networks. *IEEE Geosci. Remote. Sens. Lett.* **2019**, *16*, 1665–1669. [CrossRef]
22. Xie, M.; Jean, N.; Burke, M.; Lobell, D.; Ermon, S. Transfer Learning from Deep Features for Remote Sensing and Poverty Mapping. *arXiv* **2015**, arXiv:cs.CV/1510.00098.
23. Mazzia, V.; Comba, L.; Khaliq, A.; Chiaberge, M.; Gay, P. UAV and Machine Learning Based Refinement of a Satellite-Driven Vegetation Index for Precision Agriculture. *Sensors* **2020**, *20*, 2530. [CrossRef] [PubMed]
24. Saraiva, M.; Protas, E.; Salgado, M.; Souza, C. Automatic Mapping of Center Pivot Irrigation Systems from Satellite Images Using Deep Learning. *Remote Sens.* **2020**, *12*, 558. [CrossRef]
25. Gu, Y.; Wang, Y.; Li, Y. A Survey on Deep Learning-Driven Remote Sensing Image Scene Understanding: Scene Classification, Scene Retrieval and Scene-Guided Object Detection. *Appl. Sci.* **2019**, *9*, 2110. [CrossRef]

26. Hinton, G.E. Reducing the Dimensionality of Data with Neural Networks. *Science* **2006**, *313*, 504–507. [CrossRef] [PubMed]
27. Hu, W.; Huang, Y.; Wei, L.; Zhang, F.; Li, H. Deep Convolutional Neural Networks for Hyperspectral Image Classification. *J. Sensors* **2015**, *2015*, 258619. [CrossRef]
28. Maggiori, E.; Tarabalka, Y.; Charpiat, G.; Alliez, P. Convolutional Neural Networks for Large-Scale Remote-Sensing Image Classification. *IEEE Trans. Geosci. Remote. Sens.* **2017**, *55*, 645–657. [CrossRef]
29. Persello, C.; Stein, A. Deep Fully Convolutional Networks for the Detection of Informal Settlements in VHR Images. *IEEE Geosci. Remote. Sens. Lett.* **2017**, *14*, 2325–2329. [CrossRef]
30. Ferreira, E.; Brito, M.; Alvim, M.S.; Balaniuk, R.; dos Santos, J. BrazilDAM: A Benchmark dataset for Tailings Dam Detection. In Proceedings of the Latin American GRSS ISPRS Remote Sensing Conference, Santigo, Chile, 22–26 March 2020.
31. Gorelick, N.; Hancher, M.; Dixon, M.; Ilyushchenko, S.; Thau, D.; Moore, R. Google Earth Engine: Planetary-scale geospatial analysis for everyone. *Remote. Sens. Environ.* **2017**. [CrossRef]
32. Deng, J.; Dong, W.; Socher, R.; Li, L.J.; Li, K.; Fei-Fei, L. ImageNet: A Large-Scale Hierarchical Image Database. In Proceedings of the 2009 IEEE Conference on Computer Vision and Pattern Recognition, Miami, FL, USA, 20–25 June 2009.
33. Krizhevsky, A.; Sutskever, I.; Hinton, G.E. ImageNet Classification with Deep Convolutional Neural Networks. In *Advances in Neural Information Processing Systems 25*; Pereira, F., Burges, C.J.C., Bottou, L., Weinberger, K.Q., Eds.; Curran Associates, Inc.: Red Hook, NY, USA, 2012; pp. 1097–1105.
34. Goodfellow, I.; Bengio, Y.; Courville, A. *Deep Learning*; MIT Press: Cambridge, MA, USA, 2016.
35. Ronneberger, O.; Fischer, P.; Brox, T. U-Net: Convolutional Networks for Biomedical Image Segmentation. In *Medical Image Computing and Computer-Assisted Intervention—MICCAI 2015*; Navab, N., Hornegger, J., Wells, W.M., Frangi, A.F., Eds.; Springer International Publishing: Cham, Switzerland, 2015; pp. 234–241.
36. Girshick, R.; Donahue, J.; Darrell, T.; Malik, J. Rich Feature Hierarchies for Accurate Object Detection and Semantic Segmentation. In Proceedings of the 2014 IEEE Conference on Computer Vision and Pattern Recognition, Columbus, OH, USA, 24–27 June 2014; pp. 580–587.
37. Li, H.; Ellis, J.G.; Zhang, L.; Chang, S. PatternNet: Visual Pattern Mining with Deep Neural Network. In Proceedings of the 2018 ACM on International Conference on Multimedia Retrieval, Yokohama, Japan, 11–14 June 2018; pp. 291–299. [CrossRef]
38. Carneiro, T.; Nobrega, R.V.M.D.; Nepomuceno, T.; Bian, G.B.; Albuquerque, V.H.C.D.; Filho, P.P.R. Performance Analysis of Google Colaboratory as a Tool for Accelerating Deep Learning Applications. *IEEE Access* **2018**, *6*, 61677–61685. [CrossRef]
39. Kingma, D.; Ba, J. Adam: A Method for Stochastic Optimization. International Conference on Learning Representations. *arXiv* **2014**, arXiv:1412.6980.
40. Wang, P.; Kvalheim, A. *Environmental Impact Assessment (EIA) of Development Aid Projects. Initial Environmental Assessment*; Technical Report; Norwegian Agency for Development Cooperation NORAD: Oslo, Norway, 1994.
41. Cohen, J. A Coefficient of Agreement for Nominal Scales. *Educ. Psychol. Meas.* **1960**, *20*, 37–46, [CrossRef]

3

Spatial-Temporal Features Based Sensor Network Partition in Dam Safety Monitoring System

Hao Chen [1,2], Yingchi Mao [3,*], Longbao Wang [3] and Hai Qi [3]

[1] College of Water Conservancy and Hydropower Engineering, Hohai University, Nanjing 211100, China; chenhao.hnlcj@foxmail.com
[2] Centre of Research and Development, Huaneng Lancang River Hydropower Co., Ltd., Kunming 650214, China
[3] School of Computer and Information, Hohai University, Nanjing 211100, China; wlb@hhu.edu.cn (L.W.); qihai@hhu.edu.cn (H.Q.)
* Correspondence: yingchimao@hhu.edu.cn

Abstract: Many various types of sensors have been installed to monitor the deformation and stress in the dam structure. It is difficult to directly evaluate the operation status of the dam structure based on the massive monitoring data. The sensor network is divided into multiple regions according to the design specifications, simulation data, and engineering experiences. The local results from sub-regions are integrated to achieve overall evaluation. However, it ignores the spatial distribution of sensors and the variation of time series, which cannot meet the real-time evaluation for the dam safety monitoring. If the network partitions can provide the preliminary foundation for analyzing the dynamic change laws of the dam's working conditions in a real-way, we should consider the similarity of structure and stresses in the local region of the dam and the correlation among the monitoring data. A time-series denoising autoencoder (TSDA) is proposed to represent the spatial and temporal features of the nodes by compressing high-dimensional monitoring data. Then, a network partitioning algorithm (NPA) based on spatial-temporal features based on the TSDA is presented. The NPA ensures that the partition results can support the analysis of the physical change laws by introducing the auxiliary objective variable to optimize the network partition objective function. Experimental results on the public datasets and a real dataset from an arch dam demonstrate that the proposed network partition algorithm NPA can achieve better partition performance than TSDA+K-Means and TSDA+GMM. The NPA can improve the silhouette coefficient by 45.1% and 58.4% higher than the TSDA+K-Means and TSDA+GMM, respectively. The NPA can increase the Calinski-Harabaz Index by 30.8% and 61.6%, respectively.

Keywords: network partition; spatial-temporal feature; auto-encoder; dam safety monitoring; sensor networks

---

1. Introduction

With the rapid development of Internet of Things (IoT) technologies, various types of IoT devices (e.g., sensor nodes) can be deployed in large dam engineering to measure the different physical quantities and sense their changes in various regions of the structure, such as deformation, stress, pressure, etc. [1–4]. In the dam safety monitoring systems, the massive monitoring data are generated from the deployed sensor nodes. However, it is difficult to process a large amount of data with the traditional mechanical model, so it is impossible to directly evaluate the safe operation states of the dam. In real applications, the monitoring network is usually divided into multiple regions according to the design specifications, simulation data, and engineering experience. The local evaluation results of each region are integrated to achieve the global overall evaluation [5]. Taking an arch dam as an example, based on the dam sections and elevation, the monitoring network can be divided into multiple grids adopting horizontal and vertical partitions, respectively. Figure 1a shows the gridded partition results of the dam safety monitoring network based on the mechanical models. Different color solid dots correspond to different types of sensor nodes, and the horizontal and vertical dotted lines indicate the dam sections and elevation, respectively. Due to instrument failure, environmental changes, and monitoring requirements, the spatial distribution of sensor nodes vary [6]. Meanwhile, during the dam operation period, the measured data and their temporal features dynamically change with time.

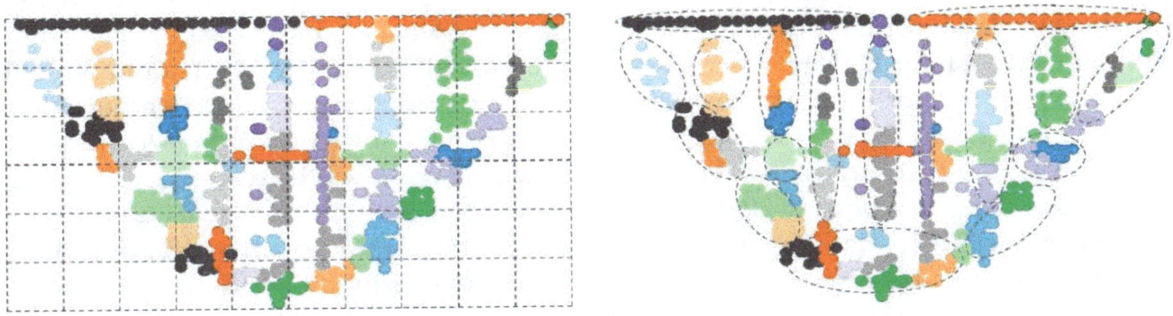

(a) (b)

Figure 1. Network partition results for dam safety monitoring system (**a**) gridded partition results with a mechanical model; (**b**) partition results with the network partitioning algorithm (NPA) based on spatial-temporal features.

Monitoring data analysis adopts the idea of "divide and conquer", which is an important algorithm based on multi-branched recursion. A divide and conquer algorithm works by recursively breaking down a problem into two or more sub-problems of the same or related types, until these sub-problems become simple enough to be solved directly. The solutions to the sub-problems are then combined to give a solution to the original problem. To evaluate the whole structure of the dam based on the monitoring data, the monitoring network of the dam can be divided into multiple regions based on the spatial-temporal features of the monitoring data from which can be obtained the local evaluation result in the regions. The local evaluation results are combined to obtain comprehensive evaluation results. Therefore, how to divide the monitoring network becomes the important problem. In this paper, the network partition problem based on the spatial-temporal features of the monitoring data is equal to the clustering problem. The clustering problem was extensively studied in various applications, e.g., image processing, feature selection, and data stream clustering. In the image features encoding, the whole image can be divided into individual blocks with the average region division method [7]. For feature selection, an unsupervised feature selection based on subspace clustering represents samples to the subspace [8], and we can obtain the data distribution through the subspace clustering. To improve the accuracy of monitoring data processing and analysis in the sensor networks, a hierarchical aggregation clustering protocol is proposed [9]. The detected events are classified in the local sensor nodes and are transmitted to the sink node. The sink node constructs the aggregation clustering tree to obtain the event classification results. In the wireless sensor networks, the network topology is divided into several regions based on the nodes' attributes and correlations in order to reduce energy consumption and balance the workload [10]. Considering the redundant deployment of nodes, the backbone nodes in the differential pattern tree are used to execute the task of events classification, which can reduce the computation cost and communication overheads [11]. Thus, the cluster head nodes can aggregate the monitoring data and make the decision quickly. In the applications of data stream clustering, considering the data distribution in each dimension, the subspace clustering method of the data stream based on the region partition was proposed to divide the data dimension based on the unit cell. Furthermore, it adopted a bottom-up strategy to gather the final clustering results [12]. Moreover, in order to decrease the congestion in the large-scale networks, the networks were partitioned into regional sub-networks. Some clustering methods for network partition were discussed [13]. However, the mentioned clustering methods did not consider spatial and temporal features.

At present, the arch beam load sharing methods are adopted to deploy the sensor nodes for the arch dam safety monitoring. The sensor network can be partitioned into the spatial arch beam monitoring grids based on the beam sections and arch monitoring base. The evaluation of each monitoring part is limited to the arch beam grid from the original section and base. For a monitoring part under the influence of internal and external loads of the dam, there is a correlation between the changing trends of different monitoring items, such as temperature, stress, deformation, osmolality, as well as the correlation of the operation states among the various parts of the arch dam after the completion of arch sealing. Unfortunately, we lack quantitative and feasible evaluation methods of correlation analysis. In addition, it is impossible to accurately simulate and predict the load and structural changes of the arch dam through the structural mechanics calculations during the design stage. Therefore, the arch beam grids are difficult to effectively reflect the dynamic change law of the physical quantities of arch dams, under the conditions of inter-annual cycle changes and long-term accumulation of various loads. The monitoring data of the arch dam varies dynamically and the corresponding data features also change with time. The monitoring network partition based on the arch beam load distribution depends on the structural mechanics calculations in the design stage. It is quite difficult to establish the relationship of the stress load between the adjacent monitoring parts in the arch dam in a real-time manner. Therefore, the existing network partition methods cannot timely reflect the dynamic changes of the dam structure, which results in the low-efficient evaluation of dam operation states.

We should consider the similarity of structure and stresses in the local region of the dam and the correlation among the monitoring data. A time-series denoising autoencoder (TSDA) is proposed to compress the high-dimension monitoring data in order to extract their spatial-temporal features. Considering the correlation among the spatial positions of the different sensor nodes and the change law of the time-series, the auxiliary distribution variables are introduced to optimize the objective function of deep clustering. Based on the spatial-temporal features of monitoring data, a network partition algorithm (NPA) is presented to cluster all sensor nodes and partition the sensor network into the different regions. Figure 1b shows the partition results based on the spatial-temporal feature of sensor nodes. The solid dots surrounded by different dotted boxes correspond to different regions clustered by the NPA. Compared with the traditional monitoring network partition based on the arch beam load sharing method, the partition results with the NPA method can reflect the relationship between the stress and load between the different monitoring parts of the arch dam based on the spatial-temporal data features. Furthermore, the divided areas are bilaterally symmetrical in the horizontal direction, and the solid dots with the same color are clustered into the same area. That is to say that the partition results reflect the high correlations among the sensor nodes within the same region based on the spatial-temporal features.

The paper is organized as follows. We introduce the related work in Section 2. The preliminary and problem statement are presented in Section 3. A time-series denoising autoencoder is proposed to compress the high-dimensional data for the spatial-temporal feature representation. In Section 5, we present our proposed NPA approach for sensor network partitioning based on spatial-temporal features. Finally, we evaluate the performance of the proposed NPA with four public datasets and large-scale real dam monitoring data in Section 6, and conclude the work in Section 7.

2. Related Work

Traditional clustering methods include partition-based clustering, fuzzy clustering, hierarchical clustering, and density-based clustering. The most widely used algorithms are K-Means clustering, Gaussian mixture model (GMM) [14–16], and fuzzy cluster analysis. However, traditional clustering methods still have a long convergence time and low clustering accuracy when processing high-dimensional data. Furthermore, an autoencoder (AE) is adopted to compress high-dimensional data (astronomy, meteorology, long time series, etc.) and represent their features [17–20]. Denoising autoencoders (DAEs) have been used to learn new representations for a wide range of machine learning tasks [21,22]. A deep clustering network (DCN) can learn the complex probability distribution functions to process high-dimensional data such as images and texts, but the complex network structures and parameter adjustments result in extra computation overheads [23].

Besides the mentioned works on the clustering with region division methods, there are also several studies regarding the combination of clustering with feature embedding learning. A kernelized K-means algorithm (DisKmeans) was embedded to a low dimensional subspace via a linear discriminant analysis (LDA), which was jointly learned with K-means clustering assignments [24]. An unsupervised feature selection method was proposed to combine the clustering and feature embedding tasks to achieve better performance [25]. However, these methods suffer from the shallow and linear embedding functions.

A deep structure, named TAGnet was introduced in [26], where two layers of sparse coding followed by a clustering algorithm are trained with an alternating learning approach. Similar work was presented that formulates a joint optimization framework for discriminative clustering and feature extraction using sparse coding [27,28]. However, the inference complexity of sparse coding forces the model to reduce the dimension of input data with Principal Components Analysis (PCA) or use an approximate solution. Moreover, hand-crafted features and dimension reduction techniques degrade the clustering performance by neglecting the distribution of input data.

To employ deep learning in graph clustering, a simple method was presented to learn a non-linear embedding of the affinity graph using a stacked autoencoder and obtain the clusters in the embedding subspace via K-means [29]. A semi non-negative matrix factorization was extended to stacked deep semi Non-negative Matrix Factorization (NMF) in order to capture the abstract information in the top layer [30]. A recurrent-based framework, named JULE, was introduced to cluster the data. In JULE, data is represented via a convolutional neural network and embedded data is iteratively clustered using an agglomerative clustering algorithm [31]. JULE achieved good results using the joint learning approach, however, it requires tuning of massive hyper-parameters, which is not practical in real-world clustering tasks.

3. Problem Statement

The monitoring data in the large dams are high-dimensional dynamic spatial-temporal data. Considering the correlation among the spatial positions of the different sensor nodes and their change rules of the time series, the network partition algorithm should cluster all sensor nodes and partition the sensor network into the different regions based on the spatial-temporal features of nodes.

Assume that there are n sensor nodes installed on a dam for the dam safety monitoring. The set of sensor nodes is $\mathbf{X} = \{x_i | i = 1, \ldots, n\}$, where each node $x_i \in \mathbb{R}^{d_x}$. $\mathbb{R}^{d_x}$ represents the original features of the observed data, including the spatial features (spatial location, the type of sensor node) and temporal features (time series of data). d_x is the dimension of the spatial-temporal features. Figure 2 illustrates the procedure of the network partition. In the dam safety monitoring networks, the spatial features are usually static, while the temporal features of the time series are high-dimensional and constantly changing. Using the embedding function with sparse coding, $\varphi : \mathbf{X} \rightarrow \mathbf{Z}$, the original samples are mapped into the latent feature space. In other words, the latent feature space is $\mathbf{Z} = \{z_i | i = 1, \ldots, n\}$, where each latent feature element $z_i \in \mathbb{R}^{d_z}$ has a much lower dimension compared to the original input data (i.e., $d_z \ll d_x$). Therefore, the function with spare coding $\varphi : \mathbf{X} \rightarrow \mathbf{Z}$ can support the dimensions reduction and features representation. Given the embedding features, we can adopt a multinomial logistic regression function (SoftMax) $f_\Theta : \mathbf{Z} \to \mathbf{P}$ to compute the probability p_{ij} of the sensor node x_i belonging to the jth cluster, where $\Theta = \{\theta_i | i = 1, \ldots, m\}$. The probability p_{ij} can be computed as follows:

$$p_{ij} = f(z_i, \Theta) = \frac{e^{\theta_j^T z_i}}{\sum_{j=1}^{m} e^{\theta_j^T z_i}} \tag{1}$$

where $\Theta = \{\theta_1, \ldots, \theta_m\} \in \mathbb{R}^{d_z \times m}$ are the SoftMax function parameters. The SoftMax function is adopted to compute the probability p_{ij} and the stochastic gradient method is used to optimize the parameters $\Theta = \{\theta_i | i = 1, \ldots, m\}$.

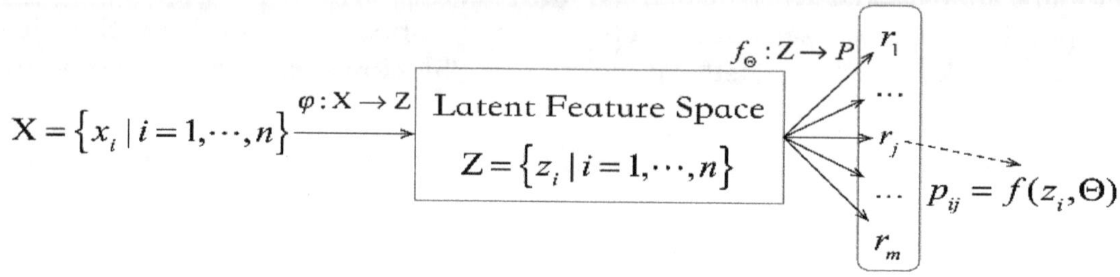

Figure 2. The procedure of network partitioning.

Therefore, the sensor network partition problem is equivalent to the sensor nodes clustering problem. In other words, n sensor nodes are clustered and partitioned into m regions, and we can have a set of regions $R = \{r_j | j = 1, \ldots, m\}$. The probability p_{ij} of sensor node x_i belonging to the jth region can be computed in Equation (1). When, $p_{ij} = \max\{p_{ik} | k = 1, \ldots, m\}$, the sensor node x_i is partitioned into the region r_j.

In this paper, we proposed a monitoring network partition method based on the spatial-temporal features of the monitoring data. We can obtain the clustering analysis results of sensor nodes with the proposed NPA algorithm. Under the influence of the internal and external loads of the dam, our approach can provide a feasible solution for correlation analysis of the different monitoring items (e.g., temperature, stress, deformation, and seepage pressure) in the same monitoring parts. In addition, our method can support the correlation analysis of the physical effects of the adjacent monitoring parts of the dam. This paper focuses on the correlation analysis of the monitoring data themselves. Our approach provides a preliminary foundation for the dam safety evaluation, rather than directly evaluate the structural safety of the dam.

4. Spatial-Temporal Feature Representation

4.1. Temporal Feature Representation

In general, the time-series data of the sensor nodes is a variable one-dimensional series with periodicity. In order to represent the change law of the temporal features of the sensor nodes, we select the monitoring data from more than one observation period (e.g., one year) as the sample. The monitoring data is not really always periodic. For the newly constructed dam, the monitoring data includes the time effect. After 5–10 years of stable operation, the time factor has little effect. In this paper, the arch dam has been operated for 10 years and the reservoir has multi-year regulation capacity. The monitoring data exhibits periodic change. Unlike the low-dimensional and discrete spatial data, the time-series of the sensor nodes are high-dimensional and continuously changing data. Due to the hardware, network conditions, the limitation in the memory, and computational capacity of the sensor nodes, the monitoring data of sensors are often misreported or lost. Therefore, feature compression and noise reduction of the spatial-temporal data from the sensor nodes are required.

To extract and represent the features of high-dimensional and noisy time-series data, a time-series denoising autoencoder (TSDA) is proposed in this paper, as shown in Figure 3. To enhance the anti-noise ability of the TSDA, random noise data is added to the sample data set during the training process. In the encoding phase, the convolutional layer (Conv2D) and the maximum pooling layer (MaxPooling2D) are applied to the compression representation of the time-series data features. In the decoding phase, the convolutional layer and the up-sampling layer (UpSampling2D) as opposed to the encoding phase, are used to reconstruct the compression representation into the original input. Due to the same input and output of the proposed TSDA autoencoder, its objective function is to minimize the reconstruction errors and optimize the encoder and decoder.

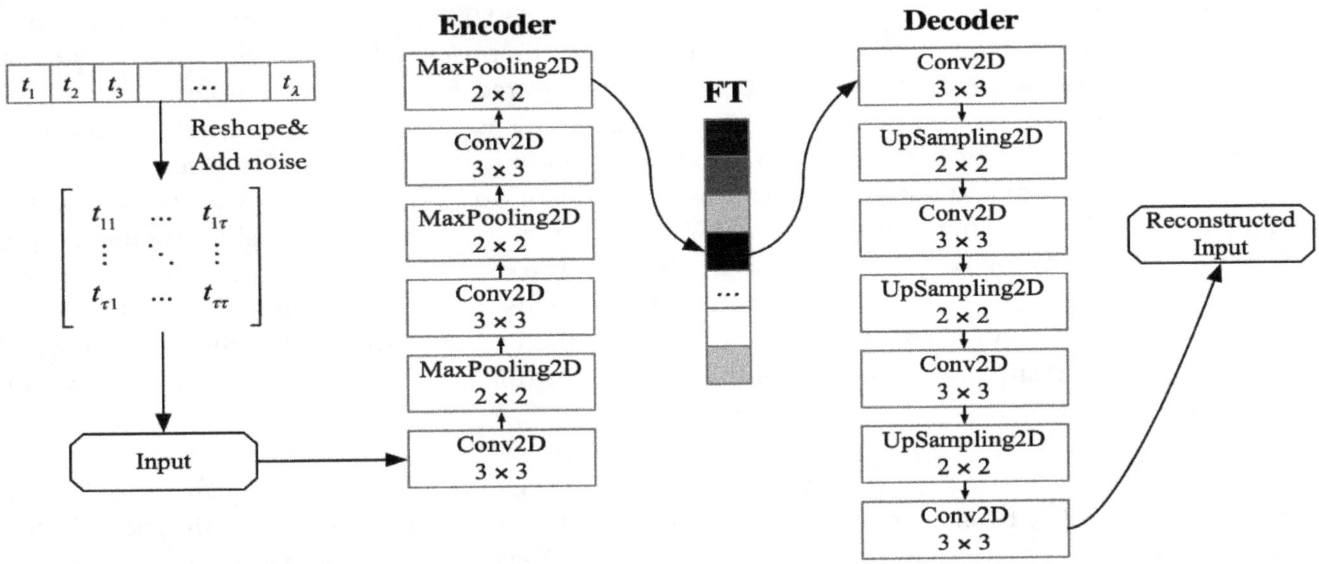

Figure 3. The structure of time-series denoising autoencoder (TSDA).

Assume the time series data of one sensor node x_i with the length λ is $T_{x_i} = [t_1, t_2, \ldots, t_\lambda]$, and its feature of time-series (FT) can be denoted as $FT_{x_i} = [ft_1, ft_2, \ldots, ft_{t_x}]$, where t_x is the number of dimensions and $t_x \ll \lambda$. The detailed procedure of TSDA is as follows.

(1) Input. To facilitate the convolution, pooling, and up-sampling operations, the incoming time-series $T_{x_i} = [t_1, t_2, \ldots, t_\lambda]$ is transformed into a matrix T'_{x_i} through the reshape operation, and then the matrix T'_{x_i} is mixed Gaussian random noise to generate the input of the TSDA.
(2) Encoding. Multiple convolutional layers and maximum pooling layers are stacked alternately to compress the input data and obtain the feature representation FT.
(3) Decoding. Multiple convolutional layers and up-sampling layers are stacked alternately to restore the data feature representation FT into the reconstructed input.
(4) Objective Function. The error of the original input and the reconstructed input is taken as a loss function. The objective is to minimize the error.

After the TSDA training is completed, the weights are saved. The feature representation FT from the encoder is the temporal feature compression representation.

4.2. Spatial Feature Representation

The spatial features of one sensor node include the node's coordinates, the importance of the node, and the instrument type, which can be processed with normalization. Suppose the spatial feature of one sensor node x_i can be denoted as $FS_{x_i} = [fs_1, fs_2, \ldots, fs_{s_x}]$, where s_x is the number of dimensions.

The spatial position of the sensor node has three attributes: longitude (e), latitude (n), and height (h). In the real measurement network systems, the value ranges of different attributes are large and the dimensional units are different. It is difficult to quantitatively describe the relative position relationship amongst the senor nodes. To eliminate the side effects of different dimensional units and the large range of values, uniform dimensions, and nonlinear normalization are adopted to calculate the relative spatial position of the sensor node in order to accelerate the convergence of training. In the coordinate system, all dimensional units are uniform in meters. To avoid negative values of attributes, the symbolic function $y = sgn(x)\frac{\ln(|x|)}{\ln(\max|x|)}$ is used for data transformation and normalization.

4.3. Examples

In this paper, the real dataset from the highest arch dam in the world is from 1 January 2017 to 31 December 2017 from 964 sensor nodes, which have 350,000 data items recording the sensors' types, the spatial coordination positions, time slots, and the observed data. The observed physical quantities include stress, pressure, deformation, opening degree, seepage, and temperature, etc. Then, the normalized time-series data of one sensor can be encoded as a 1×15 feature representation vector with the TSDA method. The similarity between the temporal feature vectors of two sensor nodes can be measured with cosine similarity. For example, the sensor node P04616 deployed in the middle of the dam is selected as a reference point, then we can analyze their similarities considering the temporal and spatial features of the sensor nodes. Note: this paper focuses on the clustering analysis of dam monitoring data, so there is no introduction to the structure of a specific dam. In addition, the so-called symmetry refers to the symmetry of the layout of the installed sensors, rather than the symmetry of the dam structure. The spatial arrangement of the sensors for this arch dam is symmetric with the arch crown beam as the center.

We will exhibit three different spatial correlations of sensors, as shown in Figures 4–6. The arch dam is a thin shell structure. A small number of sensors are installed in the Y-axis direction. In order to simplify the representation, the sensors installed in the X-axis and Z-axis directions are selected as the analysis objects.

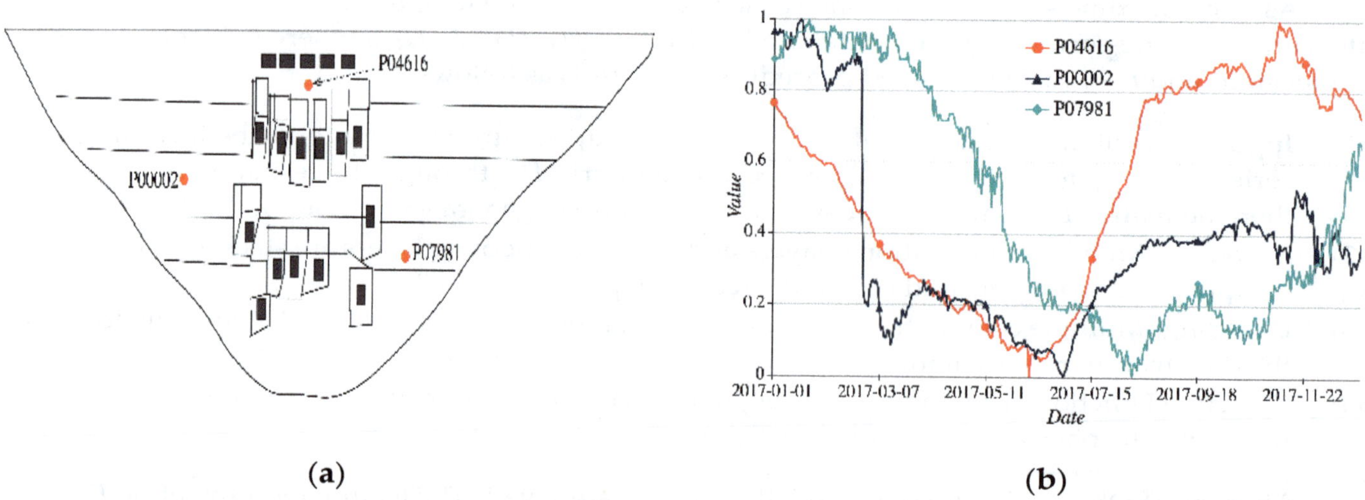

Figure 4. Spatial-temporal feature of nodes P04616, P07981, and P00002 (**a**) spatial distribution of three nodes; (**b**) sequential process lines of three nodes.

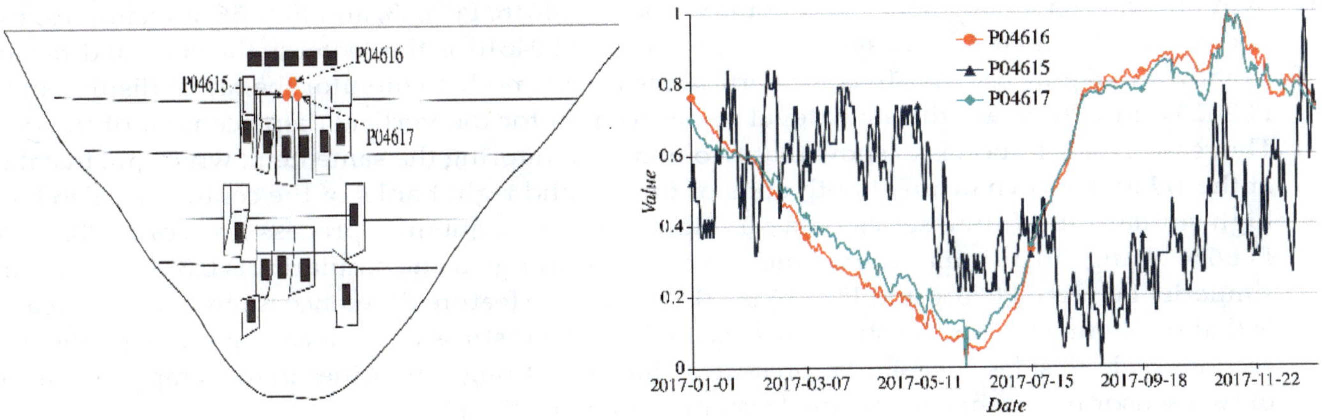

(a) (b)

Figure 5. Spatial-temporal feature of nodes P04616, P04615, and P04617 (**a**) spatial distribution of three nodes; (**b**) sequential process lines of three nodes.

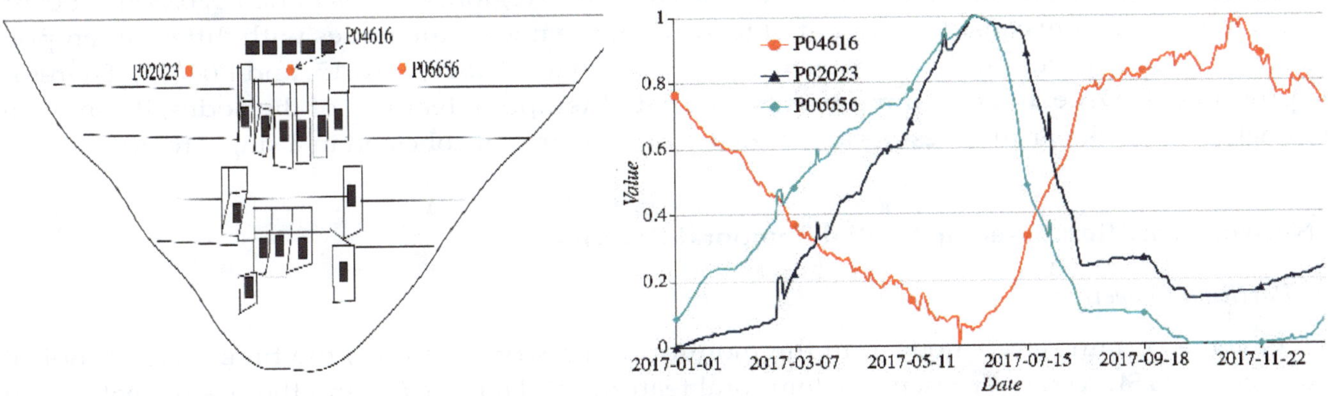

(a) (b)

Figure 6. Spatial-temporal feature of nodes P04616, P02023, and P06656 (**a**) spatial distribution of three nodes; (**b**) sequential process lines of three nodes.

(1) Asymmetrical spatial position and long-distance. Three vertical measuring nodes P04616, P00002, and P07981 are selected to monitor the radial deformation of the arch dam. The spatial distribution and the sequential process lines of three nodes are shown in Figure 4. From Figure 4, three nodes measure the same monitoring items, and their spatial positions are irrelevant. The corresponding process line trends and fluctuation ranges are completely different. The cosine similarities of temporal feature encoding of both P04616 and P00002, and P07981 are −0.017 and 0.248, respectively. The sensor nodes with different spatial features have low temporal feature encoding.

(2) Spatial proximity of different nodes. Taking three nodes P04616, P04615, and P04617 as examples, they are installed in the same dam section and height. P04616 and P04615 are vertical measuring nodes. P04616 and P04615 are to monitor the radial displacement and tangential displacement, respectively. P04617 is the stress measuring node to monitor the radial stress of the dam. The reason for selection is to illustrate that nodes monitoring different items installed in the same part of the dam may have similar and consistent trends. As shown in Figure 5, although two nodes P04616 and P04617 monitor the different physical quantities, their observed values are similar and have basically consistent trends. Their cosine similarity of feature encoding is 0.991. On the contrary, the sequential process line of P04615 is significantly different from that of P04616, and the cosine similarity is 0.266. Thus, even with two sensor nodes with close spatial distance, the similarity of the temporal feature is also low.

(3) Symmetrical spatial positions. Taking three nodes P04616, P02023, and P06656 as examples, two nodes P02023 and P06656 are symmetrical with node P04616 as the center in the vertical direction, as shown in Figure 6a. P04616 is the vertical measuring node to monitor the radial displacement. P02023 and P06656 are the static level nodes to monitor the vertical displacement of the dam. The reason for selection is to show that two nodes monitoring the same item, which are installed at the 1/4 arch crown beam at both ends of the left and right banks of the centerline, also have high similarities of time-series data. In Figure 6b, the sequential process lines of P02023 and P06656 change in a similar trend, and the corresponding cosine similarity is 0.831. The cosine similarity between P06656 and P04616 is −0.134, and the feature difference is obvious. The reason is that the symmetric spatial structures exist with similar stress in the local regions, e.g., the dam sections on both sides bear similar pressure. There are strong similarities in the temporal features of two sensor nodes that are symmetrical in spatial positions.

From Figure 4 to Figure 6, the spatial correlation of sensors is different, and the correlation of time-series characteristics of the measured values are also different. If the sensor network partition only considers the temporal features of the nodes, the sensor nodes scattered spatial positions with similar temporal features are likely clustered into the same region, such as node P02023 and P06656. Meanwhile, if we only consider the spatial features of the nodes, the nodes with different temporal features are likely partitioned into the same region, such as P04616, P04615, and P04617. Therefore, the proposed TSDA extracts and represents the spatial-temporal features of the nodes, it can obtain reasonable network partition results and realize a high degree of cohesion in a single region.

5. Network Partition Based on Spatial-Temporal Features

5.1. Partition Objective

Given a certain sensor node x_i in the monitoring network systems, the time-series denoising autoencoder (TSDA) can represent the temporal features of data as FT_{x_i}, and the spatial feature after normalization can be represented as FS_{x_i}. Using the feature mapping function with sparse coding $\varphi(x_i) = \{FT_{x_i}, FS_{x_i}\} = z_i$, the raw samples are mapped into the latent feature space.

Suppose that p_{ij} is the probability of sensor node x_i belonging to the jth cluster and $\mathbf{P}$ is the probability distribution of model predictions, q_{ij} is the target probability and $\mathbf{Q}$ is the target probability distribution. The network partition problem can be transformed into the parameter optimization problem. Expectation maximization (EM) is adopted to make the prediction distribution $\mathbf{P}$ as close to the real probability distribution $\mathbf{Q}$. In order to define the clustering objective function, we use an auxiliary target variable (that is the target probability distribution $\mathbf{Q}$) to refine the model predictions probability $\mathbf{P}$ iteratively.

Firstly, Kullback-Leibler (KL) divergence [30] is introduced to decrease the distance between the model prediction $\mathbf{P}$ and the target variable $\mathbf{Q}$.

$$\mathcal{L} = KL(\mathbf{Q} \,\|\, \mathbf{P}) = \frac{1}{n}\sum_{i=1}^{n}\sum_{j=1}^{m} q_{ij} \log \frac{q_{ij}}{p_{ij}} \tag{2}$$

The degenerate solution allocates most of the samples to a few clusters or assigns a cluster to the outlier sample. In order to avoid the degenerate solution, a regularization term is imposed on the target variable $\mathbf{Q}$. Therefore, we define the empirical label distribution of target variables $\mathbf{Q}$ as:

$$\Gamma_j = \frac{1}{n}\sum_{i=1}^{n} q_{ij} \tag{3}$$

where Γ_j can be considered as the soft frequency of cluster assignments in the target distribution. Adopting this empirical distribution Γ, we can ensure our preference for having balanced assignments by adding the following *KL* divergence to the loss function.

Assume H is the uniform prior to the empirical label distribution Γ. If there is any extra knowledge about the distribution of network partition (clustering), we can use an arbitrary distribution to initialize the uniform prior H. The objective function $\mathcal{L}$ is as follows:

$$\mathcal{L} = KL(\mathbf{Q} \parallel \mathbf{P}) + KL(\Gamma \parallel \mathrm{H}) = [\tfrac{1}{n}\sum_{i=1}^{n}\sum_{j=1}^{m} q_{ij}\log\tfrac{q_{ij}}{p_{ij}}] + [\tfrac{1}{n}\sum_{j=1}^{m}\Gamma_j\log\tfrac{\Gamma_j}{\mathrm{H}_j}] = \tfrac{1}{n}\sum_{i=1}^{n}\sum_{j=1}^{m} q_{ij}\log\tfrac{q_{ij}}{p_{ij}} + q_{ij}\log\tfrac{\Gamma_j}{\mathrm{H}_j} \tag{4}$$

where the first term $\frac{1}{n}\sum_{i=1}^{n}\sum_{j=1}^{m} q_{ij}\log\frac{q_{ij}}{p_{ij}}$ minimizes the distance between the target and model prediction distributions, the second term $q_{ij}\log\frac{\Gamma_j}{\mathrm{H}_j}$ balances the frequency of clusters in the target variables. Utilizing the balanced target variables, we can ensure the model has more balanced predictions (cluster assignments) **P** directly. It is also simple to change the prior from the uniform distribution to any arbitrary distribution in the objective function $\mathcal{L}$ if there is any extra knowledge about the probability distribution of clusters.

5.2. Iterative Solution to the Partition Problem

The objective function $\mathcal{L}$ of the network partition (clustering) contains hidden variables (target variables **Q**, empirical distribution Γ, and the prior distribution H). The EM method is utilized to optimize the objective function. The iterative solution includes two main phases: (1) target variables **Q** estimation via fixed parameters; (2) parameters update via fixed target variables **Q**.

(1) target variables **Q** estimation via fixed parameters

Using the alternative learning approach, we can estimate the target variables **Q** via fixed parameters and update the parameters while the target variables **Q** are assumed to be known (maximization step). The problem to infer the target variable **Q** has the following objective:

$$\min_{\mathbf{Q}} = \frac{1}{n}\sum_{i=1}^{n}\sum_{j=1}^{m} q_{ij}\log\frac{q_{ij}}{p_{ij}} + q_{ij}\log\frac{\varphi_j}{h_j} \tag{5}$$

where the target variables are constrained to $\sum_{j=1}^{m} q_{ij} = 1$, m is the number of clusters. All sensor nodes are clustered into their own regions through the network partition algorithm; thus, we have $\sum_{j=1}^{m} p_{ij} = 1$. This problem can be solved using the first order method, which only requires the objective function value and its gradient at each iteration. We use the partial derivative of the objective function with respect to the target variables as follows:

$$\frac{\partial\mathcal{L}}{\partial q_{ij}} = \frac{q_{ij}}{\sum_{i=1}^{n} q_{ij}} + \log\left(\frac{q_{ij}\Gamma_j}{p_{ij}}\right) + 1 \tag{6}$$

There are a large number of sensor nodes deployed in the monitoring systems when the number of sensor nodes n is often big enough, we have $\lim_{n\to\infty}\left(q_{ij}/\sum_{i'=1}^{n} q_{i'j}\right) = 0$. To investigate the problem more carefully, the gradient can be approximately computed by removing the first item in Equation (7). Setting the gradient equal to zero, we compute the closed-form solution and update the target variables **Q**.

$$q_{ij} = \psi(p_{ij}, \Theta) = \frac{p_{ij}\left(\sum_{i'=1}^{n} p_{i'j}\right)^{-\frac{1}{2}}}{\sum_{j'=1}^{m} p_{ij'}\left(\sum_{i'=1}^{n} p_{i'j'}\right)^{-\frac{1}{2}}} \tag{7}$$

(2) parameters update via fixed target variables **Q**

When the target variables are known, we update the network parameters Θ using the estimated target variables with the following objective function.

$$\min_{\Theta} = -\frac{1}{n}\sum_{i=1}^{n}\sum_{j=1}^{m} q_{ij}\log p_{ij} \tag{8}$$

To minimize the parameters Θ of the soft max layer in the deep neural network, the objective function can be converted into a standard cross-entropy loss function for clustering. The parameters of the soft max layer Θ can be updated by backpropagating the error. The above two steps are executed alternately during the iteration until convergence.

5.3. *Network Partitioning Algorithm*

To obtain reasonable network partition results in the dam safety monitoring networks, one network partition algorithm based on the TSDA (NPA) is presented to cluster all sensor nodes and partition the network into the different regions based on the spatial-temporal features. In order to establish the relationship between the stress and load among the adjacent dam space elements, the network is partitioned into regions based on the spatial-temporal of the monitoring data.

Assume those n sensor nodes deployed in the monitoring network are clustered into m regions. The set of sensor nodes $\mathbf{X} = \{x_i | i = 1, \ldots, n\}$ is the input, the distribution of model predictions $\mathbf{P} = \{p_{ij} | i = 1, \ldots, n; j = 1, \ldots, m\}$ is the output. The NPA algorithm is designed for the unsupervised clustering task and does not require a set number of regions m, however, the NPA needs to manually set the hyperparameters including training batches, maximum iterations, and iteration error thresholds. Algorithm 1 shows a brief description of the NPA algorithm.

Algorithm 1. Network Partition Algorithm based on Spatial-Temporal Features (NPA)

Input: The set of sensor nodes $\mathbf{X} = \{x_i | i = 1, \ldots, n\}$;
Output: The probability distribution of model prediction $\mathbf{P} = \{p_{ij} | i = 1, \ldots, n; j = 1, \ldots, m\}$;
1: Using the embedding function $\varphi : \mathbf{X} \to \mathbf{Z}$, and the original samples are mapped into the latent feature space
2: K-Means cluster algorithm is used to initialize the objective distribution $\mathbf{Q}$
3: **While not** convergence
4: fix the objective distribution $\mathbf{Q}$ to compute $p_{ij} = f(z_i, \Theta)$
5: update the prediction distribution $\mathbf{P}$
6: fix the parameters Θ to compute $q_{ij} = \psi(p_{ij}, \Theta)$
7: update the objective distribution $\mathbf{Q}$
8: minimize $\min\limits_{\Theta} - \frac{1}{n}\sum_{i=1}^{n}\sum_{j=1}^{m} q_{ij}\log p_{ij}$ through fixing the objective distribution $\mathbf{Q}$
9: update the parameters Θ
10: **End while**
11: **Return** $\mathbf{P} = \{p_{ij} | i = 1, \ldots, n; j = 1, \ldots, m\}$

The proposed TSDA uses the embedding function with sparse coding $\varphi : \mathbf{X} \to \mathbf{Z}$ for the temporal feature compressed representation. The uniform dimension and nonlinear normalization are adopted to represent the spatial feature. The original samples can be mapped into the latent feature space. To avoid the uncertainty of the random initialization $\mathbf{Q}$, K-Means, or GMM is selected to initialize the objective distribution $\mathbf{Q}$. In order to speed up the convergence of objective distribution, EM is used to make the prediction distribution $\mathbf{P}$ close to the real probability distribution $\mathbf{Q}$. An auxiliary target variable (that is the target probability distribution $\mathbf{Q}$) is introduced to refine the model predictions probability $\mathbf{P}$ iteratively. The iterative solution includes two main phases: (1) target variables $\mathbf{Q}$ estimation via fixed parameters; (2) parameters update via fixed target variables $\mathbf{Q}$.

6. Performance Evaluation

6.1. Experimental Setup

6.1.1. Datasets

In this paper, the network partition method NPA based on the spatial and temporal features of data was presented. In the section of the problem statement, the sensor network partition problem is equivalent to the sensor nodes clustering problem. In order to show that the NPA works well with various datasets, we have chosen the real dam safety monitoring dataset and public datasets. The real dataset from the highest arch dam in the world is the same as the dataset mentioned in Section 4.3. The public datasets come from the Keras Framework [32] and the UCI Machine Learning Repository [33], which are widely selected to evaluate the clustering performance. Table 1 provides a brief description of each dataset.

Table 1. Public Dataset Descriptions.

Dataset	Samples	Dimensions	Classes	Description
MNIST	70,000	$1 \times 28 \times 28$	10	handwritten digits images
Fashion-MNIST	70,000	$1 \times 28 \times 28$	10	fashion items images
CIFAR-10	60,000	$3 \times 32 \times 32$	10	colorful images
Reuters	11,228	\	46	News texts

6.1.2. Baselines

We evaluate the proposed NPA in comparison with state-of-the-art clustering methods on the above datasets. K-Means clustering and the Gaussian mixture model (GMM) are widely used clustering methods. The autoencoder (AE) is usually adopted to compress the high-dimensional data and represent their features. Therefore, K-Means and GMM are added to the AE for improvement as a benchmark method in the experiments, respectively. On the public datasets, AE+K-Means, AE+GMM, and the deep clustering network (DNN) are used for comparison, meanwhile, the experimental results of K-Means and GMM are retained. On the dam monitoring datasets, K-Means and GMM are further improved into TSDA+K-Means and TSDA+GMM by using the time-series denoising autoencoder (TSDA) to extract features.

6.1.3. Evaluation Metrics

Experiments on public datasets: We use three popular evaluation metrics for clustering algorithms, accuracy (ACC), normalized mutual information (NMI), and adjusted Rand index (ARI). ACC measures the extent to which the data are properly classified into the corresponding clusters, and NMI represents that the clustering results contain the amount of real information, and ARI can reflect the accuracy and purity of the clustering results. The value range of clustering accuracy (ACC) and the normalized mutual information is [0, 1]. The larger the value, the higher the clustering accuracy. The value range of the adjusted Rand index is $[-1, 1]$, where negative values indicate poor clustering performance, and 1 indicates optimal clustering performance.

Experiments on real datasets: Three methods, TSDA +K-Means, TSDA+GMM, and the NPA, are performed to conduct experiments on the dam's monitoring datasets. The silhouette coefficient (SC) and Calinski-Harabaz Index (CH) are used to evaluate the network partitioning performance.

The SC quantifies the distribution of the sensor nodes. $SC \in [-1, 0)$ represents the basic error of the network partitioning performance, 0 represents the local optimal solution. If $SC \in (0, 1]$, the higher the value, the more reasonable distribution of the sensor nodes. The CH measures the degree of local cohesiveness. The higher the value, the higher the degree of cohesiveness.

6.2. Experimental Results on Public Datasets

We compare our NPA partition algorithm, with several baselines and state-of-the-art algorithms, including K-Means, AE+K-Means, GMM, AE+GMM, on the public datasets MNIST, Fashion-MNIST, STL-10, and Reuters. The evaluation metrics are clustering accuracy (ACC), normalized mutual information (NMI), and the adjusted Rand index (ARI). Note that the labeled data are used to tune the number of hyperparameters for each algorithm. This number only shows the number of hyperparameters, which are set differently for various datasets for more efficient performance. The experimental results are an average of 10 times. The details are shown in Table 2.

Table 2. Clustering performance of different algorithms on public datasets on clustering accuracy (ACC), normalized mutual information (NMI) and adjusted Rand index (ARI).

Dataset	Metrics	K-Means	AE+K-Means	GMM	AE+GMM	DCN	NPA
MNIST	ACC	0.732	0.887	0.716	0.863	0.813	0.910
	NMI	0.627	0.812	0.637	0.827	0.729	0.876
	ARI	0.541	0.791	0.537	0.762	0.614	0.837
Fashion-MNIST	ACC	0.587	0.613	0.495	0.601	0.590	0.635
	NMI	0.567	0.607	0.494	0.582	0.576	0.652
	ARI	0.405	0.524	0.321	0.478	0.453	0.563
CIFAR-10	ACC	0.489	0.557	0.472	0.523	0.521	0.832
	NMI	0.432	0.531	0.462	0.505	0.492	0.778
	ARI	0.378	0.470	0.417	0.489	0.436	0.691
Reuters	ACC	0.521	0.762	0.534	0.638	0.607	0.782
	NMI	0.312	0.537	0.371	0.482	0.446	0.674
	ARI	0.218	0.438	0.339	0.443	0.386	0.591

As shown in Table 2, the clustering accuracy (ACC) of K-Means on MNIST, Fashion-MNIST, CIFAR-10, and Reuters are 0.732, 0.587, 0.489, and 0.521, respectively. The ACC values of AE+K-Means are 0.887, 0.613, 0.557, and 0.762, respectively. The autoencoder (AE) can improve the clustering accuracy of K-Means by 21.2%, 4.4%, 13.9%, and 46.3%, respectively. Overall, AE+K-Means outperforms K-Means. GMM on the same four public datasets can achieve the clustering accuracy ACC 0.716, 0.495, 0.472, and 0.534, respectively, while the accuracy of AE+GMM are 0.863, 0.601, 0.523, and 0.638, respectively. The AE can increase the accuracy of GMM by 20.5%, 21.4%, 10.8%, and 19.5%, respectively. Deep clustering network (DCN) is compared with AE+K-Means and AE+GMM using the autoencoder on four datasets. DCN is inferior to the other three algorithms on the whole performance. On the Fashion-MNIST dataset, DCN can only exhibit similar performance with AG+GMM. Autoencoder can significantly improve clustering performance. AE+K-Means and AE+GMM algorithms employ autoencoder structures and use the reconstruction loss function as a data-dependent regularization for training the parameters, which can avoid undesirable local minima during training. Therefore, AE+K-Means and AE+GMM can outperform K-Means and GMM on clustering performance, respectively. MNIST, Fashion-MNIST, CIFAR-10, and Reuters are high-dimensional data. AE has the ability to effectively compress high-dimensional feature data and avoid various errors in the iteration and accelerate the convergence process.

On the MNIST dataset, clustering accuracy ACC of the NPA, AE+K-Means, and AE+GMM are 0.877, 0.863, and 0.910, respectively; NMI of the three algorithms are 0.812, 0.827, and 0.876; the ARI of the three algorithms are 0.791, 0.762, and 0.837, respectively. The NPA, AE+K-Means, and AE+GMM algorithms all adopt autoencoder for feature compression representation to obtain efficient clustering performance.

Furthermore, we compare the NPA, AE+K-Means, and AE+GMM algorithms on the four public datasets, as shown in Table 3. The NPA can achieve better clustering performance than AE+K-Means and AE+GMM in the terms of ACC, NMI, and the ARI on the four public datasets. Especially on the CIFAR-10 dataset, the NPA can improve the clustering accuracy by about 50%, compared with AE+K-Means and AE+GMM. In general, the NPA is superior to AE+K-Means and AE+GMM in three aspects: clustering accuracy, clustering real information amount, and cluster purity.

Table 3. Clustering performance of different algorithms on public datasets.

Dataset	Metrics	NPA vs. AE+K-Means	NPA vs. AE+GMM
MNIST	ACC	2.6%	5.4%
	NMI	7.9%	5.9%
	ARI	5.8%	9.8%
Fashion-MNIST	ACC	3.6%	5.7%
	NMI	7.4%	12.0%
	ARI	7.4%	17.8%
CIFAR-10	ACC	49.4%	59.1%
	NMI	46.5%	54.1%
	ARI	47.0%	41.3%
Reuters	ACC	2.6%	22.6%
	NMI	25.5%	39.8%
	ARI	34.9%	33.4%

6.3. Experimental Results on Real Datasets

We evaluated the proposed NPA network partition algorithm based on the real dam safety monitoring dataset, as mentioned in Section 4.3. The layout of the sensor nodes is shown in Figure 7, where the solid dots represent the sensor nodes. As shown in Figure 8, the sensor nodes are evenly deployed at the top of the dam and the number is relatively small, while the distribution of sensor nodes on the abutments is sparse and asymmetrical due to the different stresses on both abutments of the dam. The vertical dashed lines divide the dam into multiple sections. The sensor nodes are densely distributed along the dam section joints. From both sides to the middle of the dam foundation, the distribution of the sensor nodes changes from sparse to dense. The stress of the dam heel plays a key role in the overall structural safety of the dam. The sensor nodes are densely installed at the dam heel. In addition, the so-called symmetry in this paper refers to the symmetry of the layout of the monitoring sensors, rather than the symmetry of the dam structure. The spatial arrangement of monitoring sensors for the arch dam is symmetrical with the arch crown beam dam section as the center.

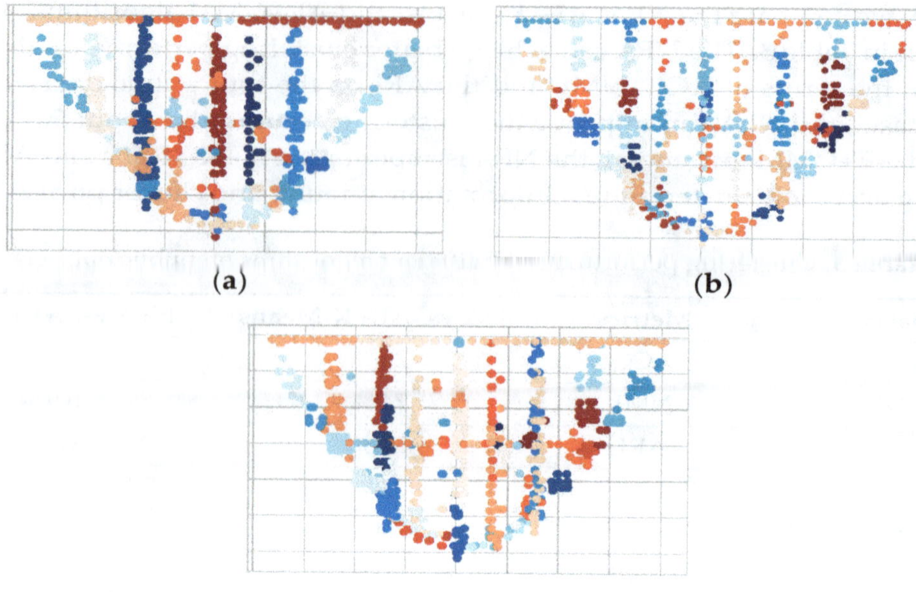

Figure 7. Network partition results for dam safety monitoring system (**a**) NPA; (**b**) TSDA+K-Means; (**c**) TSDA+ Gaussian mixture model (GMM).

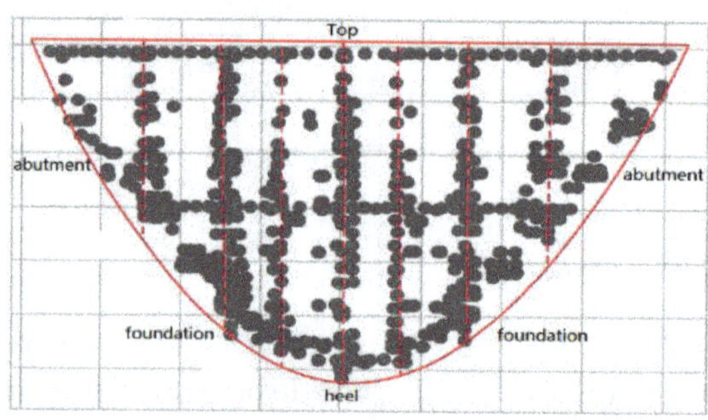

Figure 8. The spatial distribution of sensor nodes in the dam safety monitoring system.

According to the design specifications of dam safety monitoring, the sensor nodes installed in the dam foundation and heel within the same dam section are dense. Due to the strong correlation among the sensor nodes at the same dam section, the cohesiveness of the network partition results should be as high as possible. On the other hand, the deployment of sensor nodes at the top of the dam and on both sides of the dam abutments are scattered, the range of the partitioned region is quite large. Besides, the overall structure of the dam is symmetrical, and the network partition of dam safety monitoring should also exhibit a symmetrical effect.

The proposed NPA adopts the TSDA to extract the temporal features of the sensor nodes. To evaluate the partition results of the proposed NPA algorithm, we compare the NPA with the modified AE+K-Means (TSDA+K-Means) and modified AE+GMM (TSDA+GMM), as illustrated in Figure 8. The NPA is the unsupervised clustering algorithm, and there is no need to set the number of regions m in advance. TSDA+K-Means and TSDA+GMM need to set the number of regions. The number of regions m is estimated to be 51 and 53 using the canopy and affinity propagation clustering algorithm, respectively, while the experts estimate the number of regions to be 48. In the experiments, the median is taken, so $m = 51$.

As shown in Figure 8, the different colors of sensor nodes indicate different partitioned regions and the sensor nodes in the same region have the same color. Overall, the partitioned regions with TSDA+NPA exhibit the symmetrical distribution at the center of the dam. The sensor nodes in the same dam section, foundation, and heel are highly cohesive, while the nodes located at the top of the dam and both sides of the dam abutments are relatively scattered, and the corresponding regions are large. The TSDA+NPA can obtain reasonable partitioning results. On the contrary, the partitioning results with TSDA+K-Means and TSDA+GMM are unsatisfactory. The sensor nodes in the same dam section and dam heel are often partitioned into different regions with poor cohesiveness. The partition results of sensor nodes with TSDA+K-Means and TSDA+GMM are unreasonable.

Two metrics are introduced to evaluate the partition performance of the dam safety monitoring network. They are the silhouette coefficient (SC) and the Calinski-Harabaz Index (CH). The SC can quantify the distribution of sensor nodes. The higher the value, the more reasonable distribution. The CH can measure the degree of cohesiveness of network partitioning. The higher the value, the higher the degree of cohesiveness. Table 4 shows the average SC and CH values of three algorithms. The SC and CH of the NPA are significantly higher than those of TSDA+K-Means and TSDA+GMM. The SC of the three algorithms are all in the range of (0, 1]. Compared with TSDA+K-Means and TSDA+GMM, the SC of the NPA increases to 45.1% and 58.4%, respectively. As to the CH, the NPA can increase by 30.8% and 61.6%, respectively. The cohesiveness of the sensor nodes in the same region is high and the correlation between different regions is low. The NPA exhibits better performance on the network partitioning than the two other algorithms. The experimental results in Table 4 are consistent with those shown in Figure 7.

Table 4. Silhouette coefficient (SC) and Calinski-Harabaz Index (CH) with TSDA+K-Means, TSDA+GMM, and the NPA.

Metric	TSDA+K-Means	TSDA+GMM	NPA
SC	0.502	0.548	0.795
CH	580.7	717.1	938.3

7. Conclusions

In order to describe the dynamics of the physical quantities for the dam safety monitoring through the network partition in a real-time way, a network partitioning algorithm based on the spatial-temporal features of sensor nodes (NPA) was proposed. A time-series denoising autoencoder is presented to compress the high-dimension data in order to extract the spatial-temporal features. Adopting the TSDA model, the NPA introduces an auxiliary objective variable to optimize the partition objective functions to cluster the sensor nodes, so that the partition results of the sensor network can reflect the physical laws of the dam safety operation effectively. The experimental results on the public datasets show that the proposed NPA can improve the partition accuracy and cohesiveness compared with AE+K-Means, AE+GMM, and DCN in the terms of ACC, NMI, and the ARI. For the real dataset from the dam safety monitoring system, the experimental results illustrate that the NPA can also outperform TSDA+K-Means and TSDA+GMM in the terms of network partition accuracy, the SC, and the partition distribution CH. This paper focuses on the correlation analysis of the monitoring data themselves. Our approach can provide a preliminary foundation for dam safety evaluation in a real-time way, rather than directly evaluate the structural safety of the dam. In future work, we will consider the dam safety evaluation integration of artificial intelligence and domain knowledge from the dam structure.

Author Contributions: All four authors have contributed to the work presented in this paper. Methodology, Y.M.; validation, H.Q.; formal analysis Y.M. and H.Q.; resources, H.C.; dfunding acquisition, and resources, H.C.; data curation, H.C.; writing—original darft preparation, Y.M. and H.Q.; writing—review and editing, Y.M. and H.C.; visualization, L.W. and H.Q.; Software, L.W. All four authors have equally contributed to the work presented in this paper. All authors have read and agreed to the published version of the manuscript.

References

1. Su, H.; Wen, Z.; Sun, X.; Yan, X. Multisource information fusion-based approach diagnosing structural behavior of dam engineering. *Struct. Control Health Monit.* **2018**, *25*, 1–17. [CrossRef]
2. Sharma, R.; Vashisht, V.; Singh, U. Soft Computing Paradigms Based Clustering in Wireless Sensor Networks: A Survey. In *Advances in Data Sciences, Security and Applications*; Springer: Singapore, 2020; pp. 133–159.
3. Saia, R.; Carta, S.; Recupero, D.R.; Fenu, G. Internet of entities (IoE): A blockchain-based distributed paradigm for data exchange between wireless-based devices. In Proceedings of the 8th International Conference on Sensor Networks, Prague, Czech Republic, 26–27 February 2019; pp. 77–84. [CrossRef]
4. Guerreiro, J.; Rodrigues, L.; Correia, N. Resource Allocation Model for Sensor Clouds under the Sensing as a Service Paradigm. *Computers* **2019**, *8*, 18. [CrossRef]
5. Zhu, Z.; Liu, J.; Li, X. Statistical model analysis of dam deformation monitoring based on robust estimation. *J. China Inst. Water Resour. Hydropower Res.* **2018**, *16*, 28–35.
6. Mao, Y.C.; Zhang, J.H.; Qi, H.; Wang, L.B. DNN-MVL: DNN-Multi-View-Learning-Based Recover Block Missing Data in a Dam Safety Monitoring System. *Sensors* **2019**, *19*, 2895. [CrossRef] [PubMed]
7. Wan, Y.; Shi, Y.; Wu, K.; Chen, X. Laplacian sparse coding by incorporating locality and non-negativity for image classification. *Pattern Recognit. Artif. Intell.* **2017**, *30*, 481–488.
8. Zhu, P.; Zhu, W.; Hu, Q.; Zhang, C.; Zuo, W. Subspace clustering guided unsupervised feature selection. *Pattern Recognit.* **2017**, *66*, 364–374. [CrossRef]
9. Jiang, W.; Miao, C.; Su, L.; Li, Q.; Hu, S. Towards quality aware information integration in distributed sensing systems. *IEEE Trans. Parallel Distrib. Syst.* **2018**, *29*, 198–211. [CrossRef]
10. Yang, H.; Wang, X. Data Gathering Based on Regionalized Compressive Sensing in WSN. *Chin. J. Comput.* **2017**, *40*, 1933–1945.
11. Bhuiyan, M.Z.A.; Wu, J.; Weiss, G.M.; Hayajneh, T.; Wang, T.; Wang, G. Event detection through differential pattern mining in cyber-physical systems. *IEEE Trans. Big Data* **2018**. [CrossRef]
12. Yu, X.; Yin, G.; Xu, X.; Wang, J. A data stream subspace clustering algorithm based on region partition. *J. Comput. Res. Dev.* **2014**, *51*, 88–95.
13. Menendez, M.; Ambühl, L.; Loder, A.; Zheng, N.; Axhausen, K.W. Approximative network partitioning for MFDs from stationary sensor data. *Transp. Res. Rec. J. Transp. Res. Board* **2019**. [CrossRef]
14. Zivkovic, Z. Improved Adaptive gaussian mixture model for background Subtraction. In Proceedings of the 17th International Conference on Pattern Recognition (ICPR), Cambridge, UK, 26 August 2004; pp. 1–4.
15. Parthasarathy, M.; Ramya, R.; Vijaya, A. An adaptive segmentation method based on Gaussian Mixture Model (GMM) clustering for DNA microarray. In Proceedings of the 2014 International Conference on Intelligent Computing Applications (ICICA), Coimbatore, India, 6–7 March 2014.
16. Huang, W.; Wang, X.; Li, G.; Kang, H. Semi-supervised artificial immune mixture models clustering. *Chin. J. Electron.* **2016**, *25*, 249–255. [CrossRef]
17. Klampanos, I.A.; Davvetas, A.; Andronopoulos, S. Autoencoder-driven weather clustering for source estimation during nuclear events. *Environ. Model. Softw.* **2018**, *102*, 84–93. [CrossRef]
18. Fang, L.; Yin, C.; Zhou, L.; Li, Y.; Su, C.; Xia, J. A physiological and behavioral feature authentication scheme for medical cloud based on fuzzy-rough core vector machine. *Inf. Sci.* **2020**, *507*, 143–160. [CrossRef]
19. Dizaji, K.G.; Herandi, A.; Deng, C.; Cai, W.; Huang, H. Deep clustering via joint convolutional autoencoder embedding and relative entropy minimization. In Proceedings of the 2017 IEEE International Conference on Computer Vision (ICCV), Venice, Italy, 22–29 October 2017; pp. 1–13.
20. Mikalsen, K.Ø.; Bianchi, F.M.; Soguero-Ruiz, C.; Jenssen, R. Time series cluster kernel for learning similarities between multivariate time series with missing data. *Pattern Recognit.* **2018**, *76*, 569–581. [CrossRef]
21. Vincent, P.; Larochelle, H.; Lajoie, I.; Bengio, Y.; Manzagol, P.-A. Stacked denoising autoencoders: Learning useful representations in a deep network with a local denoising criterion. *J. Mach. Learn. Res.* **2010**, *11*, 3371–3408.
22. Chen, M.; Xu, Z.; Winberger, K.Q.; Sha, F. Marginalized denoising autoencoders for domain adaptation. In Proceedings of the 29th International Conference on Machine Learning (ICML), Edinburgh, UK, 16 June–1 July 2012.
23. Hershey, J.R.; Chen, Z.; le Roux, J. Deep clustering: Discriminative embeddings for segmentation and separation. In Proceedings of the 2016 IEEE International Conference on Acoustics, Speech and Signal Processing (ICASSP), Shanghai, China, 20–25 March 2016; pp. 31–35.

24. Ye, J.; Zhao, Z.; Wu, M. Discriminative k-means for clustering. In *Advances in Neural Information Processing Systems*; The MIT Press: Cambridge, MA, USA, 2008; pp. 1649–1656.
25. Wang, D.; Nie, F.; Huang, H. Unsupervised feature selection via unified trace ratio formulation and k-means clustering (track). In Proceedings of the European Conference on Machine Learning and Principles of Knowledge Discovery in Databases (ECML PKDD 2014), Nancy, France, 15–19 September 2014; pp. 306–321.
26. Wang, Z.; Chang, S.; Zhou, J.; Wang, M.; Huang, T.S. Learning a task-specific deep architecture for clustering. In Proceedings of the 2016 SIAM International Conference on Data Mining, Miami, FL, USA, 5–7 May 2016; pp. 369–377. [CrossRef]
27. Wang, Z.; Yang, Y.; Chang, S.; Li, J.; Fong, S.; Huang, T.S. A joint optimization framework of sparse coding and discriminative clustering. In Proceedings of the International Joint Conference on Artificial Intelligence (IJCAI), Buenos Aires, Argentina, 28 July–1 August 2015.
28. Guo, J.; Xie, X.; Bie, R.; Sun, L. Structural health monitoring by using a sparse coding-based deep learning algorithm with wireless sensor networks. *Pers. Ubiquitous Comput.* **2014**, *18*, 1977–1987. [CrossRef]
29. Tian, F.; Gao, B.; Cui, Q.; Chen, E.; Liu, T.Y. Learning deep representations for graph clustering. In Proceedings of the 28th AAAI Conference on Artificial Intelligence (AAAI), Québec City, QC, Canada, 21–31 July 2014; pp. 1293–1299.
30. Trigerogis, G.; Bousmailis, K.; Zafeiriou, S.; Schuller, B. A deep semi-NMF model for learning hidden representations. In Proceedings of the 31st International Conference on Machine Learning (ICML), Beijing, China, 21–26 June 2014; pp. 1692–1700.
31. Yang, J.; Parikh, D.; Batra, D. Joint unsupervised learning of deep representations and image clusters. In Proceedings of the 29th IEEE Conference on Computer Vision and Pattern Recognition, Las Vegas, NV, USA, 26 June–1 July 2016; pp. 5147–5156.
32. Chollet, F. Keras. Available online: https://keras.io (accessed on 6 October 2019).
33. Dua, D.; Graff, C. *UCI Machine Learning Repository*; School of Information and Computer Science, University of California: Irvine, CA, USA, 2019. Available online: http://archive.ics.uci.edu/ml (accessed on 6 October 2019).

4

On the Dynamic Capacity of Concrete Dams

L. Furgani [1], M. A. Hariri-Ardebili [2,3,*], M. Meghella [4] and S. M. Seyed-Kolbadi [3]

1 Mott MacDonald, Croydon CR0 2EE, UK
2 Department of Civil Engineering, University of Colorado, Boulder, CO 80309, USA
3 X-Elastica, LLC, Boulder, CO 80303, USA
4 Ricerca Sistema Energetico (RSE), 20134 Milan, Italy
* Correspondence: mohammad.haririardebili@colorado.edu.

Abstract: The purpose of this joint contribution is to study the maximum dynamic load concrete dams can withstand. The so-called "dynamic capacity functions" for these infrastructures seems now technically and commercially feasible thanks to the modern finite element techniques, hardware capabilities, and positive experiences collected so far. The key topics faced during the dynamic assessment of dams are also discussed using different point of view and examples, which include: the selection of dynamic parameters, the progressive level of detail for the numerical simulations, the implementation of nonlinear behaviors, and the concept of the service and collapse limit states. The approaches adopted by local institutions and engineers on the subject of dam capacity functions are discussed using the authors' experiences, and an overview of time and resources is outlined to help decision makers. Three different concrete dam types (i.e., gravity, buttress, and arch) are used as case studies with different complexities. Finally, the paper is wrapped up with a list of suggestions for analysts, the procedure limitations, and future research needs.

Keywords: dams; endurance time analysis; nonlinear; dynamic capacity; failure

---

1. Introduction

This paper is an extended and modified version of the contribution presented during the "3rd Meeting of EWG Dams and Earthquakes", which was held May 2019 in Lisbon, Portugal [1]. While the skeleton of the presentation was kept, many new references, literature review, simulations, and discussion were added to the paper.

Special efforts have been devoted in recent years to the dynamic assessment of large dams. High seismicity countries went through (or are still facing) re-assessment programs for their dams [2–5]. These extensive studies are required for the lack of seismic checks on older dams, updated seismic hazard parameters, and new seismic standards [6]. Despite the opportunities offered by modern technologies, it is sometimes difficult to select a cost-effective (yet high-confidence) procedure for the seismic assessment of dams. It is understandable that complex analyses make stakeholders feel uncomfortable making risk-related decisions. For this reason, in the authors' opinion, it is vital to find new procedures to balance the numerical simulations with comprehensible outcomes (i.e., the capacity curves' appreciation is easier than the judgment of stress map outputs qualitatively).

The seismic analyses of large dams must consider the following features [7,8]: specific seismic hazard assessment [9–11], step-by-step analysis approach [12], soil-structure interaction [13], fluid–structure interaction [14], and the system's nonlinear behavior [15,16]. It is a common opinion that concrete dams (especially arch dams) are unique prototypes, each having their own behavior. Even more so, it is generally expected that the seismic response of the concrete dams should be considered with an opportune degree of distinction between different case studies. Previous experiences on the application of capacity functions [17,18] demonstrated that using certain classification rules, the overall seismic behavior of concrete dams may be similar, especially towards the failure point. This aspect should exhort the scientific community and the best practice of the sector to consider capacity functions as a way forward.

While the precise time history analysis for concrete dams with dam-specific ground motion records has been the subject of much research, this paper provides a framework to estimate the capacity of dams with a novel technique. Therefore, the objective is not to validate a numerical procedure, but to provide the best estimate of the nonlinear seismic response at different shaking intensities, as well as the failure capacity. This will be accomplished using some examples showing the advantages and limitations of this approach. A summary of conventional procedures available to extract the capacity functions is also discussed. Special attention is offered to the Endurance Time Analysis (ETA) method [19]. Finally, a technical discussion is presented on the benefits that capacity functions can add to the dam industry and public interests.

1.1. International Experiences in the Seismic Assessment of Dams

Following a period in which the lack of regulations permitted designing and building dams without or with too simplified seismic design principles, the necessity to cope with the safety reassessment of these dams and to select the right method to evaluate their performance during the earthquake event is now well recognized. Despite the variety of seismic safety approaches in dam engineering, the concept of "risk assessment" is, implicitly or explicitly, present in all technical codes used in most of the developed countries, e.g., [2–5].

In Switzerland, which recently has undertaken the task of re-assessing the dams [20], some outlines are provided on the framework and efforts required to manage the dams' portfolios, referring to their risk classification. After 10 years required to assess 208 dams, important outcomes and lessons learned have been collected by the Swiss experience. It is worth noting that the cases appeared to be non-compliant with the standards where generally dams had problems primarily in static conditions or unusual design concepts. According to OFEG [21], a lower occurrence probability should be expected for earthquakes, when the failure consequences are very high at the downstream. However, it does not provide any indication on the earthquake level leading to dam failure.

Other European countries followed the example of Switzerland for seismic assessment of their dams. For instance, recently adopted dam safety regulations introduced risk concepts such as the residual safety margin of the dam against seismic actions, a kind of qualitative evaluation of the dam's capacity. Despite this, it is still necessary to bridge the gap between the seismic response determined for the target seismic levels, mainly collapse and serviceability Limit States (LSs), and the dams' ultimate capacity. Considering this, it appears clear where the so-called "capacity functions" may play their role.

The proposed engineering guide for seismic risk of dams in the U.K. guidelines is based on the dam category and has three main features [22]: the probability of occurrence of an earthquake larger than the Maximum Scenario Earthquake (MSE), the vulnerability of the dam, and the consequences of the failure. The French guide [23] is mainly focused on gravity dams. It covers the fundamental stability analysis for sliding LS with a focus on material strength. There is no new recommendation concerning seismic loads [24]. Similar experiences can be found in other countries such as India [25], Japan [26], and Turkey [27].

1.2. An Overview of Seismic Assessment Procedures

To evaluate what may be called the "failure" or "critical earthquake" of a concrete dam, it is necessary to put in place a numerical model and a seismic assessment procedure to consider the most important aspects involved. Considering recent advancements in computer science, the idea to investigate the nonlinear behavior of concrete dams towards collapse is becoming more common among researchers and engineers. Despite the excellent work done or planned in the future, the greatest obstacle for nonlinear analysis and capacity functions is the formal definition of acceptable damages against the different LSs used in the modern structural codes. At the current stage, we can simulate the structural performance decay of these structures, but we have no practical guidelines to indicate if their expected behavior is acceptable other than subjective engineering judgment.

In the buildings sector, the capacity functions are used to demonstrate that the design target and the actual structural ductility are met [28,29]. A similar route can be expected for dams, with appropriate procedures for the definition of "failure modes" and "limit states". An important aspect to be considered is the confidence with the methods used to produce the capacity functions. In the buildings sector, the use of nonlinear static pushover analysis is considered common practice [30]. Pre-defined load patterns are used to reproduce the effects produced by earthquakes. This loading pattern is then increased through a scale factor to produce the damages of the main structural elements until the point of collapse. Based on the authors' experience, it is not possible to use the same procedure for dams. This is due to the different way the dam and the surrounding domains, foundation and water, react against the seismic actions. For this reason, it is more reasonable to use a bespoke dams sector approach. This may consist of nonlinear dynamic analysis procedures where all the mechanism involved in the response can be included and properly captured without limits associated with standardized pseudo-static forces (pre-defined load pattern shapes increased monotonically).

A capacity function determined for a concrete dam under dynamic/seismic forces is always associated with concurrent actions, namely dead weight, thermal loading, and hydrostatic pressure, applied to the dam before the seismic excitation. In addition, the seismic actions are characterized by their spatial variation using three directions of the seismic action and concurrency factors. The numerosity of the combinations suggests that capacity functions cannot be developed for all of them. A procedure is then necessary to perform a screening of the most critical combinations. The proposed approach is to start with simplified linear pseudo-static procedures, move to linear dynamic analysis, and finally nonlinear dynamic analysis. This approach is also giving assurance on the validity of the analysis methods while more detailed aspects are introduced, a concept already introduced within technical guidelines. In this case, the capacity function represents the last step of a series of analyses that are code compliant.

Developing a capacity function requires that the numerical model is capable of accounting for the main nonlinearities in the dam-foundation-reservoir system during an earthquake. For gravity dams, the sliding at the base, or at the weaker lift joints, is one of the main failure modes expected [31]. This phenomenon can cut off the seismic excitation acting as a seismic isolator. The reduction of stresses above the sliding plane is obtained at the cost of a residual displacement, captured by jumps in the force-displacement diagrams. Slips occurring over the joint are also important for arch dams, but are not governing the global response of these structures. On the other hand, the right representation of joint openings and closures are essential for arch dams to represent the different behaviors of these dams under seismic actions in different directions [32,33].

In addition, according to ICOLD [34], a large portion of the concrete dam failures can be attributed to foundation issues (either geological or geotechnical). Whereas dams have generic potential failure modes, foundation failure modes are very specific [35]. Depending on the initiator (i.e., static, hydrologic, or seismic), there might be different failure modes for the foundation. These kinds of geotechnical problems are not included in this paper, but must be considered for seismic assessments of dams. It is reasonable to think that capacity functions evaluated without considering the effects of foundation failure modes must be interpreted in conjunction with the geotechnical results.

For both gravity and arch dams, cracking behavior is required to consider the effects produced by the reduction of concrete properties along the dam body and the associated changes in the way the dam deforms. These effects are represented by the slope change in the capacity function. The nonlinear model used should be also selected based on the data necessary to evaluate damages and the associated relationship with LSs [36].

Another aspect to be considered for large models is the management of computation time as already mentioned. This depends highly on modeling techniques and assumptions. If a numerical model is considered in its most effective condition (mesh and time steps are key aspects in this sense), the only way to reduce the computational time required for the analysis is the selection of alternative seismic inputs. Among the dynamic analysis techniques, the ETA is one of the most efficient methods that significantly reduces the computational cost [17,18].

2. Capacity Estimation Techniques

In this section, a summary of the procedures available for the dynamic capacity estimation of concrete dams is given. The "dynamic capacity function" is defined as the relationship between an external parameter affecting the capacity of the structure (also referred to as a "stressor"), and the "response" of the system at the macro (or structural) level [37]. The stressor, S, can be: (1) an incrementally-increasing monotonic, cyclic (pseudo-dynamic, quasi-dynamic, or dynamic), or time-dependent load (or displacement, acceleration, pressure); (2) an incrementally-decreasing resistance parameter or degradation in strength properties; or (3) a discrete increasing/decreasing critical parameter in a system leading to failure. In earthquake engineering, the stressor is typically called an intensity measurement (IM) parameter [38].

The response, R, on the other hand, is representative of the system behavior under the varying stressor. It is depicted in either an absolute or relative sense. R may be: (1) a single damage variable (DV), such as drift or energy dissipation; (2) a combination of several DVs in terms of the damage index (DI); or (3) any safety monitoring index. In the field of earthquake engineering, R is typically called an engineering demand parameter (EDP) [39]. In the field of uncertainty quantification, it is called the quantity of interest (QoI) [40]. Several dam-specific EDPs, DVs, and DIs can be found in Zhang et al. [41], Hariri-Ardebili and Saouma [42], Ansari and Agarwal [43], and Hariri-Ardebili et al. [44].

2.1. Incremental Dynamic Analysis

Incremental dynamic analysis (IDA) is a technique to estimate the seismic capacity of dams using a large set of ground motions. IDA can be decomposed into the following major branches [37]:

- Single-record IDA (SR-IDA) vs. multi-record IDA (MR-IDA)
- Single-intensity IDA (SI-IDA) vs. multi-intensity IDA (MI-IDA)
- Single-EDP IDA (SE-IDA) vs. multi-EDP IDA (ME-IDA)

Any of these IDA types might be combined together. SR-IDA is the most basic technique, in which a single record is used for nonlinear time history analysis of a dam. Let u_g be the "as-recorded" (unscaled) ground motion time history. To consider both the stronger and weaker scenarios, it can be scaled uniformly using a scale factor, λ_{SF}. The scaled ground motion can be written as $\lambda_{SF} u_g$. This should be represented by a monotonic scalable IM (e.g., peak ground acceleration (PGA), or the 5% damped spectral acceleration at the first-mode period ($S_a(T_1, = 5\%)$)). Several nonlinear analyses are then performed with the scaled ground motions. Let us assume that the response of the nonlinear system to the scaled record is R^λ. The response parameter (or EDP or QoI) can be deformation, stress, uplift, or DI.

MR-IDA is a collection of n SR-IDA for the same structural system. The required ground motion records are usually obtained from a probabilistic seismic hazard analysis (PSHA) conducted for the dam site. The number of required records typically varies given the dispersion among them and the code requirements. The order of n for framed structures is about 30–50, while for concrete dams (which are computationally expensive), it is around 12–20.

SI-IDA uses only one intensity measure parameter (e.g., PGA) to characterize the ground motion record-to-record (RTR) variability; however, MI-IDA adopts additional intensity measure parameters to enhance the uncertainty quantification and to reduce the dispersion. On the other hand, SE-IDA uses only one EDP value (e.g., relative drift), while ME-IDA accounts for the joints' EDPs (e.g., joint opening and sliding).

The combined multi-record single-intensity IDA (MR-SI-IDA) is the most common form of the IDA curves adopted by many researchers for different structures [45]. It is an IDA plot in the IM-EDP coordinate system. Figure 1 presents concrete dam-specific IDA curves developed by different researchers under various assumptions. Considering the RTR variability in the IDA capacity functions, they should be summarized using some central values like the mean, median, and 16% and 84% fractiles.

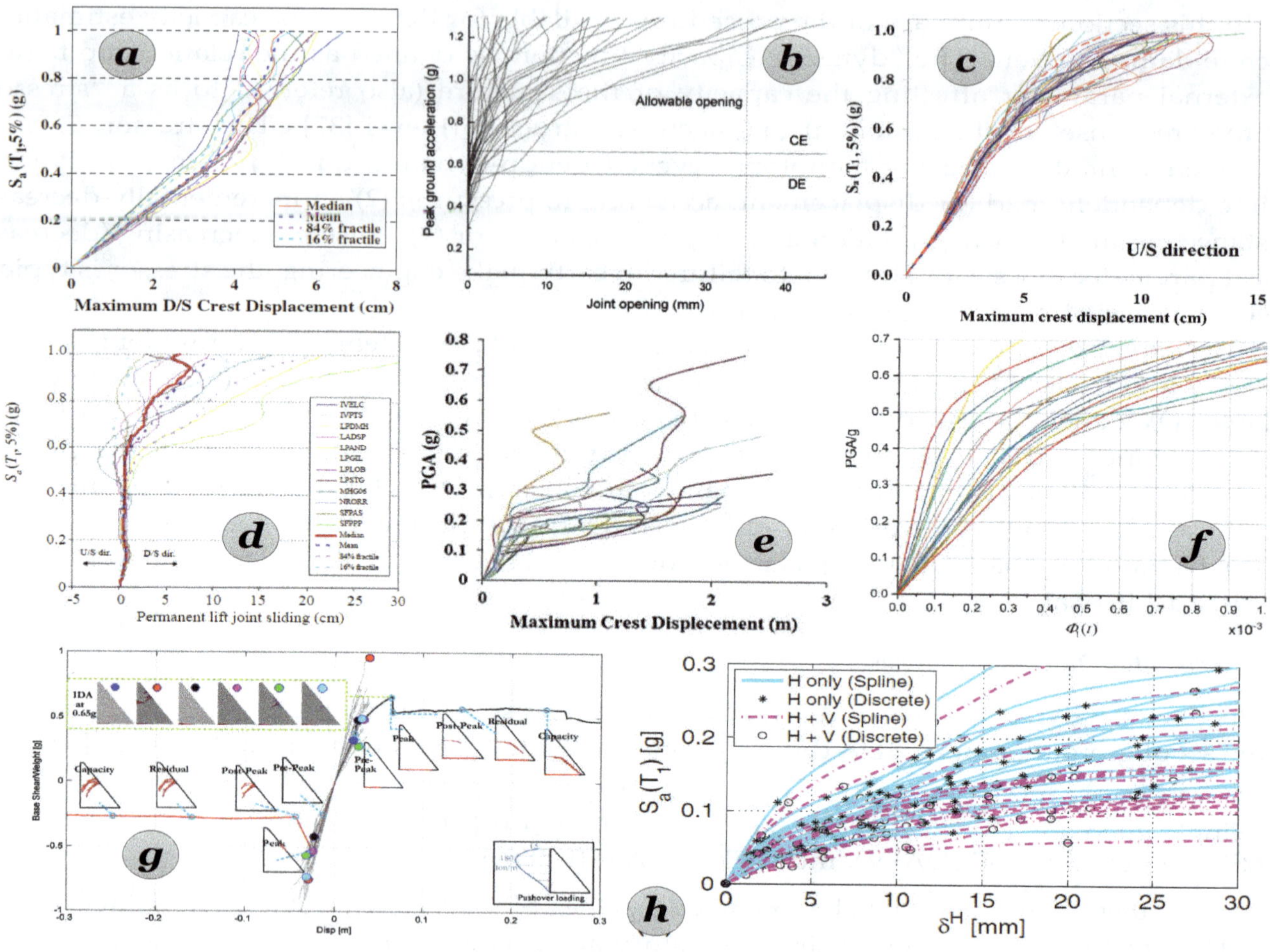

Figure 1. Sample incremental dynamic analysis (IDA) capacity functions for concrete dams: (**a**) gravity dam [46], (**b**) arch dam [47], (**c**) gravity dam [48], (**d**) gravity dam [49], (**e**) gravity dam [43], (**f**) arch dam [50], (**g**) gravity dam [51], and (**h**) gravity dam [52]. PGA, peak ground acceleration.

Therefore, different IDA versions come from different interpretations: MR-IDA is a multi-record of SR-IDA. Each of those two can be presented based on (a) single intensity (SI), or multi-intensity (MI), and/or (b) single EDP (SE), or multi EDP (ME). The simplest combination will be SR-SI-SE-IDA (i.e., single-record, single-intensity, single-EDP), which is just a curve in 2D space. The most complex one will be MR-MI-ME-IDA (i.e., multi-record, multi-intensity, multi-EDP), which consists of n curves in $m \times p$ space (where m and p are the numbers of IMs and EDPs, respectively).

2.2. Cloud Analysis

In cloud analysis (CLA) (as opposed to IDA), the dam is subjected to a set of unscaled (or as-recorded) ground motions and analyzed numerically. If the ground motion records are extracted from a specific hazard scenario, they can be represented by (M_{bin}, R_{bin}), i.e., the magnitude and distance representative of the bin. Therefore, the first n records are selected (or generated [53]). Then, they are applied to the finite element model, and subsequently, n linear or nonlinear transient simulations are performed. From these analyses, EDP vs. IM is determined and forms the so-called cloud response. The CLA method is usually used in the context of the probabilistic seismic demand model (PSDM) [54].

If the CLA is performed with a linear elastic model, the data points will have a linear trend in the arithmetic scale. Figure 2a shows the crest displacement vs. PGA of a tall gravity dam with linear elastic materials using about 2000 different ground motion records all based on a single seismic scenario. Although the results are based on linear analysis, a considerable dispersion exists at higher PGA levels. On the other hand, it is well accepted that the discrete data points resulting from nonlinear CLA have a linear trend on the logarithmic scale (Figure 2b), implying a power curve on the arithmetic scale. This figure shows the nonlinear response of another gravity dam with only 150 ground motions from six different seismic scenarios (i.e., S1–S6). As seen, the general trend is quite linear.

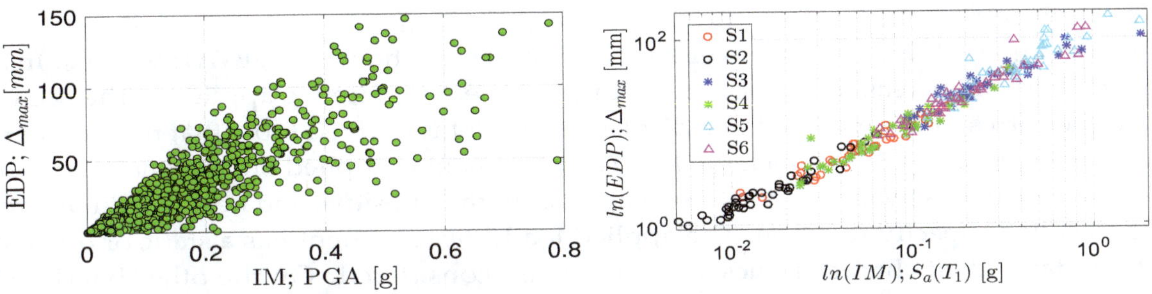

(a) Linear analysis **(b)** Nonlinear analysis

Figure 2. Sample cloud analysis (CLA) capacity functions for concrete dams. (**a**) Linear analysis; (**b**) nonlinear analysis. IM, intensity measurement; S, scenario.

2.3. Endurance Time Analysis

Endurance time analysis (ETA) is a dynamic pushover procedure used to estimate the seismic performance of structures when subjected to pre-designed intensifying excitation [55,56]. The simulated acceleration functions are intended to shake the structure from a low excitation level (with a structural response in the elastic range) to a medium excitation level (where the structure experiences some nonlinearity) and ultimately to a high excitation level causing failure [57]. All these response ranges are experienced in a single time history analysis.

The challenging part of this method lies in generating the endurance time excitation functions (ETEFs), Figure 3a. Initial generation of ETEFs were only based on acceleration and displacement response spectra [58], while the recent advances in the simulation of ETEFs can be found in Mashayekhi and Estekanchi [59], and Mashayekhi et al. [60–62], where the functions were optimized in the time and frequency domains accounting for the acceleration and displacement response spectra, as well as the nonlinear displacement, absorbed hysteretic energy, number of effective cycles, etc. Unconstrained nonlinear optimization is used to simulate ETEFs. The objective function of simulating ETEFs problems is to compute differences between the dynamic characteristics of ETEFs and their intended target values. The objective function covers a range of pre-defined times, periods, and ductility ratios.

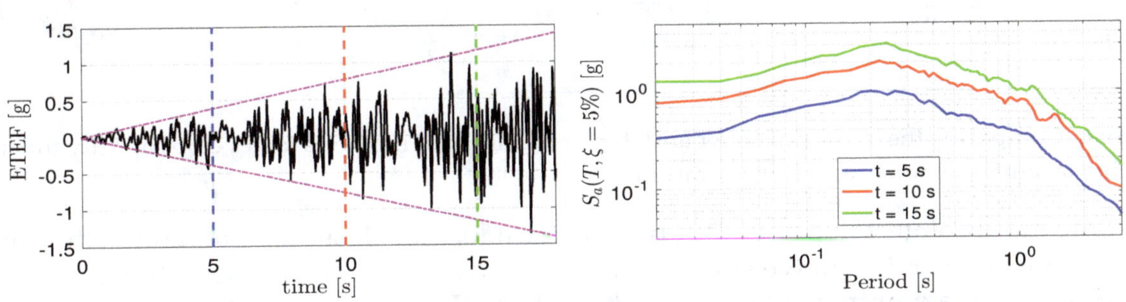

(a) Acceleration time history **(b)** Time-dependent response spectra

Figure 3. Sample endurance time excitation function (ETEF). (**a**) Acceleration time history; (**b**) time-dependent response spectra.

The ETEF is an artificially-designed intensifying acceleration time history, where the response spectra of the ETEF linearly increases with time; see Figure 3b. Ideally, the profile of the acceleration time history and response spectrum increase linearly with time. Figure 3 shows a sample ETEF and its response spectra at three different times (i.e., 5, 10, and 15 s). Note that the ETA and associated spectra are always evaluated from $t = 0$. Therefore, a particular time t_p refers to time window $[0, t_p]$. As seen, the spectrum at $t = 10$ s is nearly twice the one at $t = 5$ s, and the spectrum one at $t = 15$ s is three-times the one at $t = 5$ s. In this technique, the seismic performance is determined by the duration the structure can endure the dynamic input with increasing input intensity (i.e., acceleration). The longer the structure can endure the imposed stressor, the better the performance is. This is why it is named "endurance time analysis".

Figure 4 compares the cyclic pushover and the dynamic pushover (i.e., ETA) methods. In the cyclic pushover method, an increasing cyclic load or displacement function is applied to the finite element model, and the corresponding cyclic force-displacement plot is derived (note that this is different from monotonic pushover analysis in which the load/displacement is applied only in one direction) [63,64]. Next, the envelope of this hysteretic curve is computed (both in positive and negative directions), which forms the so-called capacity function. The applied load/displacement has a static or quasi-dynamic nature. Rate-dependent characteristics are not usually considered. On the other hand, a dynamic pushover method applies a time-dependent load/displacement and directly incorporates the material's dynamic behavior. A capacity function in the ETA method can be obtained as:

- Develop the finite element model of the dam, and apply an appropriate ETEF.
- Record the time history of any desired EDP (or QoI), e.g., crest displacement in the horizontal direction. In nonlinear models, the analysis fails at some time, t_{fail} (e.g., $t_{\text{fail}} = 5.9$ s in Figure 4). Failure is interpreted either as a numerical non-convergence or physical failure (drift beyond a displacement threshold).

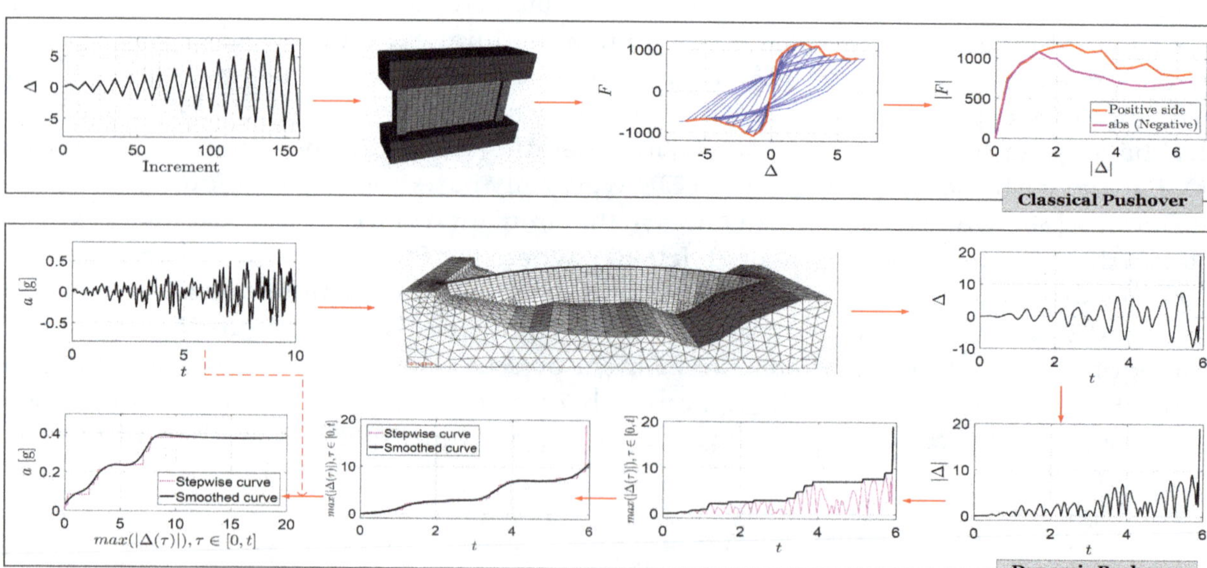

Figure 4. Comparing the cyclic pushover and dynamic pushover methods to develop capacity functions; Δ is displacement, F force, and a acceleration.

- Compute the absolute value of EDP at time interval $[0, t_{\text{fail}}]$. This step guarantees that all the peak values in positive and negative sides are incorporated in the final capacity function.
- Compute the maximum value of the absolute EDP at τ for any time interval $[0, \tau]$ using $\Omega(t) = max(abs(EDP(\tau))); \tau \in [0, t]$. The product of this step is a step-wise increasing function.
- Since the ETEFs have a stochastic nature, it is recommended to smooth them for any practical purposes. The resulting curve is called the ETA curve and is in the EDP-time coordinate system.

- Finally, the EDP-time coordinate system is flipped to make it time-EDP. Moreover, using the relationship between time and acceleration (from the original ETEF), one can convert the time parameter to its equivalent IM parameter. The final product will be a smoothed (or step-wise) curve in the IM-EDP coordinate system, which is the capacity function.
- Although this procedure is applicable only with a single ETEF, in order to reduce the uncertainty (due to the random nature of initial white noise in ETEF generation), the mean of three ETEFs typically is recommended.

Sample application of ETA method and capacity functions for concrete (arch and gravity) dams can be found in Meghella and Furgani [17], Hariri-Ardebili et al. [18], Valamanesh et al. [56], Hariri-Ardebili and Mirzabozorg [65], Hariri-Ardebili and Saouma [42], Hariri-Ardebili et al. [44].

3. Case Study Examples

According to the historical records collected by the ICOLD [66] (through its committee meetings and publications), concrete dams performed relatively well during earthquakes. The U.S. Committee on Large Dams published three comprehensive reports on the observed performance of dams during earthquakes [67–69]. In addition, Nuss et al. [70] collected a list of concrete dams that have been shaken by large earthquakes, but showed a relatively acceptable performance. Later, Hariri-Ardebili and Nuss [8] evaluated the seismic risk of these dams using a semi-quantitative approach and coupled it with finite element analyses. From previous experiences, the key role of the geometry on the seismic resistance of dams has been confirmed. For gravity dams, the use of narrowing sections at the neck is generally associated with cracking problems (e.g., Koyna dam [71,72]), while for the arch dams, a large "U" shape of the valley rift can affect negatively the overall seismic resistance of the dam [73]. These typical aspects of dams' seismic vulnerability have been confirmed by post-earthquake inspections and structural analysis. Some examples of capacity functions are reported below. The finite element simulations have not been discussed in detail deliberately, allowing the readers to focus on the concept behind the use of capacity functions.

Three case studies are considered in this paper based on their complexities:

- Gravity dam: This is a 2D model of a gravity dam including the foundation and reservoir interactions. This example explains the progressive collapse in dams.
- Buttress dam: This is a single monolith of a buttress dam without a foundation and an empty reservoir. This simple example shows the impact of material uncertainties and also introduces the aspects associated with 3D blocks.
- Arch dams: This is a full nonlinear analysis of several arch dam-foundation-reservoir systems. It includes four dams presenting a portfolio of dams.

3.1. Gravity Dam

Koyna gravity dam has a 103-m height; the thicknesses at the base and at the crest are 70.2 m and 14.8 m, respectively, for the central non-overflow monoliths; see Figure 5. The mesh was refined around the neck and dam-foundation interface to capture all the potential failure modes precisely. The fluid–structure interaction (FSI) was applied according to the pressure-based fluid elements. The massless foundation was used, which might lead to slightly different response distribution in dams, particularly for relatively soft rocks, because the radiation damping is ignored. Applied loads on the system were: dam's self-weight, hydrostatic pressure, bottom sediment, and seismic loads. Nonlinearity was originated from a concrete smeared crack model. The modulus of elasticity and compressive strength for concrete were assumed to be 26.4 GPa and 20.4 MPa, respectively. The modulus of elasticity of the foundation was 17 GPa.

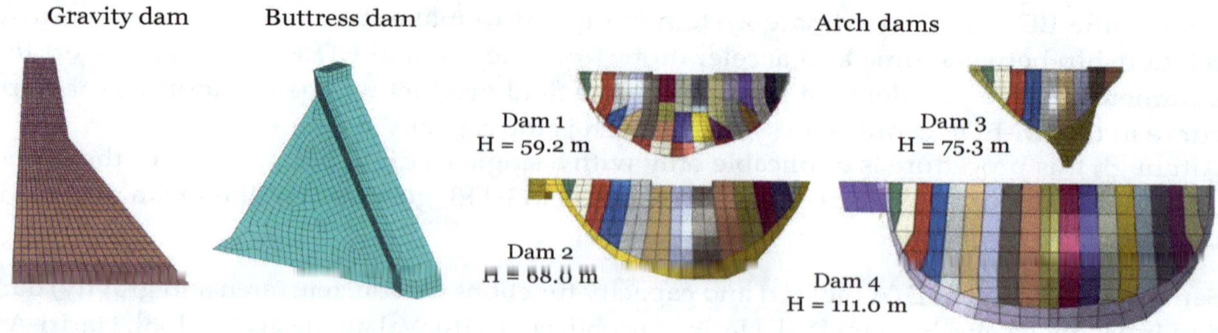

Figure 5. Finite element models for the dam body. Note: the foundation and reservoir are not shown for gravity and arch dams.

3.2. Buttress Dam

Sefidrud buttress dam has a 106-m height in the middle section and consists of 26 monoliths, each 14 m long. The slopes of the dam on the downstream and upstream faces are 1:0.6 and 1:0.4, respectively [74]. It was constructed of plain concrete. It was assumed that the dam was fixed at the base where the ETEFs were applied. Figure 5 shows the finite element mesh. A 3D co-axial rotating smeared crack model was used for the nonlinear analyses [75]. The model was capable of cracking in three orthogonal directions, as well as crushing.

3.3. Arch Dams

Four arch dams were considered in this paper to introduce the key aspects associated with their seismic response towards global failure. Together they represent a reasonable representation of concrete arch dams with heights ranging from 59–111 m [17]; see Figure 5. For each dam, the interaction with the foundation was considered using the massless approach, which for high-intensity level earthquakes represents a major limitation (as discussed later in the paper). The FSI was accounted for by direct modeling the fluid domain using acoustic elements, a non-reflecting boundary condition at the far-end domain, and a wave reflection coefficient at the reservoir bottom. The nonlinear behavior of vertical and horizontal joints separating the different blocks of the dam were modeled to reproduce both joint slips and openings. The nonlinear behavior of concrete was considered using the damage plasticity model by Lee and Fenves [16]. Applied loads were: dead weight, hydrostatic load (full reservoir), winter thermal condition, and seismic loads. The water pressure inside the contraction joints was not modeled [76].

4. Results

4.1. Progressive Collapse of a Gravity Dam

The damage response of the case study gravity dam is explained in this section subjected to a fifteen-second ETEF. Figure 6a shows the variation of the normalized crest horizontal displacement with respect to the total dam height (i.e., 103 m). The dam showed a nearly linear displacement up to a PGA of 0.6 g with $\Omega(\Delta)/H_{dam} = 0.1$. The first large normalized displacement jump occurred for a PGA of about 0.78 g (which corresponds to about 12 s of ETEF). Moreover, the variation of three cracks and the crushing is shown in Figure 6b using the concept of DI. Although many DIs have been proposed by the authors for both gravity and arch dams [42,44], a simple one was adopted. In this paper, DI is defined as a ratio of the damaged area (in a particular cracked direction or crushed situation) to the total dam cross-sectional area. Note that a similar metric was already proposed by Ghanaat [31] for linear elastic analysis and was called the percentage of overstressed regions (at different demand-capacity ratios). At PGA = 0.6 g (which was a limit to linear displacement response), the DI in Directions 1, 2, and 3 was 4%, 1%, and 0.5%, respectively. Moreover, crushing was practically zero. All four DIs showed a meaningful jump at PGA of 0.78 g. Thus, this intensity can be announced as the capacity of this dam (under current numerical assumptions).

Finally, the damage profile of the dam at two different time steps corresponding to two (relatively) big changes at the pre-failure and failure stages is shown in Figure 6c. The progressive collapse can be tracked in these two sets of plots, especially for the Crack-2 and Crack-3 directions.

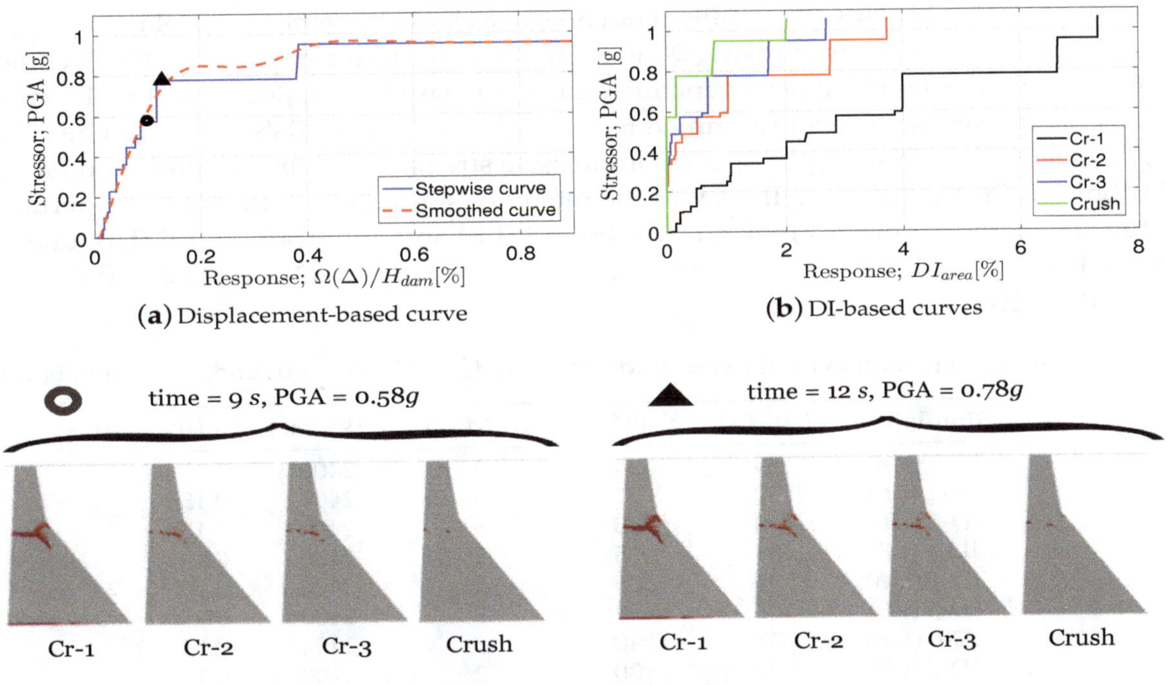

(c) Damage profiles at two different times

Figure 6. Computed capacity functions for the case study gravity dam; PGA is peak ground acceleration; Cr refers to cracking; DI is the damage index; Δ is displacement. (**a**) Displacement-based curve; (**b**) DI-based curves; (**c**) damage profiles at two different times.

The results confirmed the typical failure modes for the Koyna dam, widely investigated by different authors using traditional accelerograms (see [8] for a summary of Koyna numerical models), as well as through seismic testing on the scale model [72]. If typical concerns linked to the non-linear numerical methods are left out, the capacity function offers important information on the development of response parameters and DI over increasing seismic actions.

It is worth noting that these results were based on a single ETA simulation exploiting the advantages of this technique in capacity estimation. The advantage of computational time also unlocks opportunities over parametric analysis. With these procedures, it may be possible to establish the capacity functions associated with typical geometries of concrete gravity dams and define a classification of dams for their seismic performance. The DI curves can be defined as well, like those indicated in Figure 6b, to understand the most critical failure mode and the associated limit of acceptance.

With the global and local measure of the seismic response of dams under nonlinear behavior, it is also permitted in principle to follow the example of the buildings sector on the definition of the so-called " behavior factor" [77]. This will allow the analysts to perform linear analysis with seismic spectra properly scaled down to consider the accepted level of damages.

4.2. Uncertainty Quantification of a Buttress Dam

So far, the ETA was applied for a gravity dam with deterministic material properties. However, it is well accepted that there are different levels of uncertainties in quantifying the material properties especially for large deteriorated concrete structures [78,79]. This implies that finite element models should be analyzed multiple times, each one with different combinations of the material properties to properly account for all the potential combinations.

Again, the ETA method offers unique characteristics to provide "probabilistic capacity functions" only with a limited number of nonlinear simulations. However, the first step is to develop an efficient number of random finite element models. In general, there are two broad approaches to achieve this goal: (1) using one of the Monte Carlo family models such as Latin hypercube sampling (LHS) [80] or (2) using one of many design of experiment (DOE) methods [79]. In this paper, the DOE technique was adopted to develop a series of capacity functions, each one based on a specific material combination.

Three concrete parameters were assumed to be random variables (RVs), i.e., concrete modulus of elasticity, E, concrete tensile strength, f_t, and compressive strength, f_c. Nine different combinations are shown in Table 1. The first one, i.e., ID-0, was the reference simulation with all the properties at their mean value. Then, the properties took a lower bound (LB) and an upper bound (UB) one at a time (simulations ID-1–ID-6). Finally, two last simulations (i.e., ID-7 and ID-8) considered the combined effects of all three RVs.

Table 1. Material combination for the case study buttress dam. LB, lower bound; UB, upper bound.

Simulation	E (GPa)	f_c (MPa)	f_t (MPa)	ρ (kg/m^3)	ν (-)
ID-0 (Ref)	35	35	3	2400	0.15
ID-1 (LB)	30	35	3	2400	0.15
ID-2 (UB)	40	35	3	2400	0.15
ID-3 (LB)	35	35	2	2400	0.15
ID-4 (UB)	35	35	4	2400	0.15
ID-5 (LB)	35	30	3	2400	0.15
ID-6 (UB)	35	40	3	2400	0.15
ID-7 (LB)	30	30	2	2400	0.15
ID-8 (UB)	40	40	4	2400	0.15

Again, the buttress dam was modeled without any foundation and reservoir. The base of the dam was fixed, and an identical ETEF was applied in all cases. Subsequently, the crest displacement and the DI were recorded during the dynamic analysis. Results are summarized in Figure 7. For each of three RVs, and also the combined one, the displacement-based capacity functions are provided, which compare the UB and LB values with the reference one. Moreover, the evolution of DI is reported as a function of time. For the current purpose, only the first 13 s of the ETEF were applied, regardless of the damage level. The major observations can be summarized as follows:

- In a displacement-based capacity function (stressor-response coordinate system), the horizontal part corresponds to a large displacement with a small increase in the intensity level (i.e., numerical or physical) of failure. On the other hand, the vertical part shows the resistance of the dam (i.e., no or small drift with increasing seismic intensity).
- In general, lower and upper bound material properties decreased and increased the dynamic capacity, respectively. For example, in Figure 7b, the lower bound tensile strength, i.e., 2 MPa, reduced the dynamic capacity from 0.15 g to 0.11 g. On the other hand, the upper bound material property, i.e., 4 MPa, increased the dynamic capacity to about 0.3 g, where the analysis was not continued further.
- For the modulus of elasticity (see Figure 7a), the lower bound property reduced the dynamic capacity at pre-failure stages; however, it yielded to the same failure capacity of the reference analysis. One may note that the modulus of elasticity mainly controlled the initial stiffness of the system.
- According to Figure 7c, the compressive strength did not have any impact on the dynamic capacity of this dam. This can be attributed to the fact that the failure mode was covered by tensile damage (and not the compression damage).

- Finally, Figure 7d shows that the combined effect of all three RVs. The combined model was close to the one with only the modulus of elasticity as RV. The upper bound of the combined model did not show any damage (i.e., DI = 0) up to t = 15 s of analysis, where the simulation was terminated.

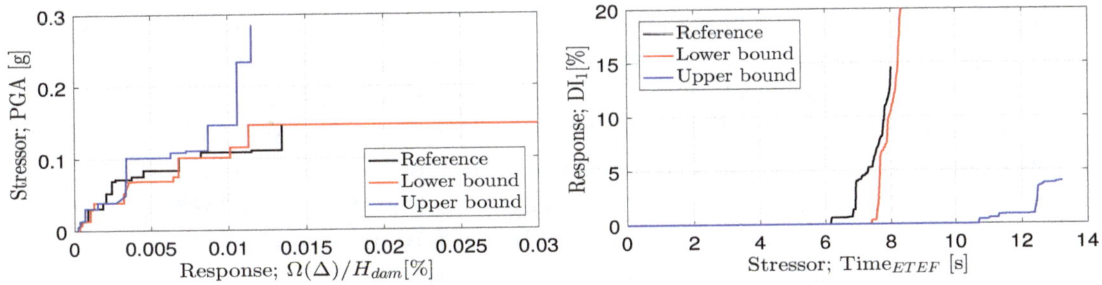

(a) E variation; ID-1, ID-2, and ID-0

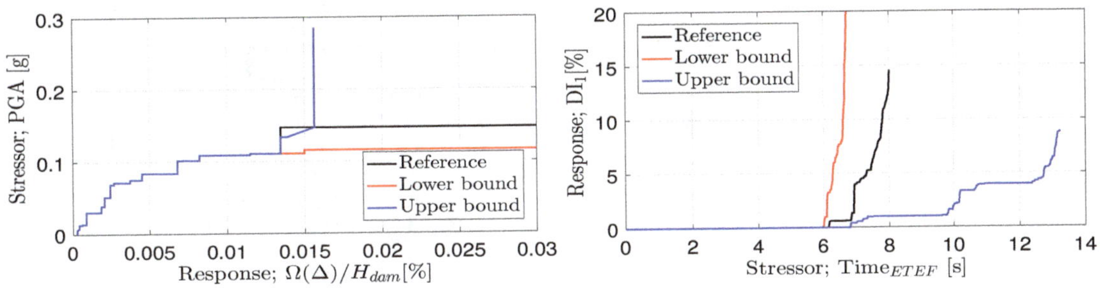

(b) f_t variation; ID-3, ID-4, and ID-0

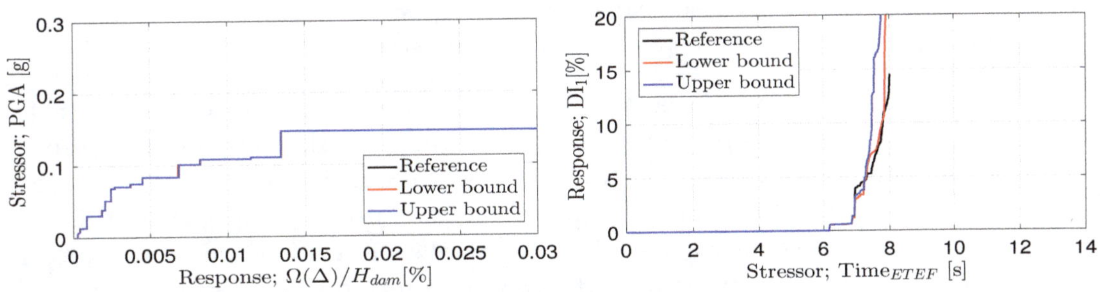

(c) f_c variation; ID-5, ID-6, and ID-0

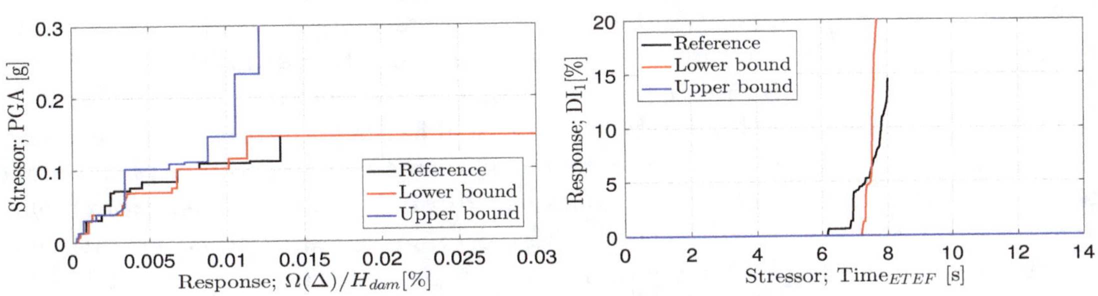

(d) $E + f_t + f_c$ variations; ID-7, ID-8, and ID-0

Figure 7. Computed capacity functions for the case study buttress dam; PGA is peak ground acceleration; DI is the damage index; Δ is displacement. (**a**) E variation; ID-1, ID-2, and ID-0; (**b**) f_t variation; ID-3, ID-4, and ID-0; (**c**) f_c variation; ID-5, ID-6, and ID-0; (**d**) $E + f_t + f_c$ variations; ID-7, ID-8, and ID-0.

These plots clearly highlight the impact of uncertain parameters in developing the dynamic capacity functions for a dam. Although it is possible to perform some extra sensitivity analysis or even develop a response surface meta-model from these data [81], they are skipped in this paper.

4.3. Assessment of an Arch Dam Portfolio

Arch and dome structures are characterized by having more key geometric parameters (i.e., curvatures). The shape of old dams has been defined during the design stage without considering explicitly the aspects related to the dynamic capacity of the structure. Figure 8 presents the relationship between the applied dynamic intensity and the response of the dams. According to the results obtained in this study, larger "U"-shaped dams appear to be the most vulnerable, while "V"-shaped dams provide higher seismic resistance. Each of the curves also shows key points associated with damages of the upstream and downstream faces of 1% and 20%, considered respectively minor and major yielding points. The relationship between these points and the associated serviceability and ultimate limit states, represented by dotted lines, gives an idea of the safety factor often translated in demand/capacity ratios. As mentioned previously, these ratios should be widely investigated and discussed to establish what can be defined as acceptable in terms of overall seismic safety.

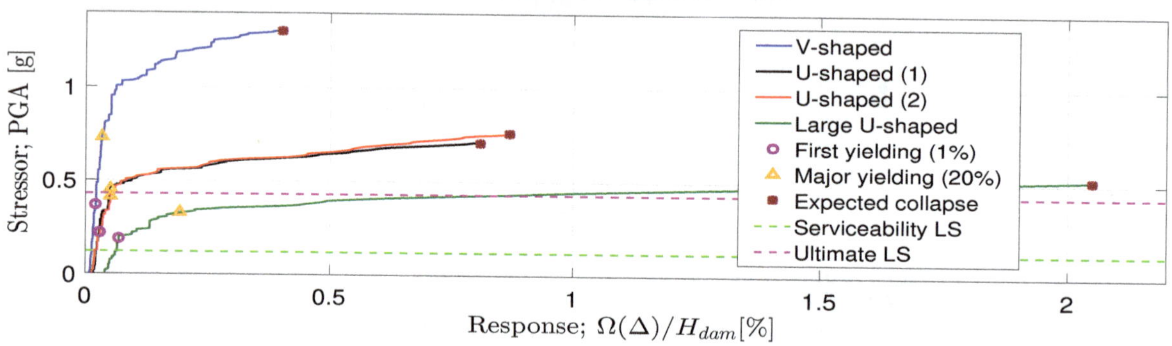

Figure 8. Computed displacement-based capacity functions for an arch dam portfolio; PGA is peak ground acceleration; Δ is displacement; LS refers to the limit state.

According to the authors, it is not reasonable to associate the serviceability limit state with the attainment of the first crack. This is particularly true for numerical models with optimized meshes not able to capture localized damages. For this reason, it is recommended to set a value that can be associated with a measurable level of damages; the 20% value is proposed in this work. For the ultimate limit state, generally associated with the uncontrolled release of water, it is recommended to use the results obtained with the numerical model paying attention to the mechanism producing the limit state.

Figure 9 shows the "V"-shaped dam behavior for the serviceability (i.e., 0.13 g) and ultimate limit states (i.e., 0.52 g). For the low-intensity seismic action, the dam did not show any damage, and the compressive stresses (minimum principal stress: blue vectors) provided an effective arch behavior. As the applied dynamic load grew, the vertical joints started to open, and the cantilever effect appeared. The cantilever effect is one of the mechanisms able to produce horizontal cracks and redistribution of stresses. According to the simulations, "through cracks" started from the downstream face propagating towards the opposite face. In these cases, it is reasonable to think that one of the mechanisms able to produce the uncontrolled release of the water (or uncontrolled leakages) is the rocking of the isolated blocks at the top of the dam [82].

The attainment of the ultimate limit state in this condition should be defined using an appropriate model or independent procedures (i.e., hand calculation based on equilibrium). In the latter case, it will be necessary to establish the acceleration able to produce the overturning of the upper blocks in the upstream direction, information to be considered while the capacity curve for the dam is used. This is an example of special considerations required to use the capacity curves properly. If general rules or procedures cannot be defined because of the bespoke nature of arch dams, the proposed ETA method can be used as a tool to compare qualitatively the dynamic response of different structures or to have a better understanding of the impact of model assumptions over the seismic response. In this last context, the capacity function can be considered self-sufficient information.

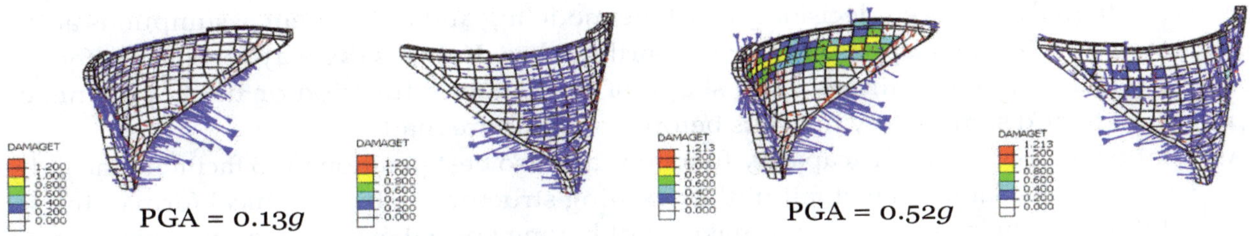

Figure 9. Damage state for dams at two different seismic intensities; PGA = 0.13 g (serviceability limit state); and PGA = 0.52 g (ultimate limit state).

On the other hand, one of the most controversial issues in dynamic analysis of dams is the concept of "massed/massless foundation". The massless foundation is a widely-adopted technique for practical seismic safety assessments of dams. Nevertheless, it has been extensively demonstrated that such a method provides over-conservative and misleading dam response [83]. To consider such effects properly, a drastic increase in the size and density of finite element domains is required [13]. Capacity functions, being able to represent the dam response in just one synthetic picture, can help to identify a critical PGA value beyond which the massless approach is no longer valid to achieve a consistent and realistic dam response; Figure 10. This critical PGA for Arch Dam #4 was about 0.2 g.

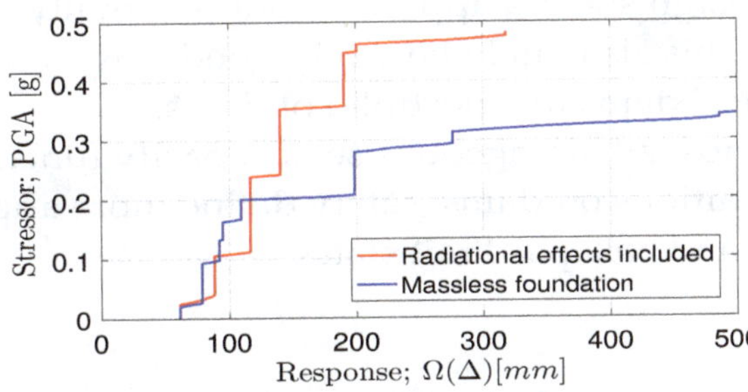

Figure 10. Capacity functions for Arch Dam #4 with two foundation models; PGA is peak ground acceleration; Δ is displacement.

5. Present and Future Opportunities Offered by Capacity Functions

During the 3rd Meeting of EWG Dams and Earthquakes held in Lisbon [1], the authors presented the results of their studies and following a constructive collaboration with other researchers and professionals agreed on the expected goals for the incoming years. It appears that the research to be done on the seismic response of dams can be divided into two fields: the interpretation of recorded response and back-analysis under low to medium earthquakes and the forecast of the response of dams for strong earthquakes able to produce the uncontrolled release of the water or the collapse of the structure. Considering the different tools required for these two goals (i.e., nonlinear behavior capabilities versus extensive input parameter calibration), it is reasonable to think that the researchers working on these two fields will develop their own procedures and will feed different practices.

The most advanced (or latest) regulations and guidelines regarding the seismic assessment of existing dams are very demanding in terms of methods, resources, and time needed for its accomplishment. In some cases, the set of numerical simulations required may be economically unsustainable. Moreover, even if properly executed, the analyses are significantly affected by uncertainty, particularly when strong earthquakes have to be considered and strong nonlinear effects on construction joints and concrete are expected to occur. Uncertainties can be handled by calibrating the models with the real response of the dam, but from the experience collected so far, few dams' experienced strong earthquakes able to activate substantially their nonlinear behavior [70]. It is worth mentioning that if something similar happened, the dam would likely be damaged enough to be dismissed or to be retrofitted significantly (i.e., new resisting concrete blocks), and this means the recorded data cannot be used. In this light, the information provided by the "capacity functions" can be very useful to make balanced decisions about the modeling and assessment assumptions addressed to the representation of the response for strong earthquakes. If this is the way forward, safety factors can be used to handle uncertainties. The shape of the capacity function or the key points can be post-processed to add some safety margins before decisions are made.

As mentioned previously, the capacity functions are also best positioned to facilitate the judgment of the safety factor available against failures for existing structures. This graphical form of the seismic response of the dams helps any decision-makers not having special expertise on the seismic response of dams to drive economic efforts for advanced studies or retrofitting works.

It is worth mentioning that standardized seismic functions such as those used in the ETA method may resolve major issues associated with the arbitrary selection of natural accelerograms. It is intuitive that it is hard to compare the seismic response of dams in the nonlinear field if the applied time histories are different. According to the experience of some of the authors, legislators are reluctant to use synthetic signals. This may represent a major obstacle for the ETA method and other similar methods. More recent studies on the generation of ETEF functions are addressing the main concerns received by the sector on their validity.

A proposal to receive the acceptance of these signals is to produce site-specific ETEF signals for different seismic regions able to represent more closely the soil and earthquake conditions. In this case, dams in the same region will be checked with the same signals defined by local codes. This appears to be a radical change from the design spectra approach, but is actually the most effective way to meet the new technical codes' goals. This also might be a very good way to compare the seismic response of various dams and identify critical dams in a portfolio of dams.

Last but not least, for the most advantageous use of capacity functions, it is fundamental that the technical codes or the key institutions on dams' safety define more appropriate acceptance limits for damages in both serviceability and collapse limit states.

6. Conclusions

Capacity functions are recognized in the building sector as a tool to judge the nonlinear behavior of framed structures. The advantages of their application seem remarkable and may be crucial within the seismic assessment procedures for concrete dams. A brief introduction to their use and the theoretical background was shown in this paper using some examples.

Following the recollection of some lessons learned on the seismic safety of dams located in high seismicity zones, the standard assessment procedures used by legislation and engineers were introduced. The new approach of "capacity functions" was then introduced for dams and the associated available methods briefly explained, in particular: incremental dynamic analysis (IDA), cloud analysis (CLA), and endurance time analysis (ETA). The last method was selected to show some advantages using capacity functions based on three examples: a concrete gravity dam, a buttress dam, and a portfolio of arch dams. The limits and the future enhancements of these seismic assessment procedures were also discussed.

Despite several applications of capacity functions in the dam industry, this procedure is still in a preliminary stage. The capacity functions may represent a powerful tool to meet some of the key requirements introduced by modern technical codes. The ETA method represents a good choice to determine the capacity functions with reasonable computational time.

The procedure introduced in the paper is focused mainly on the production of capacity curves. It is worth repeating that the modeling assumptions were not the subject of this work. Analysts doing the assessment must select the appropriate assumptions based on the best practice and the local legislation. If for the dam under assessment, it is crucial to consider special aspects such as galleries inside the dam, other structures connected to the dam (bridges, intake towers, or gates), reservoir sediments, foundation non-linearity, landslide effects, pore water pressure, dynamic uplift, lift joints, seismic input non-uniformity, or any other aspect, the analyst should select the appropriate modeling tools and adapt the procedure introduced here for the capacity function. In a certain sense, the main scope is to define a new way to add the shaking input and to get responses out of the numerical models prepared by the analysts, not a guide to set the numerical models.

It is the authors' opinion that capacity functions in the form of standardized procedures will be available in the near future. Their comparison on a large scale will highlight crucial information of the seismic response of dams characterized by similar geometries and conditions.

Author Contributions: Conceptualization, L.F. and M.A.H.A.; methodology, L.F., M.M. and M.A.H.A.; software, L.F., M.A.H.A. and S.M.S.K.; validation, L.F., M.M. and M.A.H.A.; formal analysis, L.F. and S.M.S.K.; investigation, L.F. and M.A.H.A.; resources, L.F., M.M. and M.A.H.A; data curation, L.F. and M.A.H.A; writing—original draft preparation, L.F. and M.A.H.A.; writing—review and editing, L.F., M.M. and M.A.H.A.; supervision, M.A.H.A. and M.M.; project administration, M.A.H.A.

Acknowledgments: The second author would like to express his sincere appreciation to Professor Victor E. Saouma (University of Colorado), and Dr. Jerzy W. Salamon (U.S. Bureau of Reclamation) for their support.

References

1. Furgani, L.; Hariri-Ardebili, M.; Meghella, M. Towards the Seismic Capacity Assessment of Concrete Dams. In Proceedings of the 3rd Meeting of EWG Dams and Earthquakes, Lisbon, Portugal, 6–8 May 2019; pp. 87–98.
2. USBR-Manual. *Dam Safety Risk Analysis, Best Practices Training Manual*; Version 2.2; Technical Report; U.S. Department of the Interior Bureau of Reclamation in Coorporation with the U.S. Army Corps of Engineers: Denver, CO, USA, 2011.
3. Canadian Dam Association (CDA). *Dam Safety Guidelines*; Technical Report; Canadian Dam Association (CDA): Edmonton, AB, Canada, 2007.
4. Decreto Ministeriale 26 Giugno 2014. E TRASPORTI, Ministero delle Infrastrutture. In *Norme Tecniche per la Progettazione E la Costruzione degli Sbarramenti di Ritenuta (Dighe E Traverse). Gazzetta Ufficiale 07/07/2014 n.156*; Technical Report; Gazzetta Ufficiale: Rome, Italy, 2004.
5. NSW. *Risk Management Policy Framework for Dam Safety*; Technical Report; New South Wales Government Dams Safety Committee: Parramatta, NSW, Australia, 2006.
6. Wieland, M. Features of seismic hazard in large dam projects and strong motion monitoring of large dams. *Front. Arch. Civil Eng. China* **2010**, *4*, 56–64. [CrossRef]
7. Chopra, A.K. Earthquake analysis of arch dams: Factors to be considered. *J. Struct. Eng.* **2012**, *138*, 205–214. [CrossRef]
8. Hariri-Ardebili, M.A.; Nuss, L.K. Seismic risk prioritization of a large portfolio of dams: Revisited. *Adv. Mech. Eng.* **2018**, *10*, 1687814018802531. [CrossRef]
9. ICOLD. *Selecting Seismic Parameters for Large Dams, Guidelines, Bulletin 148 (Revision of Bulletin 72)*; Technical Report; International Commission on Large Dams: Paris, France, 2010.
10. Wieland, M. Seismic hazard and seismic design and safety aspects of large dam projects. In *Perspectives on European Earthquake Engineering and Seismology*; Springer: Berlin/Heidelberg, Germany, 2014; pp. 627–650.
11. Fiorentino, G.; Furgani, L.; Magliano, S.; Nuti, C. Probabilistic evaluation of dam base sliding. In Proceedings of the 5th ECCOMAS Thematic Conference on Computational Methods in Structural Dynamics and Earthquake Engineering, Crete Island, Greece, 25–27 May 2015.
12. Hariri-Ardebili, M.A.; Kianoush, M.R. Integrative seismic safety evaluation of a high concrete arch dam. *Soil Dyn. Earthq. Eng.* **2014**, *67*, 85–101. [CrossRef]
13. Saouma, V.; Miura, F.; Lebon, G.; Yagome, Y. A simplified 3D model for soil-structure interaction with radiation damping and free field input. *Bull. Earthq. Eng.* **2011**, *9*, 1387–1402. [CrossRef]
14. Bouaanani, N.; Lu, F. Assessment of potential-based fluid finite elements for seismic analysis of dam-reservoir systems. *Comput. Struct.* **2009**, *87*, 206–224. [CrossRef]
15. Puntel, E.; Bolzon, G.; Saouma, V. Fracture Mechanics Based Model for Joints under Cyclic Loading. *ASCE J. Eng. Mech.* **2006**, *132*, 1151–1159. [CrossRef]

16. Lee, J.; Fenves, G.L. A plastic-damage concrete model for earthquake analysis of dams. *Earthq. Eng. Struct. Dyn.* **1998**, *27*, 937–956. [CrossRef]
17. Meghella, M.; Furgani, L. Application of Endurance Time Analysis Method to the NonLinear Seismic Analysis of dams: Potentialities and Limitations. In Proceedings of the 82nd Annual Meeting of ICOLD, Banff, AB, Canada, 4–5 October 2014.
18. Hariri-Ardebili, M.A.; Mirzabozorg, H.; Estekanchi, H.E. Nonlinear seismic assessment of arch dams and investigation of joint behavior using endurance time analysis method. *Arabian J. Sci. Eng.* **2014**, *39*, 3599–3615. [CrossRef]
19. Estekanchi, H.; Vafai, A.; Sadeghazar, M. Endurance time method for seismic analysis and design of structures. *Sci. Iran.* **2004**, *11*, 361–370.
20. Darbre, G.R.; Schwager, M.; Panduri, R. Seismic safety evaluation of all large dams in Switzerland: Lessons learned. In Proceedings of the 26th International Congress on Large Dams, Vienna, Austria, 1–7 July 2018.
21. OFEG. *Sécurité des Ouvrages D'Accumulation—Documentation de Base Pour la Vérification des Ouvrages D'Accumulation aux Séismes*; Technical Report; Officé Fedéral des Eaux et de la Géologie: Bern, Switzerland, 2003.
22. Gosschalk, E.; Severn, R.; Charles, J.; Hinks, J. An Engineering Guide to Seismic Risk to Dams in the United Kingdom, and its international relevance. *Soil Dyn. Earthq. Eng.* **1994**, *13*, 163–179. [CrossRef]
23. CFRB. *Guidelines for the Justification of the Stability of Gravity Dams*; Technical Report; Comité Francais des Barrages et Reservoirs: France, 2012.
24. Royet, P.; Peyras, L. French guidelines for structural safety of gravity dams in a semi-probabilistic format. In Proceedings of the 9th ICOLD European Club Symposium (ITCOLD), Venice, Italy, 10–12 April 2013; p. 8.
25. Mohan, K.J.; Kumar, R.P. Earthquakes and dams in India: An overview. *Int. J. Civil Eng. Technol.* **2013**, *4*, 6.
26. Shimamoto, K. Trial implementation of new Japanese guidelines for seismic performance evaluation of dams during large earthquakes. In Proceedings of the 75th Annual Meeting of ICOLD, St. Petersburg, Russia, 24–27 June 2007.
27. Tosun, H.; Zorluer, İ.; Orhan, A.; Seyrek, E.; Savaş, H.; Türköz, M. Seismic hazard and total risk analyses for large dams in Euphrates basin, Turkey. *Eng. Geol.* **2007**, *89*, 155–170. [CrossRef]
28. Freeman, S. The capacity spectrum method. In Proceedings of the 11th European Conference on Earthquake Engineering, Paris, France, 6–11 September 1998.
29. Fajfar, P. A nonlinear analysis method for performance based seismic design. *Earthq. Spectra* **2000**, *16*, 573–592. [CrossRef]
30. Goel, R. Evaluation of modal and FEMA pushover procedures using strong-motion records of buildings. *Earthq. Spectra* **2005**, *21*, 653—684. [CrossRef]
31. Ghanaat, Y. Failure modes approach to safety evaluation of dams. In Proceedings of the 13th World Conference on Earthquake Engineering, Vancouver, BC, Canada, 1–6 August 2004.
32. Malm, R. *Guideline for FE Analyses of Concrete Dams*; Technical Report; Energiforsk: Stockholm, Sweden, 2016.
33. Goldgruber, M.; Malm, R. Nonlinear Seismic Simulation of an Arch Dam using XFEM. In Proceedings of the Simulia Regional Users Meeting, Graz, Austria, 10 November 2014.
34. ICOLD. *Lessons from Dam Incidents, Complete Edition*; Technical Report; International Commission on Large Dams: Paris, France, 1974.
35. Boyer, D. Geologic factors influencing dam foundation failure modes. In Proceedings of the 26th Annual USSD Conference, San Antonio, TX, USA, 1–5 May 2006.
36. Malm, R.; Hassanzadeh, M.; Hellgren, R. (Eds.) *Numerical Analysis of Dams*; ATCOLD, Austrian National Committee on Large Dams: Stockholm, Sweeden, 2018; pp. 1–722.
37. Hariri-Ardebili, M.; Saouma, V. Single and multi-hazard capacity functions for concrete dams. *Soil Dyn. Earthq. Eng.* **2017**, *101*, 234–249. [CrossRef]
38. Vamvatsikos, D.; Cornell, C. Applied incremental dynamic analysis. *Earthq. Spectra* **2004**, *20*, 523–553. [CrossRef]
39. Vamvatsikos, D.; Cornell, C. Incremental dynamic analysis. *Earthq. Eng. Struct. Dyn.* **2002**, *31*, 491–514. [CrossRef]
40. Sudret, B.; Der Kiureghian, A. Comparison of finite element reliability methods. *Probab. Eng. Mech.* **2002**, *17*, 337–348. [CrossRef]

41. Zhang, Y.; Chen, G.; Zheng, L.; Li, Y.; Zhuang, X. Effects of geometries on three-dimensional slope stability. *Can. Geotech. J.* **2013**, *50*, 233–249. [CrossRef]
42. Hariri-Ardebili, M.A.; Saouma, V.E. Quantitative failure metric for gravity dams. *Earthq. Eng. Struct. Dyn.* **2015**, *44*, 461–480. [CrossRef]
43. Ansari, M.I.; Agarwal, P. Categorization of Damage Index of Concrete Gravity Dam for the Health Monitoring after Earthquake. *J. Earthq. Eng.* **2016**, *20*, 1222–1238. [CrossRef]
44. Hariri-Ardebili, M.A.; Furgani, L.; Meghella, M.; Saouma, V.E. A new class of seismic damage and performance indices for arch dams via ETA method. *Eng. Struct.* **2016**, *110*, 145–160. [CrossRef]
45. Vamvatsikos, D. Performing incremental dynamic analysis in parallel. *Comput. Struct.* **2011**, *89*, 170–180. [CrossRef]
46. Alembagheri, M.; Ghaemian, M. Seismic assessment of concrete gravity dams using capacity estimation and damage indexes. *Earthq. Eng. Struct. Dyn.* **2013**, *42*, 123–144. [CrossRef]
47. Pan, J.; Xu, Y.; Jin, F. Seismic performance assessment of arch dams using incremental nonlinear dynamic analysis. *Eur. J. Environ. Civ. Eng.* **2015**, *19*, 305–326. [CrossRef]
48. Alembagheri, M.; Seyedkazemi, M. Seismic performance sensitivity and uncertainty analysis of gravity dams. *Earthq. Eng. Struct. Dyn.* **2015**, *44*, 41–58. [CrossRef]
49. Alembagheri, M.; Ghaemian, M. Incremental dynamic analysis of concrete gravity dams including base and lift joints. *Earthq. Eng. Eng. Vib.* **2013**, *12*, 119–134. [CrossRef]
50. Chen, D.H.; Yang, Z.H.; Wang, M.; Xie, J.H. Seismic performance and failure modes of the Jin'anqiao concrete gravity dam based on incremental dynamic analysis. *Eng. Fail. Anal.* **2019**, *100*, 227–244. [CrossRef]
51. Soysal, B.; Binici, B.; Arici, Y. Investigation of the relationship of seismic intensity measures and the accumulation of damage on concrete gravity dams using incremental dynamic analysis. *Earthq. Eng. Struct. Dyn.* **2016**, *45*, 719–737. [CrossRef]
52. Hariri-Ardebili, M.; Saouma, V. Collapse Fragility Curves for Concrete Dams: Comprehensive Study. *ASCE J. Struct. Eng.* **2016**, *142*, 04016075. [CrossRef]
53. Rezaeian, S.; Der Kiureghian, A. Simulation of synthetic ground motions for specified earthquake and site characteristics. *Earthq. Eng. Struct. Dyn.* **2010**, *39*, 1155–1180. [CrossRef]
54. Shome, N. Probabilistic Seismic Demand Analysis of Nonlinear Structures. Ph.D. Thesis, Stanford University, Stanford, CA, USA, 1999.
55. Estekanchi, H.; Valamanesh, V.; Vafai, A. Application of endurance time method in linear seismic analysis. *Eng. Struct.* **2007**, *29*, 2551–2562. [CrossRef]
56. Valamanesh, V.; Estekanchi, H.; Vafai, A.; Ghaemian, M. Application of the endurance time method in seismic analysis of concrete gravity dams. *Sci. Iran.* **2011**, *18*, 326–337. [CrossRef]
57. Hariri-Ardebili, M.A.; Saouma, V.; Porter, K.A. Quantification of seismic potential failure modes in concrete dams. *Earthq. Eng. Struct. Dyn.* **2016**, *45*, 979–997. [CrossRef]
58. Nozari, A.; Estekanchi, H. Optimization of endurance time acceleration functions for seismic assessment of structures. *Int. J. Optim. Civ. Eng.* **2011**, *1*, 257–277.
59. Mashayekhi, M.; Estekanchi, H. Investigation of Non-Linear Cycles' Properties in Structures Subjected to Endurance Time Excitation Functions. *Int. J. Optim. Civ. Eng.* **2013**, *3*, 239–257.
60. Mashayekhi, M.; Estekanchi, H.E.; Vafai, H.; Mirfarhadi, S.A. Development of hysteretic energy compatible endurance time excitations and its application. *Eng. Struct.* **2018**, *177*, 753–769. [CrossRef]
61. Mashayekhi, M.; Estekanchi, H.; Vafai, H. Simulation of Endurance Time Excitations Using Increasing Sine Functions. *Int. J. Optim. Civ. Eng.* **2019**, *9*, 65–77.
62. Mashayekhi, M.; Estekanchi, H.E.; Vafai, H.; Ahmadi, G. An evolutionary optimization-based approach for simulation of endurance time load functions. *Eng. Optim.* **2019**, 1–20. [CrossRef]
63. Krawinkler, H. *Report No. ATC 24: Guidelines for Cyclic Seismic Testing of Components of Steel Structures*; Technical Report; Applied Technology Council: Redwood City, CA, USA, 1992.
64. Filiatrault, A.; Wanitkorkul, A.; Constantinou, M. *Development and Appraisal of a Numerical Cyclic Loading Protocol for Quantifying Building System Performance*; Technical Report MCEER-08-0013; University at Buffalo, State University of New York: New York, NY, USA, 2008.
65. Hariri-Ardebili, M.A.; Mirzabozorg, H. Estimation of probable damages in arch dams subjected to strong ground motions using endurance time acceleration functions. *KSCE J. Civ. Eng.* **2014**, *18*, 574–586. [CrossRef]

66. ICOLD. *Design Features of Dams to Effectively Resist Seismic Ground Motion, Bulletin 120*; Technical Report; International Commission on Large Dams: Paris, France, 2001.
67. USCOLD. *Observed Performance of Dams during Earthquakes*; Technical Report; U.S. Committee on Large Dams: Denver, CO, USA, 1992.
68. USCOLD. *Observed Performance of Dams during Earthquakes*; Technical Report; U.S. Committee on Large Dams: Denver, CO, USA, 2000; Volume 2
69. USSD. *Observed Performance of Dams during Earthquakes*; Technical Report; U.S. Society on Dams: Denver, CO, USA, 2014; Volume 3.
70. Nuss, L.; Matsumoto, N.; Hansen, K. Shaken, But Not Stirred—Earthquake Performance of Concrete Dams. In Proceedings of the 32nd USSD Annual Meeting and Conference: Innovative Dam and Levee Design and Construction for Sustainable Water Management, New Orleans, LA, USA, 23–27 April 2012.
71. Chowdhury, M.; Matheu, E.; Hall, R. Shake table experiment of a 1/20-Scale Koyna dam model. In Proceedings of the 19th International Modal Analysis Conference, IMAC XIX, Hyatt Orlando, Kissimmee, FL, USA, 5–8 February 2001.
72. Wilcoski, J.; Robert, R.; Matheu, E.; Gambill, J.; Chowdhury, M. *Seismic Testing of a 1/20 Scale Model of Koyna Dam*; Report No. ERDC TR-01-17; Technical Report; U.S. Army Corps of Engineers, Engineering Research and Development Center: Washington, DC, USA, 2001.
73. Salamon, J.; Nuss, L. *Design of Double-Curvature Arch Dams Planning, Appraisal, Feasibility Level*; Technical Report; U.S. Department of the Interior Bureau of Reclamation: Denver, CO, USA, 2012.
74. Ghaemian, M. Seismic Response of a Retrofitted Concrete Buttress Dam. In Proceedings of the 11th World Conference on Earthquake Engineering, Acapulco, Mexico, 23–28 June 1996.
75. Hariri-Ardebili, M.A.; Seyed-Kolbadi, S.M. Seismic cracking and instability of concrete dams: Smeared crack approach. *Eng. Fail. Anal.* **2015**, *52*, 45–60. [CrossRef]
76. Hariri-Ardebili, M.A.; Kianoush, M.R. Seismic analysis of a coupled dam-reservoir foundation system considering pressure effects at opened joints. *Struct. Infrastruct. Eng.* **2015**, *11*, 833–850. [CrossRef]
77. EN-1998. *Eurocode-8: Design of Structures for Earthquake Resistance*; Technical Report; The European Union Per Regulation: Brussels, Belgium, 2004.
78. Saouma, V. *Numerical Modeling of AAR*; CRC Press: Boca Raton, FL, USA, 2014.
79. Hariri-Ardebili, M.A.; Seyed-Kolbadi, S.M.; Noori, M. Response Surface Method for Material Uncertainty Quantification of Infrastructures. *Shock Vib.* **2018**, *2018*, 1784203. [CrossRef]
80. Hariri-Ardebili, M.A.; Saouma, V.E. Sensitivity and Uncertainty Quantification of the Cohesive Crack Model. *Eng. Fract. Mech.* **2016**, *155*, 18–35. [CrossRef]
81. Hariri-Ardebili, M.A. MCS-based response surface metamodels and optimal design of experiments for gravity dams. *Struct. Infrastruct. Eng.* **2018**, *14*, 1641–1663. [CrossRef]
82. Malla, S.; Wieland, M. Analysis of an arch gravity dam with a horizontal crack. *Comput. Struct.* **1999**, *72*, 267–278. [CrossRef]
83. Hariri-Ardebili, M.A.; Saouma, V.E. Impact of Near-fault vs. Far-field Ground Motions on the Seismic Response of an Arch Dam with Respect to Foundation Type. *Dam Eng.* **2014**, *24*, 19–52.

5

Lessons Learned Regarding Cracking of a Concrete Arch Dam due to Seasonal Temperature Variations

Richard Malm *, Rikard Hellgren and Jonas Enzell

KTH Royal Institute of Technology, SE 100 44 Stockholm, Sweden; rikard.hellgren@byv.kth.se (R.H.); enzell@kth.se (J.E.)

* Correspondence: richard.malm@byv.kth.se

Abstract: Dams located in cold areas are subjected to large seasonal temperature variations and many concrete dams have cracked as a result. In the 14th International Commission on Large Dams (ICOLD) Benchmark Workshop, a case study was presented where contributors should predict the cracking and displacements due to seasonal variations. In this paper, the conclusions from this case study are presented. Overall, the results from the contributors are well in line with the observations that can be made on the dam and the measurements performed. This shows that using non-linear numerical models is a suitable tool to accurately predict cracking and estimate the displacements of cracked dams. This case study also highlighted important aspects that need special consideration in order to obtain realistic results that can be used to predict the crack pattern, these being: (1) the importance of performing transient thermal analyses based on robin boundary conditions; (2) the influence of contact formulation between the concrete dam and the foundation; and (3) the use of realistic non-linear material properties. The results and conclusions presented in this paper constitute one important step in achieving best practices to estimate dam safety and better understand the potential failure modes and ageing of concrete dams.

Keywords: concrete arch dams; seasonal temperature variations; crack prediction; non-linear finite element analyses

---

1. Introduction

For many large concrete structures, the load effects that occur from variations in ambient conditions may be the dominating loads that introduce significant stresses in the structure. Due to its size and massive cross-section, significant thermal or moisture gradients may occur. This is especially the case for concrete dams where parts of the structure are subjected to water while other parts are in contact with the ambient air. Structures located in cold climates are subjected to harsh environmental conditions with large seasonal variations in temperature. The ambient temperature difference between summer and winter can be about 70 °C during a year in Northern Hemisphere countries such as Sweden [1].

Temperature variations have been shown to cause cracking in several concrete dams, [2,3]. Temperatures that deviate from the zero strain temperature in the concrete will induce either tensile or compressive strains. The zero strain temperature (or reference temperature) corresponds to the temperature that the concrete is considered stress free and hence had when first subjected to restraint. This temperature is often also denoted as the closing temperature (i.e., the temperature when the final segment is cast that causes the structure to be subjected to restraint). The induced cracks caused by cyclic seasonal temperature variations often occur close to restraints or in large, massive areas without contraction joints [4]. During winter, when the ambient air is colder than the reservoir, the downstream part of the concrete dam will shrink in relation to the concrete at the upstream part. This will induce tensile stresses in the downstream part of the dam that may result in cracking. The thermal shrinkage of the downstream part causes the dam to deform in the downstream direction. In contrast, during summer when the ambient temperature is higher than the reservoir, the dam will deform in the upstream direction and this deformation causes a risk of cracking in the upstream part of the dam.

The influence of seasonal temperature variations and its imposed deformations on concrete dams have been studied based on several different types of models in the literature. Most of the references have focused on statistical (data-based) prediction models using regression analysis. These models are based on the assumption that the dam behaviour can be described by the following effects: Hydrostatic–Seasonal-Time (HST) or Hydrostatic–Temperature–Time (HTT) (see for instance [5–7], among others). In recent years, data-based models based on machine learning techniques such as artificial neural networks (ANN), support vector machines (SVM), and boosted regression trees (BRT) are becoming more and more widely used (see [8]). All these methods have great possibilities of giving highly accurate predictions of the normal behaviour of a dam. These methods, however, need a sufficient amount of data, typically a few years of recordings, in order to train and evaluate the models. In addition, this type of model can have difficulties in detecting progressive damage evolution, as shown by [9].

Numerical models based on the finite element method have been used to assess the deformations or induced stresses from seasonal temperature variations (see [10–13], among others). In these types of studies, the predicted behaviour is also commonly calibrated or compared with measurements. This was, for instance, the case in [14,15], who performed the calibration of the thermal behaviour of the Karun III Dam with measurements, but also estimated the structural safety. There are, however, few references that have performed simulations that account for crack initiation and propagation that have aimed to predict the crack pattern and analyse how these cracks influence the dam behaviour. In the study by Malm and Ansell [3], the cracking in a concrete buttress dam due to seasonal variations was simulated and shown that the cracking that occurred on many of the concrete buttress dams in Sweden was directly caused by the seasonal thermal variation, and that the cracks found in situ could be accurately captured with the use of non-linear finite element analyses based on constitutive material models to describe cracking.

The assessment and maintenance of existing dams represent a challenge for the whole dam engineering community. The ICOLD (International Commission on Large Dams) Committee on Computational Aspects of Dam Analysis and Design has since 1991 organised 15 benchmark workshops with the aim of offering dam engineers and researchers the opportunity to share knowledge focused on modelling strategies and best praxis regarding the numerical analyses of dams. In theme A of the 14th benchmark workshop, the contributors were asked to analyse the cracking of a concrete arch dam in northern Sweden subjected to seasonal temperature variations using numerical analyses. The formulators of the theme defined and delivered the geometry, material properties, and loads, while some other aspects of the numerical modelling were intentionally undefined to allow the contributors to make their assumptions and choices [1]. The aim of this benchmark workshop was to make blind predictions of the extent of cracking and to assess its influence on the response (e.g., displacements) of the dam. In this paper, the benchmark workshop is presented along with the conclusions and lessons learned regarding the numerical analyses of the seasonal temperature variations of concrete dams.

2. Case Study

In theme A of the 14th benchmark workshop, a Swedish concrete arch dam was used as a case study to analyse the influence of cracks on the structural response of the dam. The input data can be obtained from [16].

2.1. The Reinforced Concrete Arch Dam

The dam used as a case study is located in the northern part of Sweden and is thereby subjected to significant seasonal variations in temperature between summer and winter.

The dam was taken into operation in the 1960s. The concrete arch dam is a single curvature dam with a radius of 110 m, a crest length of 170 m, and a maximum height of about 40 m and is shown in Figure 1. The dam is slender, with a crest thickness of 2.5 m and a thickness at the base of 5.7 m. The relatively low and wide geometrical properties of the dam results in a quite high length-to-height ratio of 4.25. One explanation for the slender geometry can be that it was built in a low seismicity zone and seismic loads were not considered in the design. Another unique feature is that it is built of reinforced concrete, with one reinforcement layer on the upstream surface and two layers on the downstream surface. Each reinforcement layer consisted of bars with a diameter of 25 mm and a spacing of 300 mm in both the hoop (horizontal) and vertical directions. The reinforcement bars are spliced and transmits forces over the vertical contraction joints to ensure that the dam acts as a monolithic structure.

Figure 1. (**a**) Photo of the concrete arch dam used as the reference for the benchmark workshop, (**b**) illustration of the dam geometry. From [1].

This dam, as most other dams in Sweden, was built with an insulating wall on the downstream surface to reduce the temperature gradient through the dam. However, only a few years after operation, cracks caused by the seasonal temperature variations were detected on the downstream part of the dam. These cracks continued to grow under the following years and the dam behaviour was continuously being kept under supervision with a monitoring system. The monitoring consists of measurements of the deflections of the arch dam, using a total station at the points shown in Figure 2. Initially, the measurements were made on a monthly basis, but when it was shown that no significant change in behaviour occurred, the recordings were reduced to twice a year. Eight monitoring points were placed along the crest (section 100). About 18 m below the crest, the next measuring section is located (section 200) and measuring section 300 is located about 26 m below the crest. In both of these latter sections, three measuring points are installed. This means that it is possible to evaluate the displacements along the elevation of the dam at Section 03, Section 06, and Section 09, as shown in Figure 2.

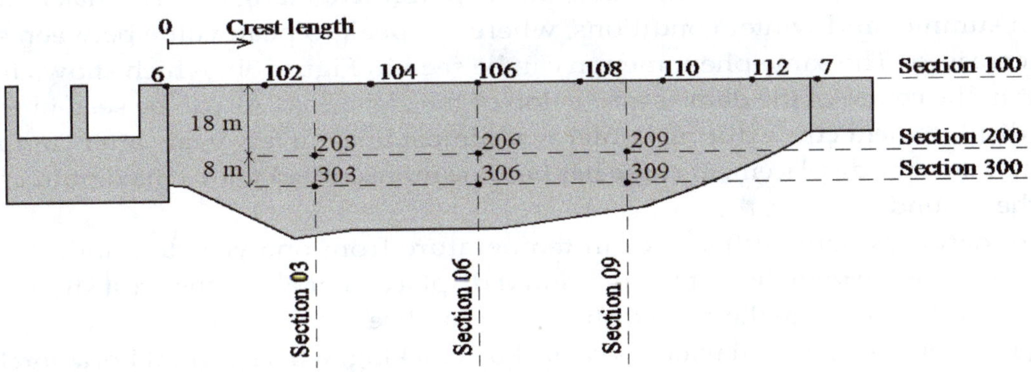

Figure 2. Downstream view of the arch dam with the spillway shown to the left and the abutment block to the right. The monitoring points are also shown where section 100 is at the crest and Section 06 is in the central part of the dam. Reproduced from [17].

In Figure 3, typical recordings from the displacement measurements are presented for the summer and winter conditions, respectively. The difference between the retention water level and the minimum water level was only 2 m for this dam, implying that the variation in water level has a small effect on its seasonal deformations. Hence, the measured displacement is primarily governed by the variations in temperature. The measured displacements are presented along the crest (see Figure 3a) where the origin of the x-axis is located at the spillway and along the elevation in the centre (see Figure 3b). The curves were created based on the recordings of the measuring points illustrated in Figure 2, and for the representation of the results along the crest, a spline curve was developed. The recorded values are illustrated in Figure 3a with squared symbols while the smooth continuous curve is a result from the spline function. Positive displacement shown in the figure corresponds to displacement in the downstream direction (x-direction) of the dam while negative displacement corresponds to upstream movement. Results are presented for two representative cases, a few years after construction (1960s) and after about 50 years of service (2010s).

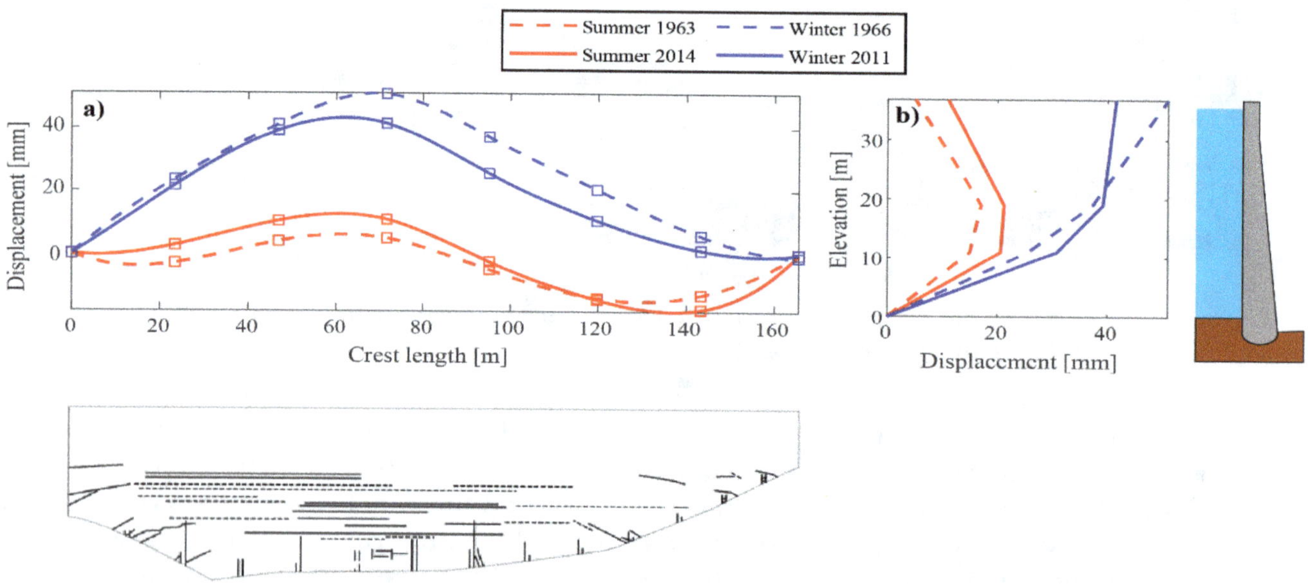

Figure 3. Measured displacement of the dam, shortly after finalisation (dashed lines) and after about 50 years of service (solid lines) for summer (red) and winter (blue) conditions, respectively, (**a**) crest displacement, (**b**) displacement along the dam height. The figure also shows the observed crack pattern on the downstream side (**bottom**) and the cross-section of the dam (**right**).

As seen in Figure 3a, the maximum displacement along the crest was in general larger in the years following the construction, and over the years, the displacement along the crest has reduced. This is valid for both summer and winter conditions, where the peak to peak value between summer and winter has decreased. The same phenomenon can be seen in Figure 3b, which shows the measured displacement in the centre of the dam as a function of the elevation. As can be seen in the figure, the shape of the displacement curve during winter was almost linear a few years after completion of the dam. Over the years, the displacement curve has become more curved with a maximum approximately 20 m above the ground.

There are naturally some differences in temperature from one year to another, which partly could explain some of the deviation of the measured displacement along the crest shown in Figure 3a. However, this could not explain the curved shape that has been measured in recent winters, as shown in Figure 3b [17]. This changed behaviour is caused by cracking, where several horizontal cracks have initiated on the downstream surface of the dam. These cracks are arching on the downstream side toward the left- and right-hand side of the dam (see Figure 3). As a result of cracking, the restraint has reduced, which resulted in a lower global temperature, inducing displacements of the dam. In Figure 3, an illustration of the observed crack pattern on the downstream surface of the dam is also illustrated. As can be seen in the figure, the cracks have mainly been initiated in the bottom two thirds of the dam height. To the right in Figure 3, the cross-section of the dam is also illustrated.

2.2. Temperature Variations

Temperature data were gathered from a weather station located about 20 km from the dam. From this weather station, measurements of the average daily temperature have been available since 1973. However, temperatures with a short duration have little effect on the structural behaviour. According to [18], daily temperature variations have minor effects since these cannot propagate in the concrete structure to a great extent. Therefore, [18] recommended that the temperature variation from one week to another is suitable to consider in numerical analyses. Therefore, the measured temperatures were averaged to give the maximum and minimum temperatures over a seven day period (i.e., weekly variations). Two extreme temperature profiles were developed based on these values, one for a warm year with the maximum temperatures and one cold year based on the minimum temperatures. These two extreme years were then combined as shown in Figure 4. The minimum weekly average temperature occurred in January and was −25.8 °C and the maximum temperature occurred in July and was 19.7 °C. This corresponds to a difference of ΔT 45.5 °C between winter and summer based on weekly average temperatures. A similar curve was also produced that started with the warm year, and the contributors were free to choose which to use (see [1]).

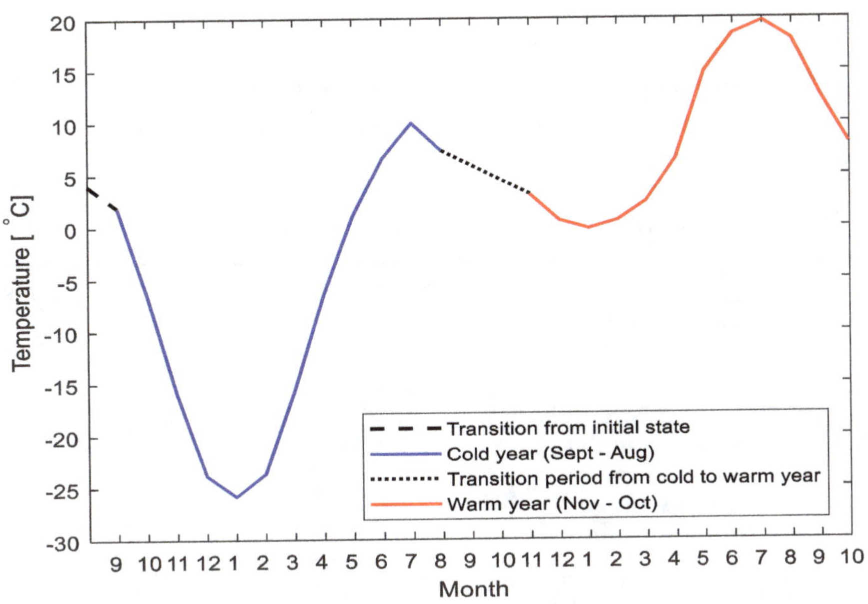

Figure 4. Illustration of the combined temperature profile for an extreme cold year followed by an extreme warm year. From [1].

The influence from solar radiation can be significant on high, thin arch dams, as shown by [14]. However, it was not considered in this case. The reason for this is that the effect from solar radiation is quite small in the Nordic countries (see [18]) and this dam has an insulation wall installed on the downstream side and thereby, the downstream part of the dam is not exposed to solar radiation.

Recordings of the water temperature were not available, but according to [18], it was defined as 70% of the ambient air temperature if the temperature was above freezing, otherwise assumed equal to zero. This assumption was based on previous measurements of water temperatures in the reservoir on other dams, which showed that the water temperature is more or less constant over the depth and just above freezing temperature (i.e., 0 °C) during the winter (see for instance [19]).

In order to include the effect of the insulation wall in the numerical models, a convective heat transfer coefficient for the downstream part of the dam was derived, which considered the thermal properties of the insulation wall and the air gap between the insulation wall and the concrete. The provided values for the convective heat coefficients are presented in Table 1.

Table 1. Conductive heat coefficients as given by [1].

Parameter	Convective Heat Coefficient (W/(m^2 K))	Comment
Downstream surface of the arch dam–air	4	The downstream surface should be considered to have lower conductive heat coefficient compared to other concrete surfaces. (the reason is that there is usually some heat insulating material installed on the downstream surface on dams in Sweden).
Concrete–air	13	For all surfaces exposed to air, except the downstream surface of the arch dam.
Concrete–water	500	
Concrete–rock	1000	
Rock–air	13	
Rock–water	500	

2.3. *Material Properties*

Both thermal and mechanical material properties are required to perform non-linear analyses regarding thermal induced cracking from seasonal variations. The material properties defined for the benchmark workshop are presented in Table 2.

Table 2. Material properties as given by [1].

	Property	Concrete	Rock	Reinforcement
Mechanical	Elastic modulus	33 GPa	40 GPa	200 GPa
	Poisson's ratio	0.2	0.15	0.3
	Density	2300 kg/m^3	2700 kg/m^3	7800 kg/m^3
	Compressive strength	38 MPa	-	-
	Tensile strength	2.9 MPa	-	-
	Yield stress	-	-	360 MPa
	Ultimate strength	-	-	600 MPa
	Ultimate strain	-	-	0.15
Thermal	Thermal expansion	1×10^{-5} K^{-1}	1×10^{-5} K^{-1}	1×10^{-5} K^{-1}
	Thermal conductivity	2 W/(m K)	3 W/(m K)	39 W/(m K)
	Stress/strain free temperature	4 °C	4 °C	4 °C
	Specific heat capacity	900 J/(kg K)	850 J/(kg K)	450 J/(kg K)

In the benchmark workshop, the fracture energy of concrete was intentionally omitted and thereby had to be defined by the contributors.

In a large concrete structure that has been built for a considerable time, the reference (stress/strain free) temperature may vary in different parts of the structure and may change over time due to ageing, creep, and cracking. This is, however, normally not considered, instead a constant reference temperature is derived for the whole structure. In this case study, the reference temperature was defined as the annual average temperature at the dam site in accordance with the recommendation of the Federal Energy Regulatory Commission (FERC) [20].

2.4. Geometrical Model and Mesh

A geometrical model had been developed by the formulators for the arch dam, its abutments, the spillway, and the rock foundation. The model was supplied to the contributors in different CAD formats and also with a suggested mesh (see Figure 5). The numerical mesh was developed in HyperMesh and the finite element program Abaqus ver. 2017. The provided mesh was based on 4-node linear tetrahedron elements (denoted C3D4), where the mesh size of the dam was defined as small enough to ensure that the element length was less than the characteristic length of the concrete material. This ensures that the elastically stored energy that is released during cracking can be captured by the surrounding elements. The characteristic length (l_{ch}) is a material property that describes the ductility of the material and is defined as

$$l_{ch} = \frac{E \cdot G_f}{f_t^2} \tag{1}$$

where E is the elastic modulus [Pa]; G_f is the fracture energy [Nm/m^2]; and f_t is the tensile strength [Pa].

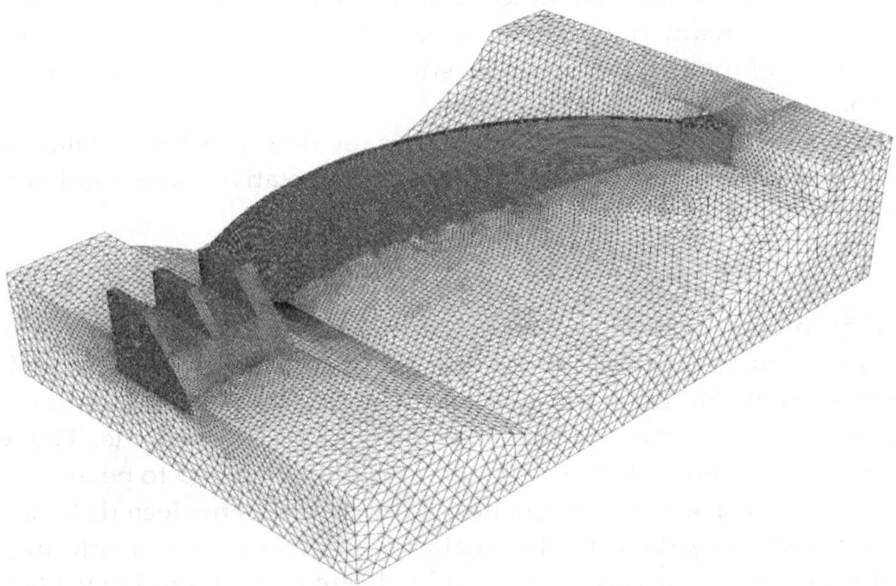

Figure 5. Illustration of the mesh provided to the contributors in the benchmark workshop.

Based on the material properties given in Table 2, the characteristic length of the concrete used in the dam was 1.1 m, and hence a smaller element length than this is required to capture the energy release during cracking. In the dam and spillway, a typical element length of about 0.75 m was used in the provided mesh. The concrete parts in the provided mesh consisted of 65,874 nodes and 292,536 elements. The provided mesh of the rock varied in element length between 0.75 and 3.0 m and the rock consisted of 43,942 nodes and 208,399 elements.

The provided mesh was highly detailed to ensure accurate results for the non-linear analyses performed to capture the crack pattern. A coarser mesh could be used, especially for the linear analyses and still obtain numerically converging results. Some contributors created their own mesh. which was slightly coarser than the one provided, but was generally smaller than the characteristic length of the concrete material.

2.4.1. Interactions and Boundary Conditions

The dam is excavated into the rock where the support is made to act like a hinge (i.e., preventing displacements but not rotation). In addition, an asphalt coating was applied at the rock surface to prevent cohesion between the dam and the rock. In the geometry of the Finite Element (FE) model, this was simplified where the excavation was neglected. Instead, a plane foundation between the rock and the concrete was defined, see Figure 6. It is crucial that the interaction between the concrete and foundation in the numerical model is defined so that it describes the real behaviour of the dam.

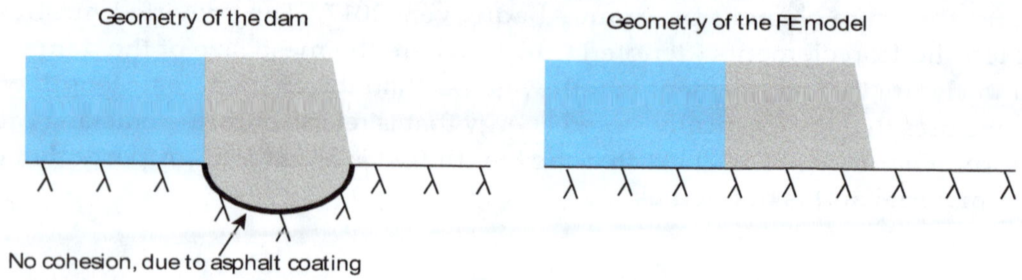

Figure 6. Illustration of the real foundation (**left**) were the dam is excavated into the rock and where cohesion is prevented by an asphalt coating. The idealised geometry in the FE model (**right**).

The contributors were free to define the boundary conditions on the rock foundation as they found suitable. All contributors defined that the bottom surface of the rock foundation was restrained in the vertical direction. In the upstream–downstream direction, the contributors either defined boundary conditions on one or both of the two parallel rock surfaces. The contributors used the same principle in the lateral direction.

Constraining two parallel rock surfaces in the upstream–downstream (or lateral) direction enforces a global restraint of the rock, which would result in a larger relative movement between the concrete and rock due to the thermal strains.

2.4.2. Loads

The main load of interest in this study was the thermal load caused by temperature gradients from the extreme temperatures described in Section 2.2. In addition to the thermal load, gravity, and water pressure were also considered in the analyses. Gravity loads were included in the analysis for the concrete based on the density given in Table 2 and the structure's volume. The water pressure was included as a hydrostatic pressure where the water level was assumed to be at the crest of the dam.

The influence from the construction sequence of the dam was neglected. In addition, other loads such as uplift pressure and ice loads were also neglected in this study. Considering the thin thickness of the arch and that the dam is relatively high, both loads are expected to have a fairly small influence on the dam behaviour, especially regarding its influence on cracking.

2.4.3. Type of Analyses and Required Results

It was mandatory for the contributors to perform two different types of analyses: thermal and mechanical. The aim was to predict both the cracking that had occurred since the dam was built in the 1960s and to assess its influence of the displacements. The contributors were asked to report the results from three steps:

1. Thermal analysis;
2. Linear mechanical analysis; and
3. Non-linear mechanical analysis.

These analyses should be described in the paper and in addition, submit results to the formulators via an Excel template file. The required results were:

1. Thermal analysis

    ○ Temperature distribution for the centre section of the dam for the months of January (minimum temperature) and July (maximum temperature).

2. Linear mechanical analysis

  - Calculated displacement on the downstream surface of the dam in the centre section as a function of the dam height, for the initial conditions (gravity load and hydrostatic water pressure) in addition to the minimum and maximum displacements due to temperature variations, respectively.
  - Calculated displacement of the downstream surface along the two lines illustrated in the figures below. In both cases, the zero on the x-axis (representing the crest length) should be defined at the side of the spillway section:
    - along the crest
    - a line ~14 m below the crest on the downstream surface
  - Contour plots illustrating the displacement of the dam;
  - Contour plots illustrating the areas which are exceeding the tensile strength (i.e., the areas that may be subjected to cracking); and
  - Vector direction plot showing the maximum and minimum principal stress directions of the downstream view of the arch dam.

3. Non-linear mechanical analysis

  - The results requested in the first three bullets of linear analyses should also be presented for the non-linear analyses.
  - Contour plots illustrating the calculated extent of cracking (i.e., showing non-linear strains, damage, crack planes, etc.).

The mechanical analyses were defined to be performed in the following sequence:

1. Gravity loads;
2. Hydrostatic pressure; and
3. Extreme temperature variations.

3. Results

In this section, the main results from the benchmark workshop are presented, where each contributor is presented with a unique number between 1 and 16. In some cases, the results from a few of the contributors were removed since they were considered as outliers, most likely caused by errors in the models. To find the results from all contributors, see [21–36].

3.1. Temperature Analyses

All contributors performed these with a one-way coupling, where the results from the thermal analysis were used as input in the mechanical analysis to impose thermal strains. First, the temperature analysis was performed by cycling the extreme thermal variations as presented in Section 2.2. These temperature distributions were then used in the mechanical to estimate the risk of cracking based on a linear elastic simulation and to simulate the extent of cracking based on non-linear analyses.

The contributors also showed that there was no significant difference between simulating the extreme temperature curves (Figure 4) based on starting with winter or summer conditions. Thereby, the calculated extent of cracking had a minor effect on which order the extreme temperatures occurred. One conclusion from the contributors was that at least two cycles of the extreme temperature curve had to be applied for the crack pattern to stabilise.

Most of the contributors performed transient thermal analyses where the temperature variations in the air and in the water was applied using Robin boundary conditions (i.e., based on convective heat transfer boundaries). No direct differences could be found between the contributors that performed analyses based on this approach. It was only possible to detect differences in temperature distribution between these contributors and the few contributors who had performed steady state analyses with Dirichlet boundary conditions (i.e., with prescribed nodal temperatures). One clear effect of using the Dirichlet boundary conditions was that the water temperature had a smaller influence on the internal temperatures in the dam compared with analyses using Robin boundary conditions. The reason for this is that water was defined with a significantly higher convective heat transfer coefficient than air (500 W/(m^2 K) between water and concrete compared to 4 W/(m^2 K) between air and concrete). It was also possible to see a clear difference between the contributors using steady state compared to transient analyses, where the latter did not reach a linear temperature gradient through the thickness.

3.2. Static Behaviour and Linear Elastic Seasonal Variations

As a first step, the contributors provided their calculated results from just the static loads (i.e., the gravity load and the hydrostatic pressure). This was introduced as a first step in the benchmark workshop to allow for easier comparison between the contributors' modelling strategies. It also provided an opportunity for the contributors to successively increase the complexity in their simulations and to be able to cross-check a simpler version of their models with the more advanced non-linear models.

In order to describe the real behaviour of the dam, it is important to prevent horizontal displacement between the concrete and rock. This can, however, be considered in several different ways, as shown by the contributors. The majority of the contributors considered some sort of interface element or springs between the concrete and rock. In the stream direction, the stiffness of the interface or the springs was defined with a high value to prevent relative movement and sliding between the concrete and rock. Some contributors defined a friction based contact formulation using a high coefficient of friction (typically higher than 20), which thereby prevented horizontal displacement at the base. Others, instead, used a built-in function where no horizontal slip could occur. The latter method is, of course, more correct and prevents horizontal movement but may result in problems finding a converging solution. The horizontal movement in the first method will be infinitely small if the coefficient of friction is high enough.

In the normal direction, most contributors defined a non-linear contact formulation, interface elements, or springs that transmits compressive forces (typically, no penetration of the master nodes into the slave nodes) but with zero stiffness for opening displacements. One of the participating teams defined a softer contact for the transition between compressive and tensile forces where a continuous function was used that also allows for some penetration of the master node into the slave nodes. This means that the axis of rotation is moved slightly in the upstream direction compared to the others that had the dam toe (downstream edge) as the axis of rotation. A softer contact is, in this aspect, more realistic in this case considering the hinge design. Other contributors used methods to prevent horizontal movement by constraining the toe (downstream edge of base) of the dam in all directions and one contributor updated the geometry of the base of the dam so that it was more similar to the real foundation.

Comparing the deformations due to static loads showed that the results were quite consistent, see Figure 7. It could be seen that the contributors obtained static displacement that could be categorised into two groups. In the first group, the maximum crest displacement varied between 16 and 18 mm, and in the second group it varied between 25 and 31 mm. The reason for this discrepancy between the two groups is that the first group used a fixed (rigid) constraint between the concrete and rock in the linear elastic analyses.

There was a very small difference between the contributions in the first group as there should be and this minor difference is believed to originate from the differences in the boundary conditions and especially in the mesh. The reason why the discrepancies were larger between the contributors that modelled non-linear contact between the concrete and rock is due to the different approaches used to model the contact formulation and likely also tolerances of the FE analyses, in addition to the reasons mentioned for the other group.

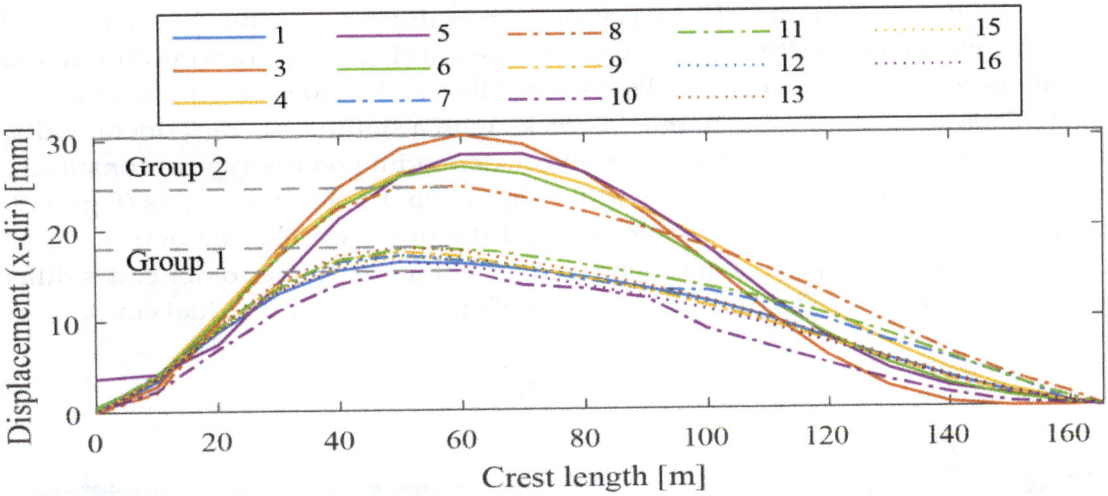

Figure 7. The calculated displacement along the dam crest for static loads based on linear elastic models, according to the contributors.

In the analysis using linear elastic material properties, the contributors were asked to provide an estimation of the areas that had a risk of cracking after application of static loads and when considering the seasonal temperature variations. The risk of cracking was in this case estimated based on the areas that were subjected to tensile stresses that exceeded the tensile strength. Most of the contributors showed a risk of cracking in the centre of the downstream surface due to static loads. When the seasonal temperatures were considered, all contributors showed significant risk of cracking on almost the entire downstream surface of the arch dam. In addition, the contributors that had modelled a fixed connection between the concrete and rock also showed a significant risk of cracking in the concrete upstream surface just above the foundation level. This is considered reasonable since the others that modelled the contact with a non-linear contact formulation showed a fairly large difference in crest displacement due to the joint opening.

3.3. Extent of Cracking

The linear elastic numerical analyses showed that the dam is already subjected to high tensile stresses on the downstream part of the dam from normal loads caused by gravity and the hydrostatic water pressure. This was later confirmed by the non-linear analyses where cracking occurred as illustrated in Figure 8. Normally, this should not be the case. However, considering that the studied dam has a high length-to-height ratio and is slender, these effects resulted in a significant risk of cracking. As mentioned previously, cracks were detected on the downstream side of the dam shortly after it was finalised. It is important to remember that this dam is reinforced, and thereby cracking does not necessarily influence the dam safety.

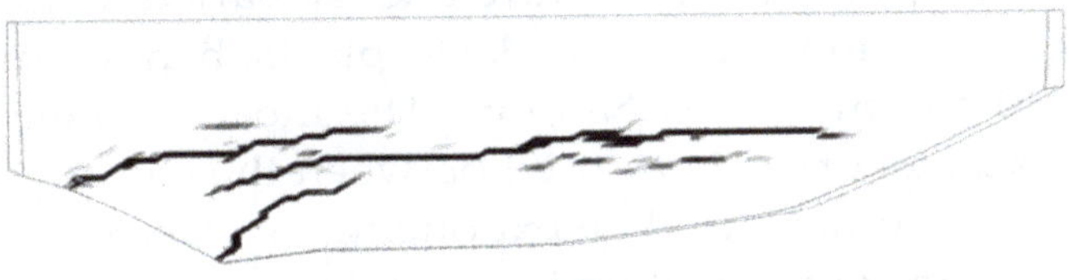

Figure 8. Example of an illustration of the predicted extent of cracking due to static loads from contributor No. 4.

In the following step, the extremely warm year and cold year were cycled to predict the cracking that had occurred. As it can be seen in Figure 9f), the contributors predicted significant cracking with several large horizontal cracks in the centre of the downstream surface of the dam and inclined cracks towards the two abutments. The crack pattern is illustrated a bit differently in the figures depending on the choice of constitutive material model, and some showed both micro and macro cracks while others only showed macro-cracks. Many of the cracks shown are micro-cracks, while a few of the horizontal cracks corresponded to macro-cracks with a significant crack opening displacement. The colour codes also varied between the contributors depending on the type of constitutive material model used, where they may present crack widths (Figure 9b,d), inelastic strains (Figure 9f), fracture planes (Figure 9e), or induced damage between 0 and 1 (Figure 9a,c). The aim of this comparison was, however, to compare the crack pattern and hence it does not matter if the colour codes differ, since the aim is to compare where the cracks occurred, number of cracks and if individual cracks were captured.

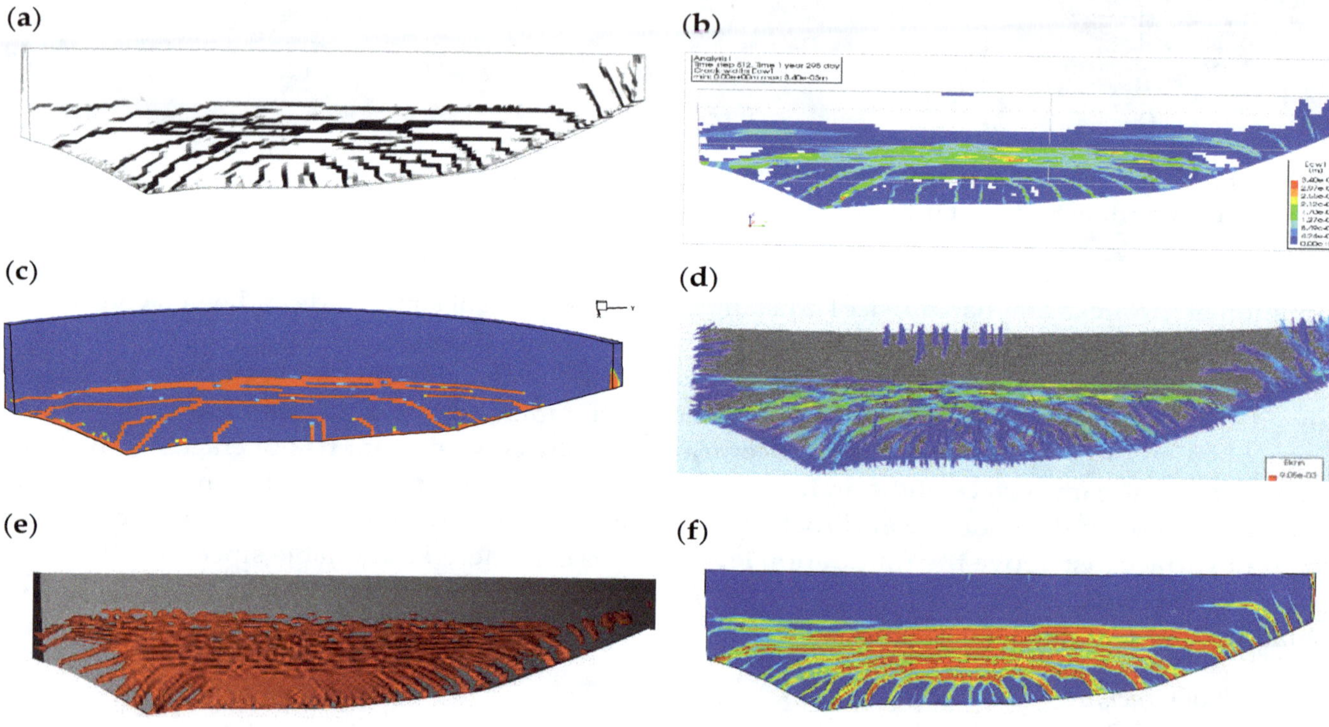

Figure 9. Example of an illustration of the predicted extent of cracking due to extreme temperatures from the following contributors: (**a**) No. 4, (**b**) No. 13, (**c**) No. 8, (**d**) No. 12, (**e**) No. 16, (**f**) No. 6.

The crack pattern obtained in the numerical analysis corresponded well with the observed crack pattern, as shown previously in Figure 3. The obtained crack pattern also corresponded well with the description given by U.S. Army Corps of Engineers (USACE) [4], who described that horizontal cracks typically occur on the downstream surface of the cantilever elements of arch dams at half the dam height.

In the benchmark workshop, the used fracture energy varied significantly, from contributors using a brittle material model to others using an elasto-plastic behaviour corresponding to infinite fracture energy between 140 Nm/m^2 to 600 Nm/m^2. The mean of their fracture energies was 250 Nm/m^2. The choice of unloading curve also varied between the contributors where some assumed a linear decaying function while others used an exponential curve. For instance, the curve defined by [37] was used to define the crack propagation in concrete.

The choice of fracture energy had a significant influence on the possibility of obtaining detailed results regarding the predicted cracking. In cases where lower fracture energy was defined, it was possible to visualize individual macro cracks. In the analyses defined with high fracture energy, it was only possible to identify larger areas that were more or less damaged/cracked. This is illustrated in Figure 10, where the top figure shows the crack pattern from one of the contributors using a fracture energy of 200 Nm/m^2 compared to a solution by the same contributor where a fracture energy of

500 Nm/m^2 was used. It should be remembered that this dam is reinforced, and the difference between using higher or lower fracture energy is expected to be significantly higher for unreinforced concrete dams. The reason for this is that the reinforcement carries the load after crack initiation and therefore, in a simplified sense, can be considered as increased fracture energy.

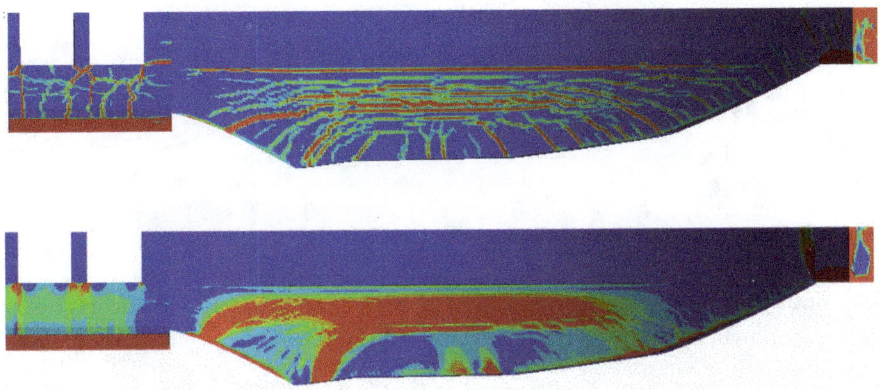

Figure 10. Example of an illustration of the predicted extent of cracking from contributor No. 5 using a lower fracture energy (200 Nm/m^2 in the top) and a higher fracture energy (500 Nm/m^2 bottom).

Overall, the contributors achieved accurate predictions of the crack pattern, regardless of whether the constitutive material model used was based on damage theory, plasticity theory, fracture mechanics, or a combination of these.

3.4. Deformation of the Arch Dam

The calculated deflection of the dam varied (as expected) significantly more than the displacements from the seasonal temperature variations based on a linear elastic material model. The maximum displacements from the contributors based on linear elastic material models varied typically between 40 mm and 65 mm, with an average of 55 mm.

In Figure 11, a comparison of the calculated displacement along the crest is presented together with the measured displacements previously shown in Figure 3. The maximum crest displacement obtained with non-linear material models varied between approximately 50 to 80 mm, with an average of 65 mm. The main reason for the discrepancy is believed to be due to the differences in fracture energy. The contributors with low fracture energy may have obtained yielding of some of the reinforcement bars, which would greatly increase the deformations compared to contributors with a higher fracture energy where the reinforcement was in the elastic regime.

The shape of the calculated displacement curves corresponded well with the measured deformed shape. The maximum crest displacement from the measurements was slightly over 50 mm for the dam after a few years (uncracked) and about 40 mm for the dam in recent years (cracked).

It should be noted, however, that the calculated displacements are performed with larger temperature variations and hence some overestimation of the displacement is expected. The difference in temperature variation is not so large, however, which means that the calculated displacement should be only slightly above the measured displacement, as shown by [17].

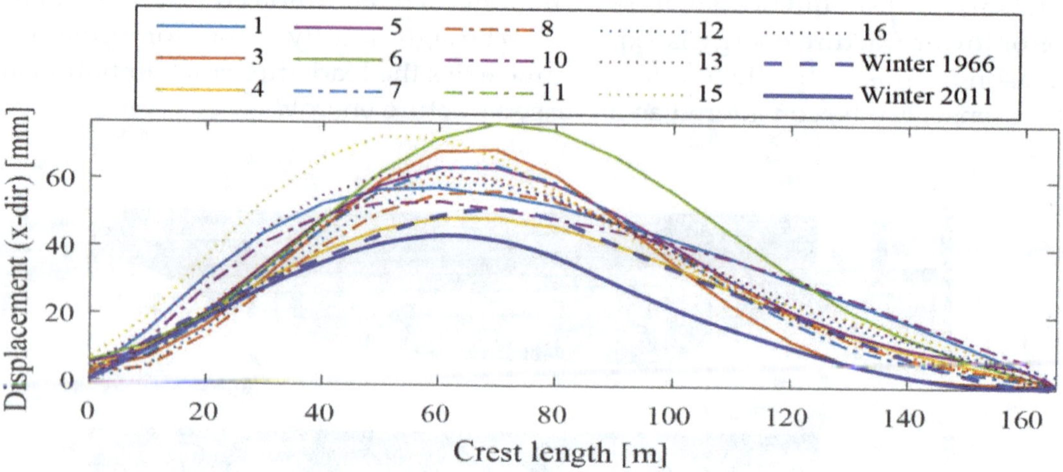

Figure 11. The calculated displacement along the dam crest during winter conditions based on non-linear analyses, according to the contributors.

4. Discussion

As shown previously, if a concrete dam is subjected to significant thermal stresses, there is a risk that cracking will occur, which may significantly alter the behaviour of the dam. When the cracks develop, the structural stiffness is reduced, and hence also the thermal forces of restraint. This results in a new equilibrium state where continued thermal variation primarily causes the opening and closing of these cracks, and lower induced stresses in the concrete as a result.

To be able to analyse this effect, detailed models are required that can capture the crack initiation and propagation in concrete such as non-linear finite element analyses based on constitutive material models. Other types of methods such as data-based (statistical) models or simplified models may capture the behaviour of an undamaged or an already damaged case, but can typically not capture the transition from undamaged to damaged or answer questions regarding the current state of a cracked dam [9]. However, the different methods have their benefits and the use of one type of model does not exclude the possibility of also using others. In the condition assessment of dams, it would be preferable if both detailed non-linear FE analyses and data-based models are used when applicable.

One simplification of the FE analyses presented in this paper is the use of a constant reference temperature. However, using a too low reference temperature typically means that the extent of cracking during a cold winter may be underestimated and vice versa. The obtained degree of cracking is typically more or less similar, even though the obtained crack width or crack length may vary to some extent [38]. Another important factor is that extreme temperature was developed to simplify the seasonal variations for over 50 years into a cold and a warm year. The approach used in this paper has, however, been shown by [39] to provide accurate results where the main types of cracks found in situ could be captured.

Overall, the results from the contributors are well in line with the observations and measurements made on the dam. All contributors indicated a significant risk of cracking due to the seasonal temperature variations based on the linear elastic analyses. In general, the contributors also achieved good agreement between the predicted crack patterns and the crack pattern observed in situ. As expected, the non-linear analyses predicted larger discrepancies regarding displacements compared to the simpler analyses with a linear elastic material model. However, for a large majority of the contributors, good agreement was obtained regarding displacements of the dam.

There are several factors that contribute to the differences between different predictions. The main identified factors that influenced the results were:

1. Temperature distributions;
2. Boundary conditions;

3. Interaction between concrete and rock; and
4. Choice of fracture energy and non-linear constitutive model.

The methodology of performing the seasonal temperature distribution analyses may have a significant influence on the results. In the benchmark workshop, most of the contributors performed transient thermal analyses with Robin boundary conditions. This is the most suitable for this application, and no direct difference in results could be found between the contributors that used this approach. A larger difference was found for the contributors that performed steady state analyses with Dirichlet boundary conditions. These analyses showed a larger influence from the ambient air temperature and obtained linear thermal gradients through the thickness of the dam.

The definition of boundary conditions had a large influence primarily regarding the possibility of obtaining converging solutions. The imposed boundary conditions on the rock could result in significant restraint of the rock mass, if all exterior surfaces were restrained in their normal direction. This prevented the rock mass from deforming and hence forced the temperature variation to primarily cause relative deformation between the arch dam and the foundation. To only restrain one of the two parallel surfaces in the stream and lateral direction, respectively, reduces the relative restraint between the dam and the rock. This is thereby considered to be more suitable in cases where thermal variations are considered. The only downside of this approach is that the rigid body motion of the rock foundation must be removed before the results are compared with the measurements or the other contributors.

One of the most influential factors is how the interaction between the concrete and rock foundation was modelled. The reason for this is that significant joint opening occurred if a suitable interaction was modelled. Preventing joint opening at the foundation level would initiate a horizontal crack along the upstream surface.

As mentioned previously, the estimated amount of fracture energy varied from very low in the brittle models up to very high in the most ductile models. Concrete used in dams differs to some extent from conventional concrete, and limited information exists about suitable concrete fracture energy for dam applications. The following characteristics are typical for concrete used for hydropower applications, according to Brühwiler [40]:

- Use of a larger maximum size of the aggregates (100 mm in this specific case);
- Slightly lower tensile strength; and
- Slightly lower elastic modulus.

The fracture energy for concrete used in dams is, due to these reasons higher, than for conventional concrete. For conventional concrete, the fracture energy can, for instance, be determined according to Model Code 2010 [41]. With the properties given in Table 2, this would result in a fracture energy of 140 Nm/m^2. This is thereby considered to be conservative in this case.

Ghaemmaghami and Ghaemian [42] showed that increasing the maximum aggregates from 30 mm to 60 mm almost doubled the fracture energy. Brühwiler [40] presented the results from the measured material properties of three arch dams, where the fracture energy (mean values) varied between 230 Nm/m^2 to 310 Nm/m^2. The mean value of the fracture energy used by the contributors is thereby within the range given by [40]. The use of too low a fracture energy often has a large influence on the possibility of achieving a converging solution, where brittle analyses are difficult to perform due to the rapid unloading that is required as a crack is initiated. The results clearly showed the importance of using realistic fracture energy for the type of concrete used in the dam. Using a too low value will result in the risk of reinforcement yielding and potentially the overestimation of the deformations. Individual cracks could not be captured by the contributors who used high value results of the fracture energy. These contributors only captured that cracking occurred in larger areas.

Overall, the contributors achieved accurate predictions of the crack pattern, regardless of whether the constitutive material model used was based on damage theory, plasticity theory, fracture mechanics, or a combination of these. Using linear unloading curves generally results in an overestimation of the structural stiffness (see [18]). However, it was not possible to draw any conclusions from the benchmark workshop regarding the choice of the unloading curve.

It is necessary to be able to estimate the root causes for cracking and to obtain accurate predictions of the extent of cracking since this is the first step of assessing the current structural safety of an existing dam or the need and type of strengthening. Therefore, the results and conclusions presented in this paper are one important step in achieving increased knowledge of the dam safety of existing structures with a better understanding of potential failure modes and ageing of concrete dams.

5. Conclusions

In this paper, the case study of a concrete arch dam subjected to cracking caused by seasonal temperature variations has been presented. This case was part of the 14th ICOLD Benchmark Workshop, where 16 teams of contributors provided their predictions of the dam's behaviour. From this, it has been possible to see the influence of different assumptions regarding modelling the crack propagation, contact formulation between the rock and the dam, and approaches to simulate the temperature variation. Some of the important findings from this benchmark were:

- Non-linear numerical models can be successfully used to predict cracking and estimate the displacements of the studied dam.
- It is important to perform transient temperature analyses with Robin boundary conditions for these applications to obtain a suitable temperature distribution.
- The interaction between concrete and rock has a large influence on the predicted displacements, especially for analyses based on linear material properties.
- The magnitude of the fracture energy of concrete influenced the accuracy of the predicted displacements of the reinforced concrete dam, but had a significantly larger influence on the possibility of capturing realistic crack patterns.

Author Contributions: Conceptualization, R.M. and R.H.; Methodology, R.M. and R.H.; Software, R.H. and R.M.; Validation, R.M., R.H., and J.E.; Formal analysis, R.M. and J.E.; Investigation, R.M.; Resources, R.M. and R.H.; Data curation, R.M.; Writing—original draft preparation, R.M.; Writing—review and editing, R.M., R.H., and J.E.; Visualization, R.M. and J.E.; Supervision, R.M.; Project administration, R.M.; Funding acquisition, R.M. All authors have read and agree to the published version of the manuscript.

Acknowledgments: The research presented was carried out as a part of the Swedish Hydropower Centre, SVC. SVC has been established by the Swedish Energy Agency, Energiforsk, and Svenska Kraftnät together with Luleå University of Technology, KTH Royal Institute of Technology, Chalmers University of Technology and Uppsala University (http://www.svc.nu).

References

1. Malm, R.; Hellgren, R.; Ekström, T.; Fu, C. Cracking of a concrete arch dam due to seasonal temperature variations—Theme A. In Proceedings of the 14th ICOLD International Benchmark Workshop on Numerical Analysis of Dams, Stockholm, Sweden, 6–8 September 2017; pp. 18–75.
2. Tarbox, G.; Charlwood, R. *Investigating the Structural Safety of Cracked Dams*; CEATI Report No. T122700-0226; Dam Safety Interest Group (DSIG): Bellevue, DC, USA, 2014; 91p.
3. Malm, R.; Ansell, A. Cracking of Concrete Buttress Dam Due to Seasonal Temperature Variation. *ACI Struct. J.* **2011**, *108*, 13–22.
4. USACE. *Arch Dam Design*; EM 1110-2-2201; US Army Corps of Engineers (USACE): Washington, DC, USA, 1994; 240p.
5. Willm, G.; Beaujoint, N. Les méthodes de surveillance des barrages au service de la production hydraulique d'Electricité de France, problèmes anciens et solutions nouvelles (The methods of surveillance of dams to serve

hydraulic Production at Électricité de France: Old problems and new solutions), Q34/R30. In Proceedings of the 9th International Congress on Large Dams (ICOLD), Istanbul, Turkey, 4–8 September 1967.

6. Chouinard, L.E.; Bennett, D.W.; Feknous, N. Statistical Analysis of Monitoring Data for Concrete Arch Dams. *J. Perform. Constr. Facil.* **1995**, *9*, 286–301. [CrossRef]
7. Léger, P.; Leclerc, M. Hydrostatic, Temperature, Time-Displacement Model for Concrete Dams. *J. Eng. Mech.* **2007**, *133*, 267–277. [CrossRef]
8. Salazar, F.; Morán, R.; Toledo, M.Á.; Oñate, E. Data-Based Models of the Prediction of Dam Behaviour: A Review and Some Methodological Considerations. *Arch. Comput. Methods Eng.* **2017**, *24*, 1–21. [CrossRef]
9. Hellgren, R.; Malm, R.; Ansell, A. Performance of data-based models for early detection of damage in concrete dams. *Struct. Infrastruct. E* **2020**, in press.
10. Léger, P.; Venturelli, J.; Bhattacharejee, S.S. Seasonal temperature and stress distributions in concrete gravity dams. Part 1 Modeling. *Can. J. Civ. Eng.* **1993**, *20*, 999–1017.
11. Léger, P.; Venturelli, J.; Bhattacharejee, S.S. Seasonal temperature and stress distributions in concrete gravity dams. Part 2 Behavior. *Can. J. Civ. Eng.* **1993**, *20*, 1018–1029. [CrossRef]
12. Léger, P.; Seydou, S. Seasonal Thermal Displacements of Gravity Dams Located in Northern Regions. *J. Perform. Constr. Facil.* **2009**, *23*, 166–174. [CrossRef]
13. Maken, D.; Léger, P.; Roth, S. Seasonal thermal cracking of concrete dams in northern regions. *J. Perform. Constr. Facil.* **2014**, *28*. [CrossRef]
14. Mizabozorg, H.; Hariri-Ardebili, M.A.; Shirkhan, M.; Seyed-Kolbadi, S.M. Mathematical modeling and numerical analysis of thermal distribution in arch dams considering solar radiation effect. *Sci. World J.* **2014**, *2014*, 597393. [CrossRef]
15. Mizabozorg, H.; Hariri-Ardebili, M.A.; Seyed-Kolbadi, S.M. Structural safety evaluation of Karun III Dam and calibration of its finite element model using instrumentation and site observation. *Case Stud. Struct. Eng.* **2014**, *1*, 6–12. [CrossRef]
16. ICOLD Benchmark Workshop 2017. Available online: https://www.icold-bw2017.conf.kth.se/workshop/ (accessed on 30 January 2020).
17. Enzell, J.; Tollsten, M. Thermal Cracking of a Concrete Arch Dam due to Seasonal Temperature Variations. Master's Thesis, KTH Royal Institute of Technology, Stockholm, Sweden, 2017.
18. Malm, R. *Guideline for FE Analyses of Concrete Dams*; Energiforsk Report 2016:270; Energiforsk AB: Stockholm, Sweden, 2016.
19. Andersson, O.; Seppälä, M. Verification of the Response of a Concrete Arch Dam Subjected to Seasonal Temperature Variations. Master's Thesis, KTH Royal Institute of Technology, Stockholm, Sweden, 2015.
20. FERC. *Engineering Guidelines for the Evaluation of Hydro Power Projects*; Federal Energy Regulatory Commission Division of Dam Safety and Inspections: Washington, DC, USA, 1999.
21. Enzell, J.; Tollsten, M. Cracking of a concrete arch dam due to seasonal temperature variations. In Proceedings of the 14th ICOLD International Benchmark Workshop on Numerical Analysis of Dams, Stockholm, Sweden, 6–8 September 2017; pp. 77–86.
22. Roth, S.N.; Dolice, D.M. Thermal cracking of a concrete arch dam. In Proceedings of the 14th ICOLD International Benchmark Workshop on Numerical Analysis of Dams, Stockholm, Sweden, 6–8 September 2017; pp. 87–96.
23. Hassanzadeh, M.; Ferreira, D. Thermo-mechanical analysis of a concrete arch dam Influence of temperature and fracture energy. In Proceedings of the 14th ICOLD International Benchmark Workshop on Numerical Analysis of Dams, Stockholm, Sweden, 6–8 September 2017; pp. 97–106.
24. Schclar Leitão, N.; Monteiro Azevedo, N.; Castilho, E.; Braga Farinha, L.; Câmara, R. Thermal Cracking Computational Analysis of a Concrete Arch Dam—Theme A: Thermal Cracking of a concrete arch dam. In Proceedings of the 14th ICOLD International Benchmark Workshop on Numerical Analysis of Dams, Stockholm, Sweden, 6–8 September 2017; pp. 107–117.
25. Lie, R.; Aasheim, E.E.; Engen, M. Thermal Cracking of a Concrete Arch Dam—FE analyses with a 3D non-linear material model for concrete. In Proceedings of the 14th ICOLD International Benchmark Workshop on Numerical Analysis of Dams, Stockholm, Sweden, 6–8 September 2017; pp. 118–127.
26. Popovici, A.; Sârghiuță, R.; Ilinca, C.; Anghel, C. Cracking of a concrete arch dam due to seasonal temperature variations. In Proceedings of the 14th ICOLD International Benchmark Workshop on Numerical Analysis of Dams, Stockholm, Sweden, 6–8 September 2017; pp. 128–136.

27. Tzenkov, A.; Frissen, C.; Santurjian, O. Cracking of a Concrete Arch Dam Subjected to Harsh Environmental Conditions—Analysis of Unusual Conditions due to Temperature Variations. In Proceedings of the 14th ICOLD International Benchmark Workshop on Numerical Analysis of Dams, Stockholm, Sweden, 6–8 September 2017; pp. 137–146.
28. de-Pouplana, I.; Gracia, L.; Salazar, F.; Oñate, E. Cracking of a concrete arch dam due to seasonal temperature variations—Theme A. In Proceedings of the 14th International Benchmark Workshop on the Numerical Analysis of Dams, Stockholm, Sweden, 6–8 September 2017; pp. 147–156.
29. Mouy, V.; Molin, X.; Fray, S.; Oukid, Y.; Noret, C. Behaviour of a thin arch dam under thermal load, with brittle behaviour—Analysis of the influence of some computation parameters. In Proceedings of the 14th ICOLD International Benchmark Workshop on Numerical Analysis of Dams, Stockholm, Sweden, 6–8 September 2017; pp. 157–166.
30. Gasch, T.; Ericsson, D. Thermally-induced cracking of a concrete arch dam using COMSOL Multiphysics. In Proceedings of the 14th ICOLD International Benchmark Workshop on Numerical Analysis of Dams, Stockholm, Sweden, 6–8 September 2017; pp. 167–176.
31. Goldgruber, M.; Lampert, R. Thermal Cracking of a Concrete Arch Dam. In Proceedings of the 14th ICOLD International Benchmark Workshop on Numerical Analysis of Dams, Stockholm, Sweden, 6–8 September 2017; pp. 177–186.
32. Shao, C.; Manso, P.A.; Gunn, R.; Pimentel, M.; Schleiss, A.J. Thermal cracking of a thin arch dam in a wide valley with skin reinforcement. In Proceedings of the 14th ICOLD International Benchmark Workshop on Numerical Analysis of Dams, Stockholm, Sweden, 6–8 September 2017; pp. 187–197.
33. Shahriari, S.; Lora, F.; Zenz, G. Thermal cracking of a concrete arch dam—Microplane approach. In Proceedings of the 14th ICOLD International Benchmark Workshop on Numerical Analysis of Dams, Stockholm, Sweden, 6–8 September 2017; pp. 198–207.
34. Hjalmarsson, F.; Pettersson, F. Finite element analysis of cracking of a concrete arch dam due to seasonal temperature variations. In Proceedings of the 14th ICOLD International Benchmark Workshop on Numerical Analysis of Dams, Stockholm, Sweden, 6–8 September 2017; pp. 208–217.
35. Frigerio, A.; Mazzà, G. Thermal cracking of a thin concrete arch dam. In Proceedings of the 14th ICOLD International Benchmark Workshop on Numerical Analysis of Dams, Stockholm, Sweden, 6–8 September 2017; pp. 218–227.
36. Varpasuo, P. Thermal cracking of a concrete dam, PVA Engineering Services contribution to the ICOLD2017 Benchmark Exercise Theme, A. In Proceedings of the 14th ICOLD International Benchmark Workshop on Numerical Analysis of Dams, Stockholm, Sweden, 6–8 September 2017; pp. 228–237.
37. Cornelissen, H.; Hordijk, D.; Reinhardt, H. Experimental determination of crack softening characteristics of normal weight and lightweight concrete. *Heron* **1986**, *31*, 45–56.
38. Malm, R.; Könönen, M.; Bernstone, C.; Persson, M. Assessing the structural safety of cracked concrete dams subjected to harsh environment. In Proceedings of the ICOLD 2019 Annual Meeting/Symposium, Ottawa, ON, Canada, 9–14 June 2019; pp. 383–397.
39. Malm, R. Predicting Shear Type Crack Initiation and Growth in Concrete with Non-Linear Finite Element Method. Ph.D. Thesis, KTH Royal Institute of Technology, Stockholm, Sweden, 2009.
40. Brühwiler, E. Fracture of mass concrete under simulated seismic action. *Dam Eng.* **1990**, *1*, 153–176.
41. *Fib Model Code 2010, Bulletins 55 and 56*; International Federation for Structural Concrete: Lausanne, Switzerland, 2012.
42. Ghaemmaghami, A.; Ghaemian, M. Large-scale testing on specific fracture energy determination of dam concrete. *Int. J. Fract.* **2006**, *141*, 247–254. [CrossRef]

6

Instrumented Health Monitoring of an Earth Dam

S.M. Seyed-Kolbadi [1], M.A. Hariri-Ardebili [2,3,*], M. Mirtaheri [1] and F. Pourkamali-Anaraki [4]

1 K. N. Toosi University of Technology, Tehran 1996715433, Iran; mahdi_kolbadi@sina.kntu.ac.ir (S.M.S.-K.); mmirtaheri@kntu.ac.ir (M.M.)

2 University of Colorado Boulder, CO 80818, USA

3 University of Maryland, College Park, MD 20743, USA

4 University of Massachusetts, Lowell, MA 02142, USA; Farhad_Pourkamali@uml.edu

* Correspondence: mohammad.haririardebili@colorado.edu.

Abstract: This work evaluates the stability of the Boostan earth dam by investigating its long-term performance and interpreting the measured data. To measure the dam response, several sensitive locations are instrumented. This process includes measuring various quantities such as pore water pressure, water level, and internal stress ratios using inspection devices such as ordinary and Casagrande piezometers, and total pressure cells. The recorded data shows that the pore pressure is in good agreement with the initial (stable) design condition. The installed piezometers show that the drainage is efficient, and the water table in the body is adequate. The instrument also shows a reasonable horizontal stress in the dam body. Overall, the condition of the case study dam is assessed to be normal. The results of this case report can be used as a guide in similar dams for instrumented health monitoring.

Keywords: stability assessment; instrumentation; earth dam; health monitoring

---

1. Introduction

The protection, maintenance, and proper operation of dams not only lead to a long life span of the structure, but they are also significant for health, safety, and environmental aspects [1]. The most significant problems that occur in the major infrastructure are those associated with the lack of proper maintenance procedures [2]. The high cost and long duration of constructing dams dictates the need for a comprehensive online maintenance and protection program. Early detection of defects and abnormalities is a critical step towards integrity assessment of dams, health monitoring, and overall risk management. To simulate the behavior of a dam during the construction time, it is necessary to precisely measure several parameters such as displacement, pore water pressure, leakage, etc. For this reason, multiple instrumental equipment are used [3].

Dam health monitoring is crucial to the project risk management and may significantly reduce the project risk exposure and future consequences, see Figure 1. A general risk management circle includes five steps:

- Identify: Establish the risk components.
- Analysis: Evaluate the likelihood of each risk element, and the associated consequences.
- Plan: Determine the general strategies to minimize the risk likelihood.
- Monitor: Continuously measure and update all variables that may indicate risk.
- Control: Take various risk reduction actions.

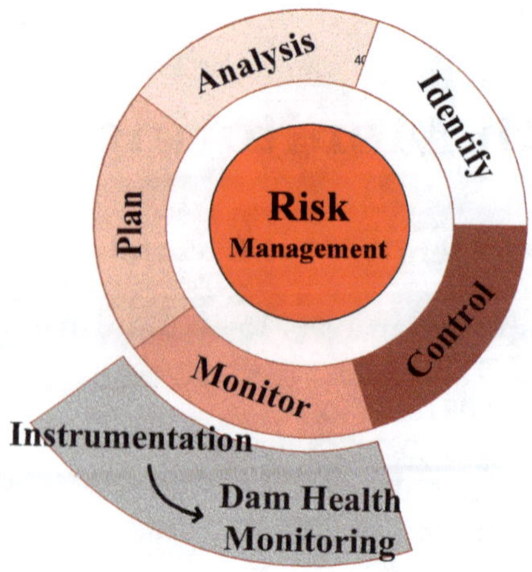

Figure 1. Interconnection of risk management with dam health monitoring and instrumentation.

The health monitoring procedure is often used in the context of risk analysis to identify the potential failure modes (PFMs) [4]. The identified failure modes coupled with health monitoring are key metrics to understand the structural, hydraulic, and geothectical conditions. Postprocessing the dam monitoring data will identify the PFMs [5].

Many researchers provided recommendations for monitoring and analyzing earth dams and levees [6,7]. Instrumentation is used for a wide spectra of measurements to identify the variations of the physical and mechanical quantities that are implemented in the area of geotechnical engineering and constructing projects [8]. The quantities variations such as displacement, and water pressure are measured using instrumented data [9,10]. Placement of these instruments at sensitive locations on the dam structure along with proper measuring devices, and finally online data registration lead to a comprehensive dam safety assessment [11].

Structural failure of earth and rockfill dams may cause irreparable damages in the downstream facilities. The dam performance should be assessed from different behavioral aspects during health monitoring period (e.g., pore water pressures, tensions, displacements, and other controlling parameters). The results from these assessments and their corresponding interpretations should be regularly reported in a proper fashion. Moreover, special consideration in the event of emergencies (e.g., flood or earthquake) must be addressed [12]. By comparing the current and past behavior, the stability of the dam is evaluated [13].

Dam risk assessment can be utilized in various scales, from different perspectives, and also subjected to multiple hazards [14]. One may focus on macro-scale studies such as dam portfolio in a region [15], or perform a series of advanced numerical simulations [16–18], experimental tests [19,20], or field investigations on a particular dam. This paper attempts to investigate the capabilities of instrumentation in controlling the design and quality of an earth dam construction. Also, the stability of the dam is evaluated by examining its current condition. The most sensitive locations within the dam body is identified. The common concerns regarding the instrumented devices are addressed for the case study dam, and their impact on the overall stability is evaluated. This paper provides a comprehensive database for an earth dam with 10+ years of the monitored data.

2. Literature Review

Visual inspection was the first safety assessment method for embankments [11]. Next, instrumentation and monitoring tools were invented [21]. The 1853 recorded data from topographic levels of the Gross Bois in France can be considered as the first application of dam data monitoring [22]. During late 19th century, the open-pipe piezometers were used in USA, India, Europe, and Asia to study the leakage water running under the dams specialized for irrigation and being constructed on alluvial areas.

Instrumentation and monitoring of structures and reservoirs are important concerns in dam engineering [23]. The installation and monitoring of appropriate instrumentation are useful tools to visually investigate the safety of an embankment dam. Furthermore, instrumentation may be used to confirm that the long-term performance, such as creep deformation, is at the desired level [24]. In 1907, this kind of instrumentation was employed by the Britain to study the water level fluctuations in earth dams. In the USA, the reservoir water level and the hydrostatic gauges were initially employed in the rehabilitation office of earth dams to measure and record the water level, and pore water pressure [5]. The installation of the hydrostatic pressure gauges in 1938–1939 for Cobol Dam in the New Mexico, USA, is an example of early instrumentation [25]. Roy Karlson designed an instrument that measured the pressure using a sensor involving a stretched wire in the concrete and earth dams [12]. Also, in 1930s the researchers began to make use of the controlled instruments for the performance evaluation of the dam constructions with planning data, and the construction methods. This attempt led to the design of the two-piped hydraulic piezometers for measuring the pore pressure [26].

Gikas and Sakellariou [27] studied the long-term settlement behavior of Mornos Dam based on a finite element analysis in comparison with the records of installed instruments. Vibration wire sensors have a long life span, high accuracy, and stability. These sensors have been proved to be appropriate for special and severe environmental conditions [28]. Its exit signal frequency makes it less sensitive to the environmental noise and deficiencies in the conducting current than other sensors. Nowadays, the vibration wire sensors are the most applicable among the implemented instrumentation [29]. During 1930s and 1940s, the use of vibration wire sensors for behavior analysis gained popularity worldwide. In recent years, the instrumentation has gone under considerable changes that are indebted to the advancements of electronics and microcomputer technology. The advancements in these fields made it easier to collect and process the data; this has caused a change in the nature of instrumentation results. Recently, it has been made feasible to receive the information from dams' instrumentation in the remotest areas of the world via satellites, and to transmit it to offices in other locations [30]. The information can be immediately simplified by computers and be put into a format that makes it possible to instantly recognize the performance deviations of the construction.

According to the statistical results from the investigation of 14,700 dams reported in ICOLD [31], Shi et al. [32], it was noticed that 7% were going to be destroyed and another 0.7% were totally destroyed. According to this report, a major part of the constructed dams was of earth and/or trench-type, and they had a very higher percentage of destruction in comparison to concrete dams (i.e., 74%). This clearly represents the necessity of the behavior analysis of earth dams [33].

The main failure modes of earth dams are piping and leakage from the dam body, foundation problems, and water overflow over the dam which can be identified by proper behavior analysis [34]. In a dam construction project, the instrumenting and stability analysis could be divided in four general phases: planning, construction, initial drainage, and operation. In each of the phases, the controlling parameters, such as leakage, displacement, and water pressure, are measured, evaluated, and controlled. According to the water level variations during the dam operation, the behavior analysis will gain high importance [35]. The rapid decrease or increase of the reservoir water level can lead to dangerous problems. Therefore, this issue must be under continuous control [36]. One may note that the dam behavior after severe events such as an earthquake or flood should be investigated individually for each dam [37].

3. Materials and Methods

3.1. *Type of Instruments*

The instruments used in dam monitoring can be divided into two main groups: [38]:

- Mechanical instruments: The main system of these instruments measures the quantity of interest directly and converts the measurement into a target parameter [39]. Their installation is easy, and in the case of failure, they can be easily replaced. Moreover, they are relatively cheap, and can be easily manufactured in a large amount. However, their application is limited in some conditions, and they may fail to properly meet the required measurement accuracy.

- Electrical instruments: They initially developed to cover the shortcomings of the mechanical instruments. Their measurement factor is mainly vibrational. Electrical instruments measure the quantity indirectly, and their reading results should be converted to the parameters via some formulas [40].

In another classification, the instruments can be categorized as follows. [41]:

- Stress measurements, e.g., Carlson soil stress meters and flat jacks.
- Pore pressure measurement, e.g., open standpipe piezometers and Casagrande piezometers.
- Flow measurement, e.g., weirs and impeller flow transducers.
- Temperature measurement, e.g., thermocouples and thermistors.
- Strong motion measurements.

3.2. Reading, Recording, and Processing the Data

The recording time of instruments should be in accordance with the instructions provided by the project instrumentation engineer. It usually requires a long enough time to capture the potential abnormalities. The time step should be dynamically adjusted based on the predicted (or recorded) variations. All other executive activities should be also recorded for future stability assessment purposes. Nearly all the response variations during the normal lifetime of dams are related to the changes in pool elevation and climate conditions (e.g., environmental temperature, amount of rainfall). Therefore, all the dam responses should be recorded in accordance to these variables, and visualized by time. Finally, the processing may require some level of corrections to the reading, as well as synchronizing multiple readings in a station.

The so-called cause and effect plots should be developed to properly address the response variations. For example, plotting the response variations, such as body and foundation deformation or pore water pressure versus to the pool level variations, is an effective way to explain the changes. Moreover, whenever the dam recording is interrupted (for example the instrument is failed), one may estimate the missing data by tracking the water level and temperature variations.

4. Case Study Dam

4.1. Boostan Dam Characteristics

Boostan Dam is constructed in Golestan province, and located ~35 km northeast of Gonbad city, and 23 km upstream of the Golestan Dam. The dam is placed on the Gorganrood River with coordination of 37° 25′ 30″ east longitude and 55° 25′ 00″ north latitude. The main purpose of this dam is flood control, irrigation. The dam is also capable of regulating 28 Mm^3 water per year. General characteristics of the dam are summarized in Table 1. Figure 2a shows a satellite photo of Boostan Dam, and the general specifications of the dam and its associated structures are shown in Figure 2b.

Table 1. General characteristics of the Boostan Dam. PMF: probable maximum flood.

Body Specification		Reservoir Specification		Design Specification	
Dam type	Homogeneous Earthen	Reservoir volume	5 Mm^3	PGA	0.15 g
Dam height	35 m	Flood control volume	18 Mm^3	Spillway capacity	785 m^3/s
Dam length	642 m	Normal level area	4.1 km^2	Spillway type	Ogee
Foundation width	265 m	Normal level high	96 m	Spillway length	40 m
Crest width	10 m	Agriculture water volume	28 Mm^3	Flood design	PMF

(a) Satellite image **(b)** Dam and reservoir

Figure 2. Boostan Dam.

4.2. Instruments in Boostan Dam

At the design stage of the Boostan Dam, six main sections were marked to be instrumented in the dam body. One of them (Section #2) is investigated in this paper. Section #2 is located exactly in line with the center of the former river axis, and the foundation has special sensitivities around there. Each of these sections comprises a set of tools such as standing and electric pipe piezometers and pressure measuring cells mounted at certain distances from the dam axis. In this dam, a total of 34 piezometers have been installed in 6 sections, 34 electrical piezometers in 3 sections, and 5 dampers in 3 different sections. Six pairs of pressure cells are also installed in a cross section (one horizontal and one vertical), as well as two triangular overflow discharge measuring tools on the left and right sides of the dam.

The drainpipes in the earth dams are required to ensure that the main body's aggregates will not be washed off. In Boostan Dam, a horizontal drainage blanket has been used as a supplement along with the vertical chimney drainpipe. Also, considering that the chimney drainpipe was in the vicinity of the dam central line, the critical conditions of the leakage forces are considerably reduced. The installed instruments in the dam body and foundation of the Boostan Dam (to screen the piezometeric level fluctuations) are cascade and vibrational piezometers. The location map of these instruments for Section #2 are shown in Figure 3. The cut-off wall is built just below the body and the dam crest. It is extended 20 m beneath the dam base for Section #2. The cut-off wall damps the energy of the water passing under the dam, and avoids extensive uplift pressure by extending water passage through the porous soil. Therefore, its proper operation should be evaluated by the instrumented devices.

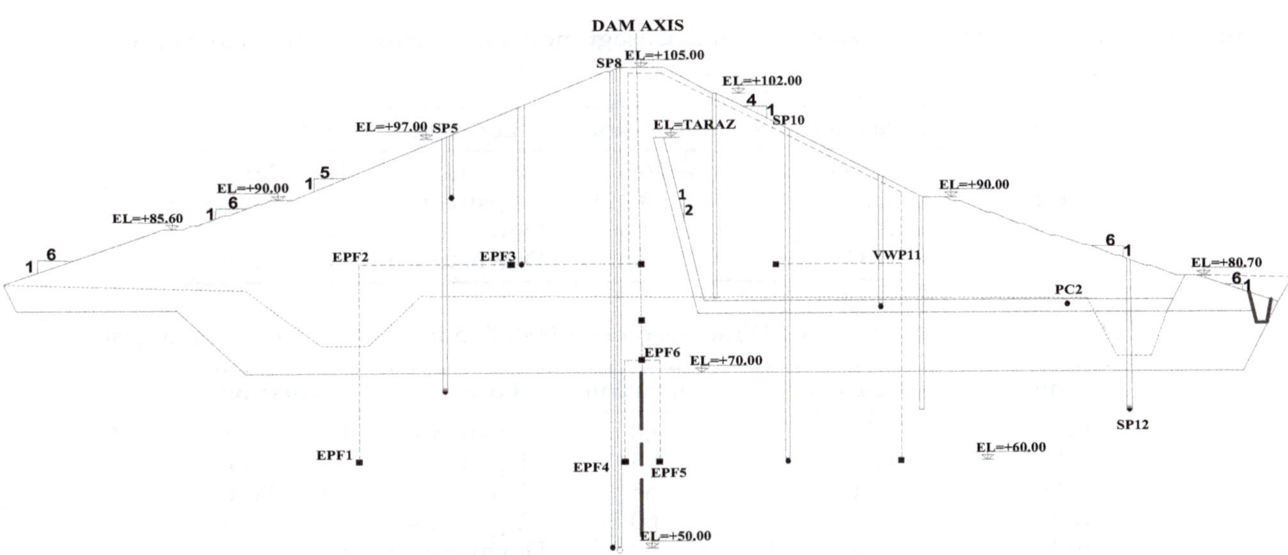

Figure 3. Instrumentation map of Boostan Dam in Section #2.

- **Piezometers:** Variation of pore water pressure is very important parameter in dam health monitoring. An increase in pore water pressure decreases the (effective) shear stresses (and thus treats the dam stability). The pore pressure leakage during the dam construction are affected by the leakage force from the water flow into the dam. The required time for downstream soil satiation and creation of a stable leakage has a direct relation to the soil type and the water regime behind the dam. In the Boostan Dam, a homogeneous earth dam, the drainage blanket is very effective, and can make the current lines vertical. Moreover, the presence of the drainage blanket reduces the pore water pressure to a great extent.
- **Casagrande Piezometer:** This is an effective and economical tool to measure the water pressure. It involves a plastic porous pipe at the bottom of the bore and a PVC pipe to record the water level. When the probe is entered into the pipe, upon its contact with water level, a continuous vocal signal is produced. The upstream level of the piezometer pipes in elongation is determined through survey operations, based on which the level and water height (total head) calculations are performed. Total head (or piezometeric level) is a summation of the installed piezometer level, z, and $\frac{U}{\gamma_w}$, where U refers to pore pressure. Figure 4a shows a sample of installed Casagrande piezometer. Table 2 shows the instrument specifications and their locations corresponding to Figure 3.
- **Vibrating Wire Piezometer:** The vibrational piezometers are reliable and stable instruments to measure the pore water pressure. The output of the piezometer includes a signal that is independent from impedance and the contact resistance, and can transmit the signal to a long distance. The sensor includes a porous tip piece often made of ceramic materials and a diaphragm. This diaphragm at its bottom is attached to a prestressed piece. When the pore pressure changes, the diaphragm moves, and alters the tension in the attached warp. The frequency of the normal vibration wire is a function of the imposed stress. Sample of the installed wire piezometer is shown in Figure 4b. Table 3 shows the instrument specifications and their locations corresponding to Figure 3.
- **Total Pressure Cell:** The effective soil pressure as well as the water pressure can be measured by total pressure cells. The instrument includes a flat cell filled with oil. The pressure imposed on the cell is imposed on the oil and is transformed into some signals by a transducer. They are connected to a digital reading device via some cables. This device measures the pressure imposed on the cell, see Figure 4c. A total of six pressure cells have been installed in Boostan Dam. Table 4 shows the instrument specifications and their locations. the location of TPC-2 can be found in Figure 3 at the lower discharge place.

Table 2. Characteristics of Piezometers and Casagrande Piezometer installed in Section #2 of Boostan Dam.

Name	Distance from Axis	Elevation	Location	Situation
SP-5	45.00	96.5	Upstream	Foundation
SP-8	7.75	50.0	Upstream	Foundation
Sp-10	33.00	61.0	Downstream	Foundation
SP-12	113.25	68.0	Downstream	Foundation

Table 3. Characteristics of Vibrating Wire Piezometer installed in Section #2 of Boostan Dam.

Name	Distance from Axis	Elevation	Location	Situation
EPF-1	65.00	60	Upstream	Foundation
EPF-2	65.00	85	Upstream	Dam Body
EPF-3	30.00	85	Upstream	Dam Body
EPF-4	3.85	60	Upstream	Foundation
EPF-5	3.85	60	Downstream	Foundation
EPF-6	0.00	72	Axis	Foundation

Table 4. Specifications of total pressure cells installed in Boostan Dam.

Name	Elevation	Direction	Location	Section
TPC-1	78.6	Vertical	Lower Discharge	# 1
TPC-2	77.6	Horizontal	Lower Discharge	# 2
TPC-3	77.1	Vertical	Lower Discharge	# 3
TPC-4	76.8	Horizontal	Lower Discharge	# 4
TPC-5	77.0	Vertical	Lower Discharge	# 5
TPC-6	75.9	Horizontal	Lower Discharge	# 6

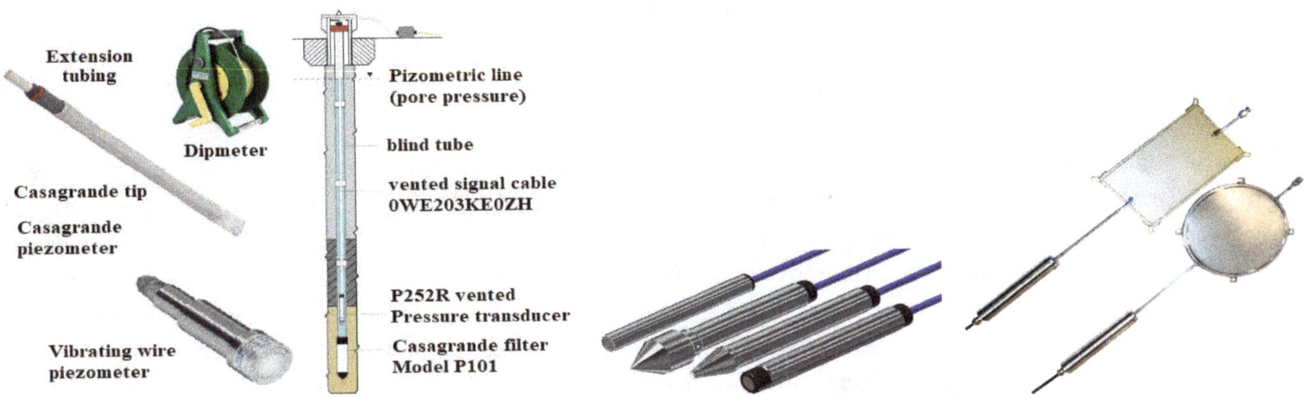

(**a**) Casagrande piezometer, and the reading gauge (**b**) Vibration wire piezometer (**c**) Total pressure cell

Figure 4. Sample of installed instruments in Boostan Dam.

5. Results and Observations

Data from instrumentation in Boostan Dam are recorded hourly, and collected monthly. Data are collected for approximately 12 years (2006 to 2018). The collected and filtered data are then used to plot several appropriate graphs, and finally to evaluate the stability of the dam.

5.1. Performance of the Casagrande Piezometers

To evaluate the stability of the dam, three quantities are tracked: (1) piezometeric level; (2) pore pressure; and (3) hydraulic loss, see Figure 5. Four Casagrande piezometers have been installed in the foundation: SP-5 and SP-8 are located upstream of the dam, and SP-10 and SP-12 are located downstream of the dam. Screening the general trend of the piezometeric level variations reveals a reduction due to water passing through the foundation layers. Variations of the piezometeric level in SP-5 are in close agreement with pool level variations (which is reasonable considering the spatial location of SP-5 at level 69.5 m). The upper level of the piezometer pipe is located at 97 m, and thus we lose the reading if the pool level reaches this value.

SP-8 is installed at the upstream dam body (and is more close to the bottom of the cut-off wall), and is less sensitive to the water level fluctuations. SP-10 and SP-12 piezometers, which are located in the dam downstream, are more affected by the water level fluctuations and show a logical hydraulic loss compared to the reservoir level. The sudden jumps can be attributed to the reading and recording errors. As it is expected, the piezometeric level in SP-12 shows a considerable difference with regard to the reservoir water level. The water level fluctuations are effective on the piezometer's behavior. According to these plots, the hydraulic loss is increased by getting away from the reservoir (e.g., SP-12). Total hydraulic loss in different piezometers is SP-5 from 0.0 to 2.18 m, SP-8 from 1.65 to 13.64 m, SP-10 from 2.43 to 15.53 m, and SP-12 from 9.23 to 21.5 m.

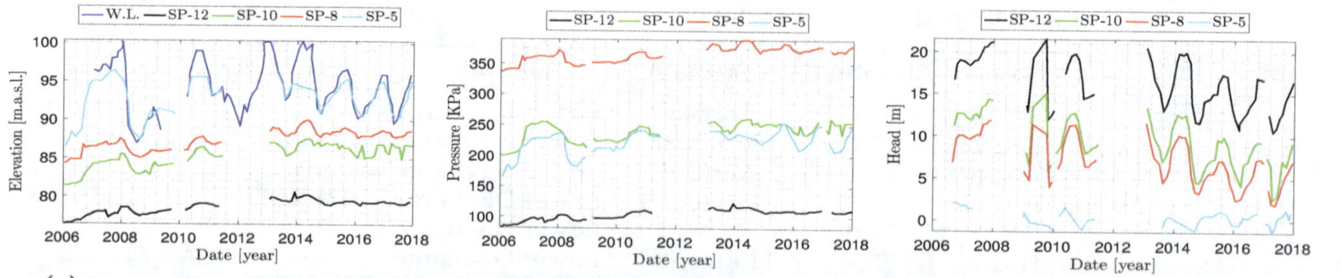

(a) Piezometric level variations **(b)** Pore pressure variations **(c)** Hydraulic loss variations

Figure 5. Recorded data by Casagrande piezometers at Section #2.

5.2. Performance of the Vibration Wire Piezometers

Similar to the previous section, the piezometeric level, pore pressure, and hydraulic loss curves are computed for the vibration wire piezometers; see Figure 6. Based on the locating map of the installed instruments, EPF-1 and EPF-4 are located in the upstream, whereas EPF-5 and EPF-10 piezometers are installed in the downstream cut-off wall at foundation level (EPF-10 is not operating anymore). According to the plots, the piezometeric level follows the same trend as the pool elevation at the upstream. For example, the piezometeric level in EPF-1 and EPF-4 is close to the reservoir level. The piezometeric level changes when the pool level exceeds 95 m.

The water head loss is increased by getting away from the reservoir. As seen, the general trend of fluctuations in the pore pressure is similar to those curves from the piezometeric level. Total hydraulic loss in different piezometers can be summarized as EPF-1 from 0 to 4.36 m, EPF-4 from 0.42 to 8.49 m, and EPF-5 from 6.91 to 15.38 m. Comparing the water head loss values in 2005 and the previous years, the general condition of the dam seems to be normal.

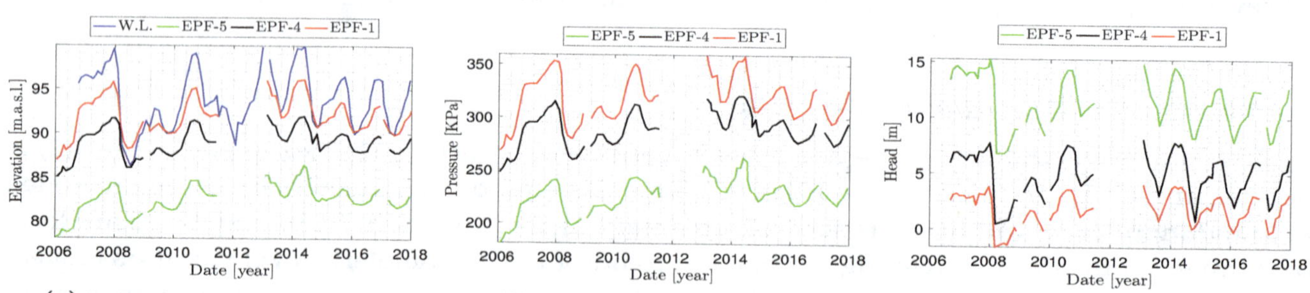

(a) Piezometric level variations **(b)** Pore pressure variations **(c)** Hydraulic loss variations

Figure 6. Recorded data by vibration wire piezometers at section #2.

5.3. Hysteresis Curve of the Foundation Piezometers

A hysteresis curve presents a graph where the horizontal axis is the reservoir water level and the vertical one is the pressure head. The hysteresis curve is plotted for different recording dates, and thus is a metric to show the stability of the foundation and performance of the cut-off wall. The temporal nature of "water level" and "pressure" curves in Figure 7a is converted to the hysteresis curve in Figure 7b for a sample piezometer. The structural health of a dam foundation (i.e., the lack of scouring in the foundation particles) has a direct relation with narrow hysteresis curve. It is also possible to fit a first-order polynomial to the data points. Such a predictive model presents the pressure as a function of the water head.

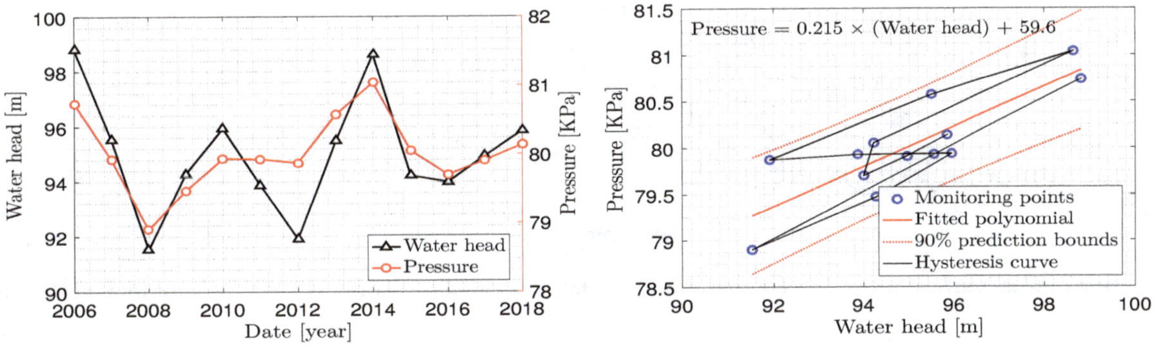

(a) Temporal variation **(b)** Generic hysteresis curve and its predictive model

Figure 7. Developing a hysteresis curve.

5.4. Excess Pore Pressure Ratio

The excess pore pressure ratio, R_U, is defined as the fraction of pore pressure from pizometric reading, $U = \gamma_w \times h_w$, to the vertical effective overburden stress, $\sigma_v = \gamma_s \times h_s$ (where h presents the height and γ is the specific density). This is a very important parameter in the stability assessment of earth dams. The dam performance is evaluated as follows.

- If $R_U < 0.4$: Dam conditions are safe and favorable
- If $0.4 < R_U < 0.6$: There is a potential risk. The considered section should be continuously evaluated for updated results.
- If $R_U > 0.6$: Dam is in a critical condition. As the pore pressure is high, the water tends to exit the soil. This may eventually lead to dam failure.
- If $U > \sigma_v$ (i.e., $R_U > 1$): A hydraulic failure occurs in the dam.

Variations of R_U at Section #2 are plotted in Figure 8 for Boostan Dam. For the majority of pizometers, this value is less than 0.4, showing the satisfactory condition with no immediate risk of uplift. EPF-1 is located at the closest distance from the reservoir, and is highly affected then. The R_U associated with this pizometer shows some values beyond 0.4 (and at some dates even beyond 0.5). Therefore, this section should be monitored regularly for any sudden changes in future.

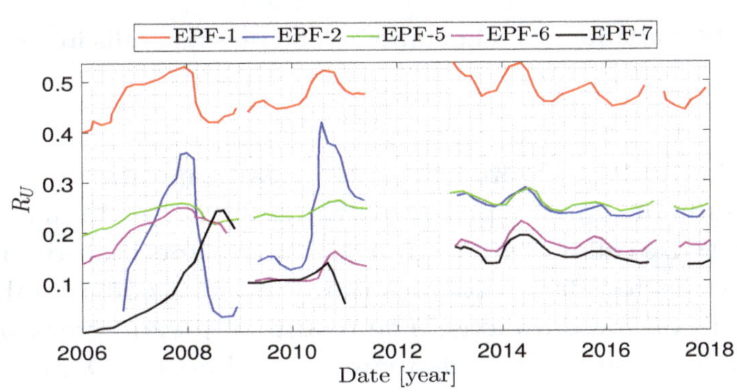

Figure 8. Variation of R_U in Section #2.

5.5. Performance of the Total Pressure Cells

The location and direction of six total pressure cells (TPC) are shown in Table 4. All of them are installed in Section #4. Variation of the vertical, horizontal, and ratio of total pressure are shown in Figure 9. In general, the fluctuations in both horizontal and vertical pressure components are consistent with the reservoir level variations. In the most of locations, the vertical pressure is higher than the horizontal one. Only in one location their ratio varies around unit. This results also show that TPC-4 is more sensitive to the water level.

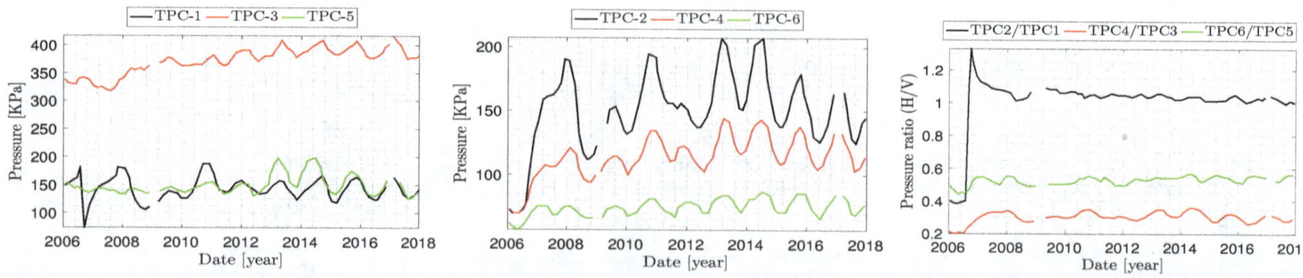

(a) Variation of total vertical pressure **(b)** Variation of total horizontal pressure **(c)** Variation of total H/V pressures

Figure 9. Variation of total pressure in Section #4.

Finally, Figure 10 presents the variations of arcing ratio in different total pressure cells. It is simply the ratio of total pressure values to the soil overburden pressure. As seen, the ratios from TPC-1, TPC-2, TPC-3, and TPC-5 range between 0.3 and 0.7. They correspond to favorable and normal conditions. The curve associate with TPC-4 and TPC-6 shows smaller values, and it can be attributed to the fact that their direction is no longer aligned with the horizon.

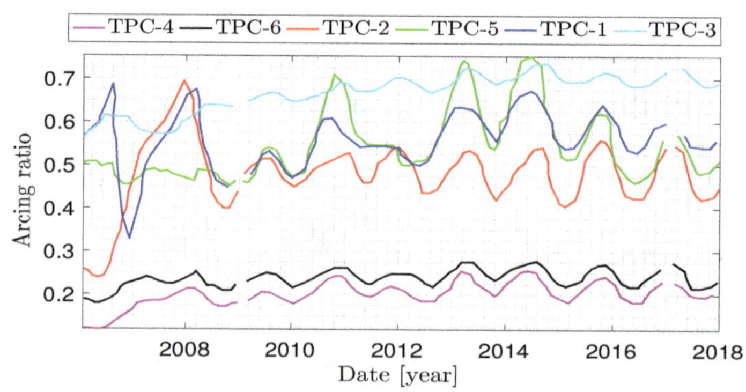

Figure 10. Variation of arcing ratio in total pressure cells in section #4.

6. Summary

This paper reports the findings of instrumented health monitoring of an earth dam, and the subsequent stability assessment. Three types of instruments (i.e., ordinary Piezometers, Casagrande piezometers, and total pressure cells) were used in this study, and their data were collected for about 12 years. Analyzing the collected data indicates that there is a considerable pressure drop due to the presence of cut-off wall on the dam axis. The results of piezometers show that the drainage is efficient, and the water table in the body is adequate. In the Boostan Dam, the excess pore pressure is acquired from the electrical piezometers data and their fluctuations over time indicate a safe operation of dam with no uplift risk. Finally, the instrument shows a reasonable horizontal stresses. Overall, the condition of the case study dam is assessed to be normal during its first 15 years of operation. Although there is a limited drainage at the downstream of the dam seal wall, it does not seem to be an immediate threat to dam safety.

Author Contributions: Conceptualization, S.M.S.-K., M.M.; methodology, S.M.S.-K.; investigation, S.M.S.-K.; resources, S.M.S.-K.; data curation, S.M.S.-K.; writing–original draft preparation, S.M.S.-K., M.M.; writing–review and editing, M.A.H.-A., F.P.-A.; visualization, S.M.S.-K., M.A.H.-A., F.P.-A.; supervision, M.A.H.-A.; funding acquisition, M.A.H.-A. All authors have read and agreed to the published version of the manuscript.

References

1. USACE. *EM 1110-2-1908: Instrumentation of Embankment Dams and Levees;* Technical Report; USACE: Washington, DC, USA, 1995.
2. Ballard, R.F. The US Army Corps of engineers seismic strong-motion instrumentation program. In *Strong Motion Instrumentation for Civil Engineering Structures*; Springer: Berlin/Heidelberg, Germany, 2001; pp. 157–166.
3. Taylor, P.W. *Engineering and Design: Instrumentation for Safety Evaluations of Civil Works Projects*; Technical Report; USACE: Washington, DC, USA, 1985.
4. On Dams, U.S.S. *Routine Instrumented and Visual Monitoring of Dams Based on Potential Failure Modes Analysis*; Technical Report; U.S. Society on Dams: Denver, CO, USA, 2013.
5. Dunnicliff, J. *Geotechnical Instrumentation for Monitoring Field Performance*; John Wiley & Sons: Hoboken, NJ, USA, 1993; pp. 97–108.
6. Robertson, K. *An Instrument to Monitor the Tilt of Large Structures;* Technical Report; Army Engineer Topographic Labs: Fort Belvoir, VA, USA, 1983.
7. US Department of the Interior, Bureau of Reclamation. *Design Standards No. 13: Embankment Dams*; Technical Report; US Department of the Interior, Bureau of Reclamation: Denver, CO, USA, 2014.
8. Lindsey, J.; Edwards, D.; Keeter, A.; Payne, T.; Malloy, R. *Instrumentation Automation for Concrete Structures; Report 1: Instrumentation Automation Techniques*; Technical Report; Wyle Labs: Hampton, VA, USA, 1986.
9. Currier, B.; Fenn, M.H. *Repair, Evaluation, Maintenance, and Rehabilitation Research Program. Instrumentation Automation for Concrete Structures. Report. 3. Available Data Collection and Reduction Software*; Technical Report; Wyle Labs: Hampton, VA, USA, 1987.
10. O'Neil, E.F. *IInstrumentation Automation for Concrete Structures. Report 4. Demonstration of Instrumentation Automation Techniques at Beaver Dam, Eureka Springs, Arkansas*; Technical Report; Army Engineer Waterways Experiment Station Vicksburg Ms Structures Lab: Vicksburg, MS, USA, 1989.
11. Bartholomew, C.L.; Murray, B.C. *Embankment Dam Instrumentation Manual*; US Department of the Interior, Bureau of Reclamation: Denver, CO, USA, 1987; pp. 71–83.
12. ASCE Task Committee on Instrumentation and Monitoring Dam Performance. *Guidelines for Instrumentation and Measurements for Monitoring Dam Performance*; American Society of Civil Engineers: Reston, VA, USA, 2000.
13. Li, X.; Li, Y.; Lu, X.; Wang, Y.; Zhang, H.; Zhang, P. An online anomaly recognition and early warning model for dam safety monitoring data. *Struct. Health Monit.* **2019**, *1*, 145–166. [CrossRef]
14. Hariri-Ardebili, M.A. Risk, Reliability, Resilience (R3) and beyond in dam engineering: A state-of-the-art review. *Int. J. Disaster Risk Reduct.* **2018**, *31*, 806–831. [CrossRef]
15. Ecorys, O. The Role of Market-Based Instruments in Achieving a Resource Efficient Economy. In *Report for the European Commission, DG Environment. Ecorys, Cambridge Econometrics and COWI, Rotterdam*; Rotterdam, The Netherlands, 2011.
16. Farias, M.M.d.; Cordão Neto, M.P. Advanced numerical simulation of collapsible earth dams. *Can. Geotech. J.* **2010**, *47*, 1351–1364. [CrossRef]
17. Seyed-Kolbadi, S.; Sadoghi-Yazdi, J.; Hariri-Ardebili, M. An Improved Strength Reduction-Based Slope Stability Analysis. *Geosciences* **2019**, *9*, 55–71. [CrossRef]
18. Guo, X.; Dias, D.; Pan, Q. Probabilistic stability analysis of an embankment dam considering soil spatial variability. *Comput. Geotech.* **2019**, *113*, 93–103. [CrossRef]
19. Matsushima, K.; Mohri, Y.; Yamazaki, S.; Hori, T.; Ariyoshi, M.; Tatsuoka, F. Design of earth dams allowing temporary overtopping based on hydraulic failure experiments and flood analysis. In *Geosynthetics in Civil and Environmental Engineering*; Springer: Berlin/Heidelberg, Germany, 2008; Volume 1; pp. 757–762.
20. Sui, W.; Zheng, G. An experimental investigation on slope stability under drawdown conditions using transparent soils. *Bull. Eng. Geol. Environ.* **2018**, *77*, 977–985. [CrossRef]
21. Piqueras, J.M.; Pérez, E.S.; Menéndez-Pidal, I. Water seepage beneath dams on soluble evaporite deposits: A laboratory and field study (Caspe Dam, Spain). *Bull. Eng. Geol. Environ.* **2012**, *71*, 201–213. [CrossRef]
22. Frémion, F.; Bordas, F.; Mourier, B.; Lenain, J.F.; Kestens, T.; Courtin-Nomade, A. Influence of dams on sediment continuity: a study case of a natural metallic contamination. *Sci. Total. Environ.* **2016**, *547*, 282–294. [CrossRef]

23. Mizuno, M.; Hirose, T. Instrumentation and monitoring of dams and reservoirs. *Water Storage Transp. Distrib.* **2009**, *1*, 1–8.
24. Charles, J.; Tedd, P.; Watts, K. The role of instrumentation and monitoring in safety procedures for embankment dams. *Water Resour. Reserv. Eng.* **1992**, *1*, 311–320.
25. Majoros, M.; Sneed, H.M. The softest program test system. *J. Syst. Softw.* **1981**, *2*, 289–296. [CrossRef]
26. Jansen, R.B. *Advanced Dam Engineering for Design, Construction, and Rehabilitation*; Springer Science & Business Media: Berlin/Heidelberg, Germany, 2012.
27. Gikas, V.; Sakellariou, M. Settlement analysis of the Mornos earth dam (Greece): Evidence from numerical modeling and geodetic monitoring. *Eng. Struct.* **2008**, *30*, 3074–3081. [CrossRef]
28. Zhu, P.; Leng, Y.; Zhou, Y.; Jiang, G. Safety inspection strategy for earth embankment dams using fully distributed sensing. *Procedia Eng.* **2011**, *8*, 520–526. [CrossRef]
29. Masoumi, I.; Ahangari, K.; Noorzad, A. Integrated fuzzy decision approach for reliability improvement of dam instrumentation and monitoring. *J. Struct. Integr. Maint.* **2018**, *3*, 114–125. [CrossRef]
30. Chugh, A.K. Stability assessment of a circular earth dam. *Int. J. Numer. Anal. Methods Geomech.* **2013**, *37*, 2833–2859. [CrossRef]
31. ICOLD. *Deterioration of Dams and Reservoirs*; Laboratorio Nationai de Engenharia Civil, LNEC: Lisboa, Portugal, 1983; pp. 1–312.
32. Shi, Z.M.; Wang, Y.Q.; Peng, M.; Chen, J.F.; Yuan, J. Characteristics of the landslide dams induced by the 2008 Wenchuan earthquake and dynamic behavior analysis using large-scale shaking table tests. *Eng. Geol.* **2015**, *194*, 25–37. [CrossRef]
33. Rashidi, M.; Haeri, S.M. Evaluation of behaviors of earth and rockfill dams during construction and initial impounding using instrumentation data and numerical modeling. *J. Rock Mech. Geotech. Eng.* **2017**, *9*, 709–725. [CrossRef]
34. Hui, S.; Charlebois, L.; Sun, C. Real-time monitoring for structural health, public safety, and risk management of mine tailings dams. *Can. J. Earth Sci.* **2018**, *55*, 221–229. [CrossRef]
35. Masoumi, I.; Ahangari, K.; Noorzad, A. Optimal monitoring instruments selection using innovative decision support system framework. *Smart Struct. Syst.* **2018**, *21*, 123–137.
36. Masoumi, I.; Naraghi, S.; Rashidi-nejad, F.; Masoumi, S. Application of fuzzy multi-attribute decision-making to select and to rank the post-mining land-use. *Environ. Earth Sci.* **2014**, *72*, 221–231. [CrossRef]
37. Meixner, O. Fuzzy AHP group decision analysis and its application for the evaluation of energy sources. In Proceedings of the 10th International Symposium on the Analytic Hierarchy/Network Process, Pittsburgh, PA, USA, 29 July–1 August 2009; Volume 29, pp. 2–16.
38. Wieland, M.; Kirchen, G. Long-term dam safety monitoring of Punt dal Gall arch dam in Switzerland. *Front. Struct. Civ. Eng.* **2012**, *6*, 76–83. [CrossRef]
39. Li, F.; Wang, Z.; Liu, G.; Fu, C.; Wang, J. Hydrostatic seasonal state model for monitoring data analysis of concrete dams. *Struct. Infrastruct. Eng.* **2015**, *11*, 1616–1631. [CrossRef]
40. Machan, G.; Bennett, V.G. Use of inclinometers for geotechnical instrumentation on transportation projects: State of the practice. *Transp. Res. Circ.* **2008**, *1*, 1–92.
41. Zhao, J.; Jin, J.; Zhu, J.; Xu, J.; Hang, Q.; Chen, Y.; Han, D. Water resources risk assessment model based on the subjective and objective combination weighting methods. *Water Resour. Manag.* **2016**, *30*, 3027–3042. [CrossRef]

7

Probabilistic Identification of Seismic Response Mechanism in a Class of Similar Arch Dams

M. A. Hariri-Ardebili [1,2,*], M. Heshmati [2], P. Boodagh [1] and J. W. Salamon [3]

1 Department of Civil Environmental and Architectural Engineering, University of Colorado, Boulder, CO 80302, USA
2 X-Elastica LLC, Boulder, CO 80303, USA
3 U.S. Bureau of Reclamation, Denver, CO 80225, USA
* Correspondence: mohammad.haririardebili@colorado.edu.

Abstract: Different numerical models have been proposed for seismic analysis of concrete dams by taking into account the nonlinear behavior of concrete and joints; interaction between the dam, foundation, and reservoir; and other seismic hazard considerations. Less focus, however, has been placed on the real seismic performance of the dams and their relative correlation. This paper investigates the linear and nonlinear seismic performance of two similar high arch dams with relatively different response mechanisms. The response correlation is performed from statistical and probabilistic points of view. Similarities and differences are highlighted, and the best practice to compare the responses in a class of dams is presented. It is found that some demand parameters and seismic intensity measures can reduce the dispersion of the results and increase the correlation. In general, the dam geometry has a direct relation with the deformation and spatial distribution of potential damaged area. However, it is not related to the localized damage at the most critical location. Moreover, the real crack pattern (from nonlinear analysis) is more discrete compared to the continuous overstressed/overstrained regions (from linear analysis).

Keywords: arch dams; probabilistic; nonlinear; seismic; response correlation

---

1. Introduction

Safety assessment of the existing dams is an important task in risk-based management of infrastructures [1,2]. Risk analysis might be performed on different scales depending on the identified hazard level and importance of the asset. Methods ranging from pure qualitative assessment, to semi-quantitative approaches, to fully quantitative ones might be implemented [3]. Nearly all the quantitative methods depend (to some extent) on a precise numerical model (which reflects the physics-based problem) [4].

In the context of structural engineering, developing an advanced finite element model of the dam and its calibration with site measurements is a first step towards next-generation quantitative risk analysis. Many of the current legislation recognize numerical evaluation as a standard technique for safety assessment of dams [5–9]. The risk-based safety evaluation of dams should consider all the existing hazard scenarios (e.g., flooding, earthquake, aging) and their interactions.

Once the risk associated with a class of dams is evaluated, the repair and rehabilitation can be focused on the high-hazard dams [10]. Therefore, it is important to evaluate the relative capacity of dams subjected to the potential regional hazard scenarios. In the case of co-located portfolio of dams [11], one may evaluate their relative capacity under similar hazard scenarios (whenever possible).

Of the various single- and multi-hazard capacity functions for different dam types, this paper focuses only on seismic hazard for high arch dams. Seismic response of dams is studied from a probabilistic point of view [12]. Various seismic intensity levels (SILs) are also compared, corresponding to different earthquake return periods. Several linear and nonlinear transient simulations are performed, and the response mechanism of high arch dams is compared. This study helps to identify the relative seismic capacity of different dams when they are subjected to similar hazard scenarios. The results can be used to prioritize the investments on strengthening the low capacity dams.

2. Probabilistic Analysis and Seismic Intensity Levels

Depending on the type of probabilistic analysis, single or multiple SILs might be considered. A single SIL aims to evaluate only a specific annual frequency of exceedance, λ. A multiple SIL, however, tries to cover a broad range of λ. Based on the Poisson probability model [13]:

$$P_E = 1 - e^{-\lambda.t} \tag{1}$$

where P_E is the occurrence probability during the time life t of structure (generally assumed to be 100 years for dams), and λ is inverse of the return period, T_R.

International Committee on Large Dams (ICOLD) recommendations are usually adapted for a limited number of SILs, while the Applied Technology Council (ATC) recommendations (for building) are adapted for multiple SIL-based analyses of concrete dams. The following subsections of this paper summarize these recommendations which are mainly adopted from Hariri-Ardebili [14].

2.1. ICOLD Recommendations

There are two basic seismic loads for the design of new dams or the safety evaluation of existing dams [15–17]:

- Operating Basis Earthquake (OBE) represents the SIL at the dam site for which only minor (easily repairable) damage is acceptable, and the dam should remain functional. OBE corresponds to the return period of 145 years (50% probability of exceedance in 100 years).
- Safety Evaluation Earthquake (SEE) represents the SIL at the dam site for which a dam must be able to resist without uncontrolled release of the reservoir water. The SEE ground motion can be obtained from a probabilistic seismic hazard analysis (PSHA) and/or a deterministic seismic hazard analysis (DSHA). For large and high consequence dams, SEE is defined as:
    - Maximum Credible Earthquake (MCE): produces the largest expected ground motion at the dam site and is estimated based on DSHA. According to ICOLD [15], the ground motion parameters should be estimated at the 84th percentile level.
    - Maximum Design Earthquake (MDE): corresponds to return period of 10,000 years (1% probability of exceedance in 100 years) and is estimated based on PSHA.

    Note that for moderate consequence dams, the SEE ground motion parameters should be estimated at the 50 to 84th percentile level (based on DSHA) and need not have a λ smaller than $\frac{1}{3000}$ (based on PSHA). For low consequence dams, the SEE ground motion parameters should be estimated at the 50th percentile level (based on DSHA) and need not have a λ smaller than $\frac{1}{1000}$ (based on PSHA).

    In addition, the following aspects must be considered [18]:

- The three components of the spectrum-matched acceleration time histories must be statistically independent. One cannot scale one of the acceleration components and use in other direction.
- The duration of strong ground shaking shall be selected in such a way that aftershocks are also covered.
- For the safety check of a dam, at least three different earthquakes shall be considered for the SEE ground motion.

2.2. ATC Recommendations

The Applied Technology Council [19] proposes three types of performance assessments (i.e., intensity-, scenario-, and time-based). Time-based performance assessment (TBPA) evaluates a dam's performance over a period of time, considering all earthquakes that may occur in that period of time, and the probability that each will occur. TBPA considers uncertainty in the magnitude and location of future earthquakes, as well as the intensity of motion resulting from these earthquakes. SILs and the corresponding ground motions are defined as follows:

- Generate a seismic hazard curve, λ *vs.* $S_a(T_1)$, for the dam site.
- Compute seismic intensity range which covers the dam response from no (or negligible) damage to collapse. As a recommendation, the minimum and maximum spectral acceleration can be assumed as: $S_a^{min}(T_1) = 0.05g$, and $S_a^{max}(T_1) = S_a(T_1)|_{\lambda=0.00002/yr}$, where $\lambda = 0.00002/yr$ corresponds to $T_R = 50,000$ years.
- Split the $[S_a^{min}, S_a^{max}]$ range into N_{IM} equal intervals; calculate and record $\Delta\lambda_i$ in each interval; identify the midpoint spectral acceleration in each interval and the corresponding λ_i. For 2D model of gravity dams, N_{IM} is recommended to be 8; while for 3D arch dams it may be reduced to 4.
- Develop a target response spectrum, $S_a^{trg}(T)$, based on data collected from each midpoint. Three types of response spectra are acceptable: (1) uniform hazard spectra, (2) conditional mean spectra, and (3) conditional spectra.
- For each target response spectrum, select and scale suites of n ground motion triplets as follows:
    - Select a candidate suite of ground motion triplets from available recorded motions (e.g., PEER [20]).
    - For each ground motion triplet, construct the geomean spectrum for the horizontal components over a period range of (T_{min}, T_{max}) as $S_a^{geo}(T) = \sqrt{S_a^{H_1}(T) \times S_a^{H_2}(T)}$, where T_{min} and T_{max} can be selected as $0.2T_1$ and $2.0T_1$, respectively. T_1 is the fundamental period of dam-reservoir-foundation system.
    - Compare $S_a^{geo}(T)$ and $S_a^{trg}(T)$, and select those ground motion horizontal pairs which are similar in shape to the target response spectrum within the period range of (T_{min}, T_{max}).
    - Amplitude-scale all three components of each ground motion triplet by the ratio of $\frac{S_a^{trg}(T_1)}{S_a^{geo}(T_1)}$.

There is no magic rule for the number of selected ground motions in each level; however, when there is a significant scatter in spectral shape of the selected records or a poor fit to the target spectrum, n = 11 or more triplets of motions may be needed. The use of fewer than n = 7 motion pairs is not recommended [21].

3. Criteria for Performance Evaluation

Since two sets of linear elastic and nonlinear damage analyses are performed for the dams, two sets of different criteria are required to interpret the results. This paper adopts the criteria proposed by Ghanaat [22], USACE [23] for the linear elastic analysis, and by Hariri-Ardebili et al. [24] for nonlinear simulations.

3.1. Linear Analysis

Results of the linear elastic analysis should be interpreted with respect to some predefined criteria; therefore, some indices are introduced first [14]:

- Demand Capacity Ratio (DCR): This local index refers to the ratio of the calculated stresses or strains in a dam body to the tensile strength of mass concrete or its equivalent strain.
- Cumulative Inelastic Duration (CID): This local index refers to the total duration of stress (or strain) excursions above a stress (or strain) level associated with a certain DCR.
- Damage Spatial Distribution Ratio (DSDR): This global index refers to the ratio of the overstressed (or overstrained) region to total dam area at the specific DCR.

The aforementioned local indices should be computed for at least one critical node. Then, the dam performance should be evaluated using the plots shown in Figure 1, which presents a threshold surface for local indices. The vertical axis, CID (or DSDR), varies between zero and CID* (DSDR*). For arch dams, a CID* = 0.4 and DSDR* = 15% are recommended by USACE [23]. Finally, the evaluation strategy can be summarized as follows:

- If DCR $\leq$ 1.0 for all N_{SIL}, the dam response is in linear elastic range. No (or minor) damage is expected.
- If 1.0 < DCR < 2.0, the dam response is in nonlinear phase. The status of the critical nodes should be checked as follows:
  - If even one of the critical nodes (completely or partially) exceeds the threshold surface, being in zone B, significant damage is expected. In this condition, performing detailed nonlinear analysis is required.
  - If all the critical nodes are in zone A, application of the linear elastic procedure is allowed only if the DSDR does not exceed the threshold, zone A (of right plot).
  - If all the critical nodes fall below the threshold curve in CID–DCR plot (i.e., zone A), but DSDR exceeds the threshold in DSDR–DCR plot (i.e., zone B), severe damage is expected in the dam body. Performing detailed nonlinear analysis is required.
- If DCR $\geq$ 2.0, severe damage (at least localized damage) is expected. If it is accompanied by considerable spatial extension, zone B in the right plot, global damage is also expected. In this case, performing detailed nonlinear analysis is required.

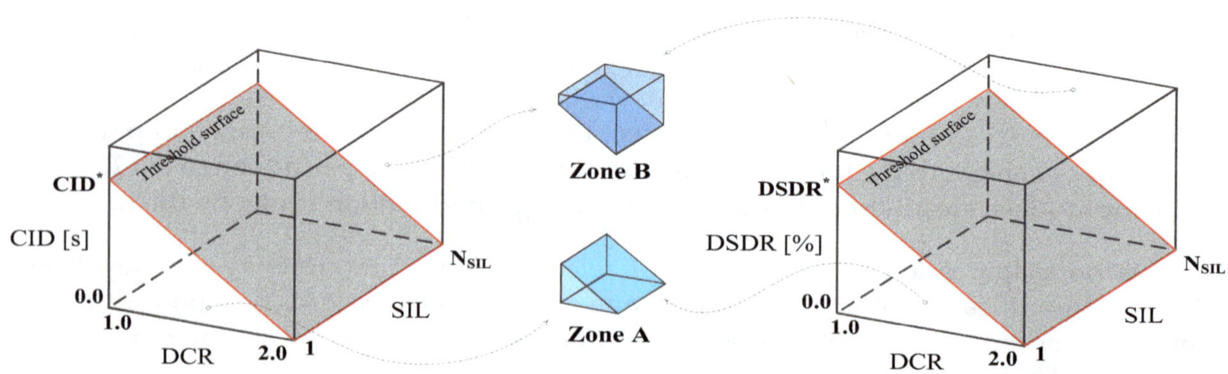

Figure 1. Seismic performance evaluation of arch dams based on linear analysis.

Application of these linear analysis based criteria on concrete dams (gravity, arch, and buttress) and their partial calibration with nonlinear models can be found in [24–32].

3.2. *Nonlinear Analysis*

Results of nonlinear analyses can be directly correlated with safety of the dam [33]. There is no established criteria/metric for nonlinear evaluation of concrete arch dams. This means that there is no limit state (LS) function from nonlinear analyses to correlate its performance with observed damage. However, multiple indices can be considered, such as concrete cracking (and crushing), joint opening/sliding, and the dissipated energy during damage. Different researchers combined these indices with the damage index (DI) concept to present the expected damage in a more quantitative form [34–37]. This paper does not directly use the concept of DI for arch dams. The objective is to compare the critical value of nonlinear damage metrics in two dams which are subjected to similar seismic hazard scenarios. Moreover, a correlation among different damage metrics will be investigated.

4. Case Study Dams

Two high arch dams are used as case studies in this paper. The first one is Dez Dam (hereafter referred to as Dam-1), and the second one is Karun-III Dam (hereafter referred to as Dam-2). The heights of Dam-1 and Dam-2 are 203 and 205 m, respectively; while their crest lengths are 240 and 462 m. General information, as well as the finite element model calibration for these dams, can be found in [38,39].

Dam-1 includes a Pulvino (and peripheral joint), while Dam-2 has massive concrete blocks on its sides. The finite element program ANSYS [40] is used for all simulations. The dams are modeled using six- and eight-node structural elements, Figure 2. Foundation rock is modeled as a mass-less medium, and the reservoir water is included using the pressure-based fluid elements. In the nonlinear analyses, the concrete is modeled based on the extended rotating smeared crack model [41], and the contraction (and peripheral) joints are simulated using the node-to-node contact elements. Lift joints were not considered in this paper. The total numbers of elements in Dam-1 are 792, 3770, and 3660 for the body, foundation, and reservoir, respectively. The numbers of elements in Dam-2 are: 3958, 21,848, and 23,022.

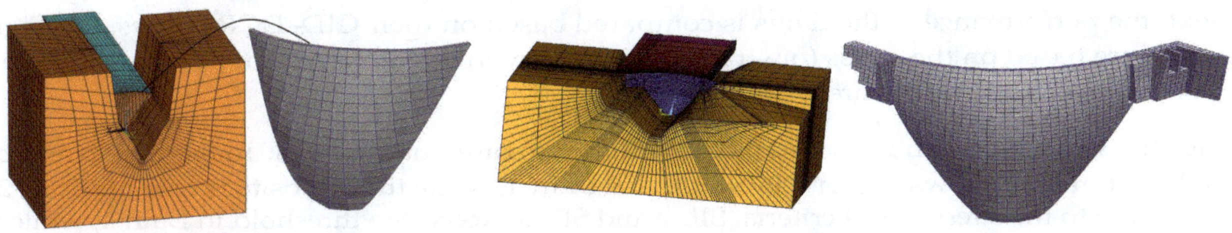

(a) Dam-1 **(b)** Dam-2

Figure 2. **(a,b)** Finite element mesh of two case study dams.

The applied loads on the system are: dam body dead load (based on several construction stages), hydrostatic load in normal water level (with gradually impounding the reservoir), silt pressure, thermal loads (with summer temperature condition; Figure 3), and, finally, seismic load (with three component ground motion records). Up to 10 and 4 construction stages were considered in Dam-1, and Dam-2, respectively. In the applied method, the concrete weight for ith stage is first applied, followed by grouting of the joints in ith stage, and then, concrete weight again for $i+1$th stage. This is continued to construct the entire dam, and to grout all the joints. Material properties can be summarized as follows: Modulus of elasticity in concrete is 40 and 30 GPa for Dam-1 and Dam-2, respectively; Poisson's ratio is 0.2; mass density for concrete is 2400 kg/m^3; and concrete compressive strength is 35 and $[25-35]$ (inner and outer parts) for Dam-1 and Dam-2, respectively.

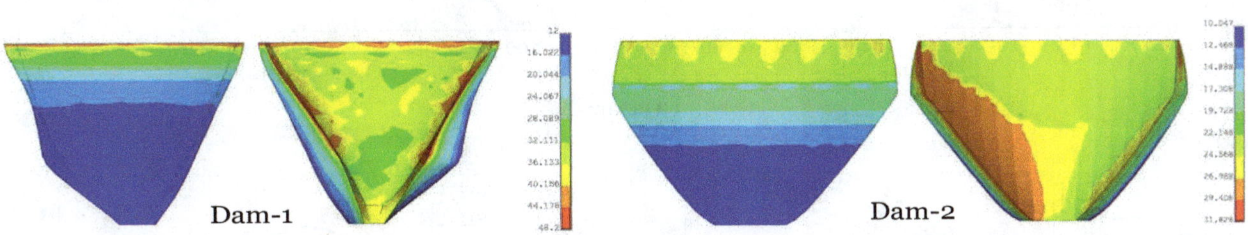

Figure 3. Summer temperature distribution on the dam face resulted from thermal transient analysis.

According to the PSHA of Dam-1, three SILs are identified, and nine site-specific real ground motion records are also selected from NGA-WEST2 list [20]. Each of nine ground motions are scaled to three SILs. Subsequently, $3 \times 9 = 27$ three-component scaled records are produced for each dam. Finally, 27 records are applied to 3 types of finite element models: linear elastic model, nonlinear model with joints, and nonlinear model with concrete damage. This requires a total of $27 \times 3 = 81$ (linear and nonlinear) transient analyses for one dam. The results presented in Section 5 summarize $81 \times 2 = 162$

simulations. Considering the complexity of models, each analysis took from 6 to 60 h to complete in a standard 8 GB-RAM workstation.

5. Results and Discussions

Results are presented in two sections. Section 5.1 presents the response of the linear and nonlinear systems from a statistical point of view. The outputs are presented for each dam separately and compared as a function of SIL. In Section 5.2, the outputs from both dams are paired with different ground motion intensity measure (IM) parameters and compared from a probabilistic point of view.

5.1. Statistical Response Comparison

Figure 4 presents the maximum displacement of dams along the crest and height of the crown cantilever. For each SIL, the mean, μ, value of nine ground motion records, as well as their lower and upper bounds (i.e., $\mu \pm \eta$, where η is standard deviation) are presented. Clearly, increasing the SIL increases the drift response. In general, the drift response of Dam-2 is about 2.5 times that of Dam-1. Their height is about the same; however, the crest length in Dam-2 (including the massive concrete blocks) is about twice that of Dam-1.

Next, the performance of the dams is compared based on their CID–DCR curves, See Figure 5. These plots are based on the stress (or strain) time history in the most critical point within the dam body. Some major observations are:

- The stress-based and strain-based CID–DCR metrics are not identical. In fact, the stress-based CID–DCR metric shows higher CID values for Dam-1, while the opposite is true for Dam-2.
- According to the stress-based criteria, SIL-2 and SIL-3 exceed the threshold in Dam-1, while SIL-3 exceeds the threshold only at DCR = 1 and 2 in Dam-2.
- The strain-based CID–DCR curves for both dams share more similarities. In both cases, the mean curve exceeds the threshold extensively for SIL-3, partially for SIL-2 (only DCR = 1), and never for SIL-1.
- Moreover, the curves associated with Dam-2 decay faster than the curves belonging to Dam-1. This means that the localized damage risk at higher DCR values is lower for Dam-2.

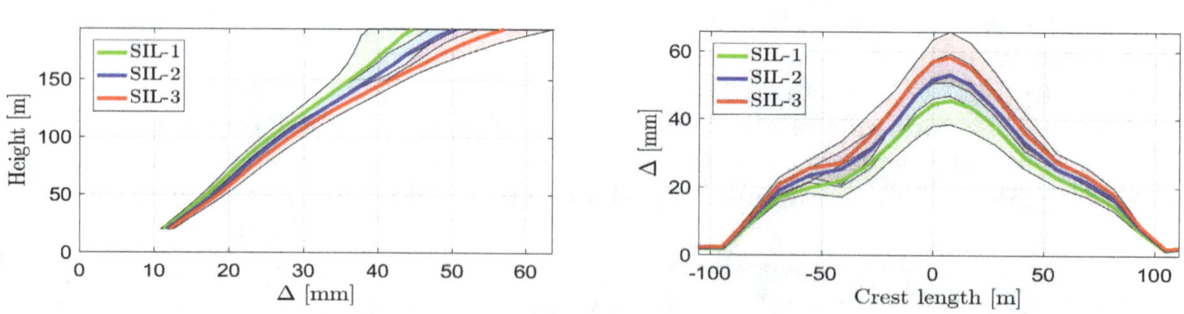

(a) Dam-1

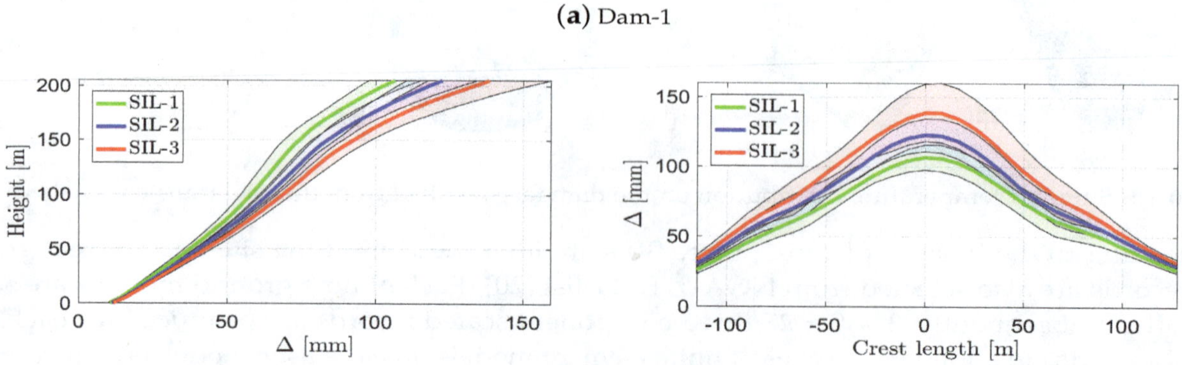

(b) Dam-2

Figure 4. (**a**,**b**) Maximum displacement response of arch dams subjected to three seismic intensity levels.

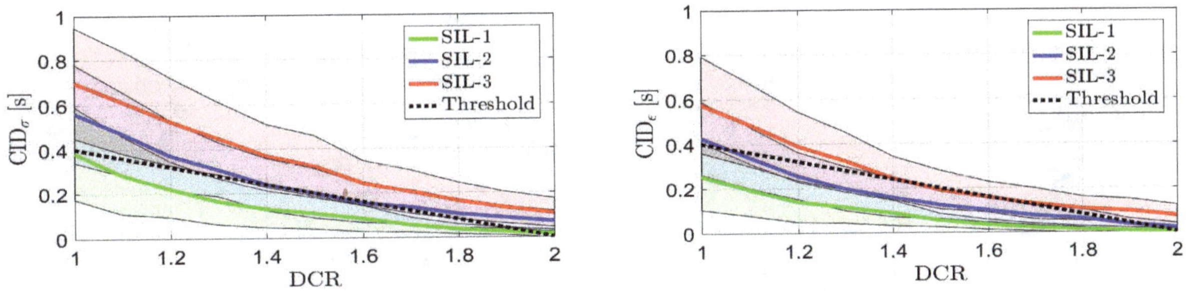

(**a**) Dam-1

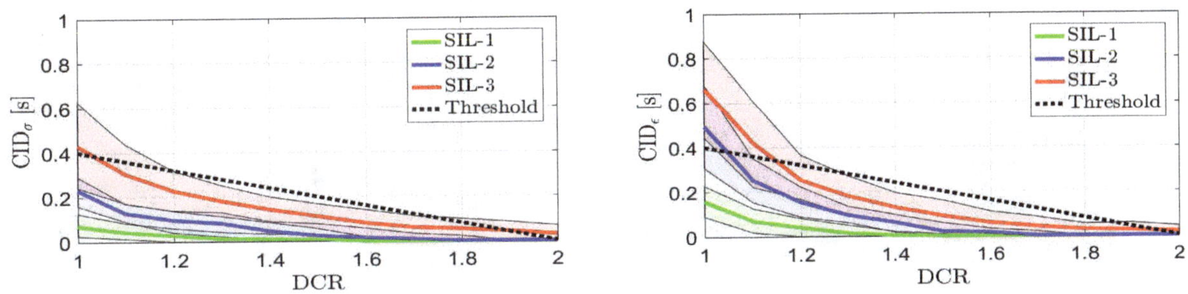

(**b**) Dam-2

Figure 5. (**a**,**b**) Cumulative Inelastic Duration–Demand Capacity Ratio (CID–DCR) of arch dams subjected to three seismic intensity levels.

So far, the localized potential damage is studied in the context of CID–DCR curves. The global behavior can be investigated in the context of DSDR–DCR plots, See Figure 6. Again, results are presented in terms of overstressed and overstrained regions on upstream (US) and downstream (DS) faces of the dams. Major findings can be summarized as follows:

- The stress-based and strain-based responses have a similar pattern for Dam-1, while they are different for Dam-2.
- In Dam-2, the strain-based DSRD is higher than stress-based one for lower DCR values.
- For Dam-1, the US face is a bit more critical than DS one. However, Dam-2 has a different mechanism and the DSDR in DS is much higher.
- In Dam-1, only SIL-3 exceeds the threshold at DCR = 1. In Dam-2, both SIL-3 and SIL-2 exceed the threshold at several DCRs (especially at the DS face).

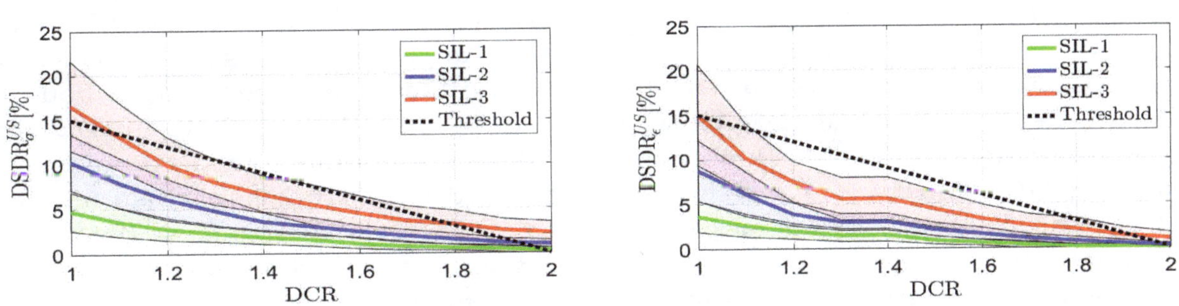

(**a**) Dam-1; US

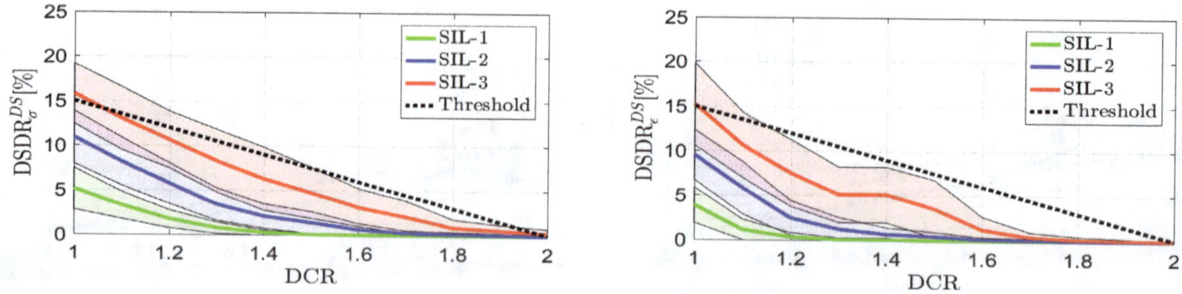

(**b**) Dam-1; DS

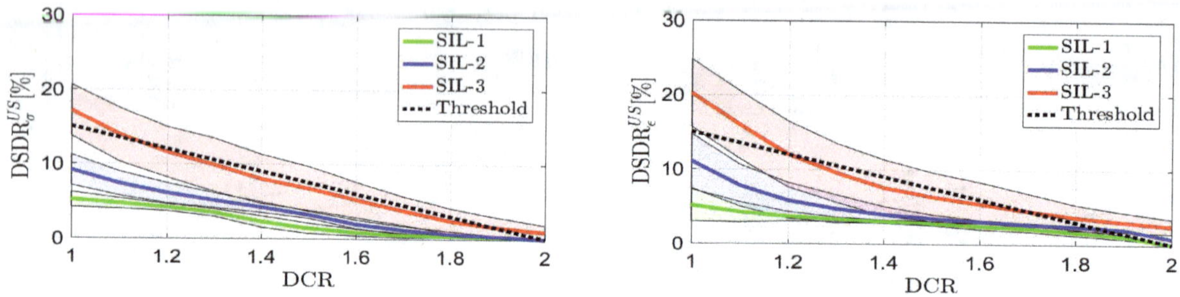

(**c**) Dam-2; US

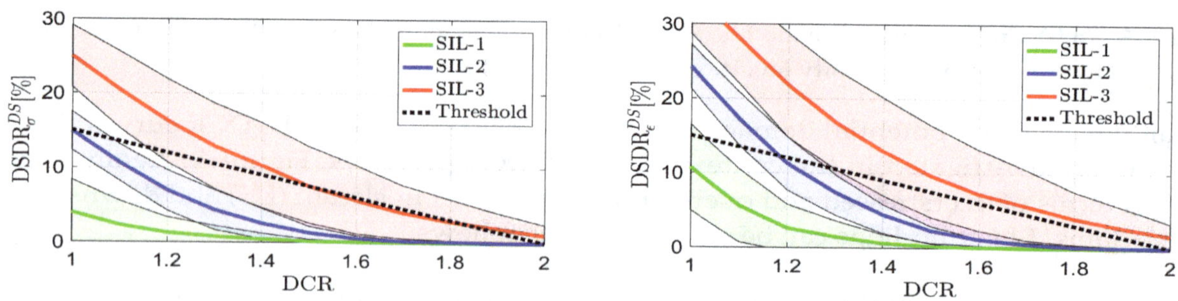

(**d**) Dam-2; DS

Figure 6. (**a–d**) Damage Spatial Distribution Ratio (DSDR)–DCR of arch dams subjected to three seismic intensity levels.

Figures A1 and A2 (in Appendix A) provide a visualization of the DSDR in Dam-1 and Dam-2, respectively. As opposed to Figure 6, which presents the statistics of DSDR at various DCRs, Figures A1 and A2 present the individual DSDR only at DCR = 1. Moreover, the graphical presentation is only provided for the US face of Dam-1 and the DS face of Dam-2 (which are critical face). In Dam-1, the central parts of the body near the crest are the most vulnerable regions during the seismic excitation. In Dam-2, the left and right quarter points in the upper half of the dam are the most vulnerable regions.

The predicted potential damaged area from linear analyses in Figures A1 and A2 are further validated based on a series of nonlinear simulations with the concrete smeared crack model, see Figure 7. Only the results of SIL-3 (which is the critical level) are provided for both the US and DS faces. In general, there is an acceptable consistency between the potential damaged area and the real cracked element for Dam-1. Nonlinear simulations show some limited extra damage next to the dam-foundation interface. The resulting damaged area in Dam-2 is less than the expected damage from linear analysis. The crack pattern is more discrete compared to the continuous overstressed/overstrained regions. The nonlinear simulations, however, show a considerable amount of damaged area at the dam-foundation interface. This is partially identified by linear analysis. Note that the concrete near the foundation is reinforced (to some degrees), and these peripheral cracks may not appear in the real dam.

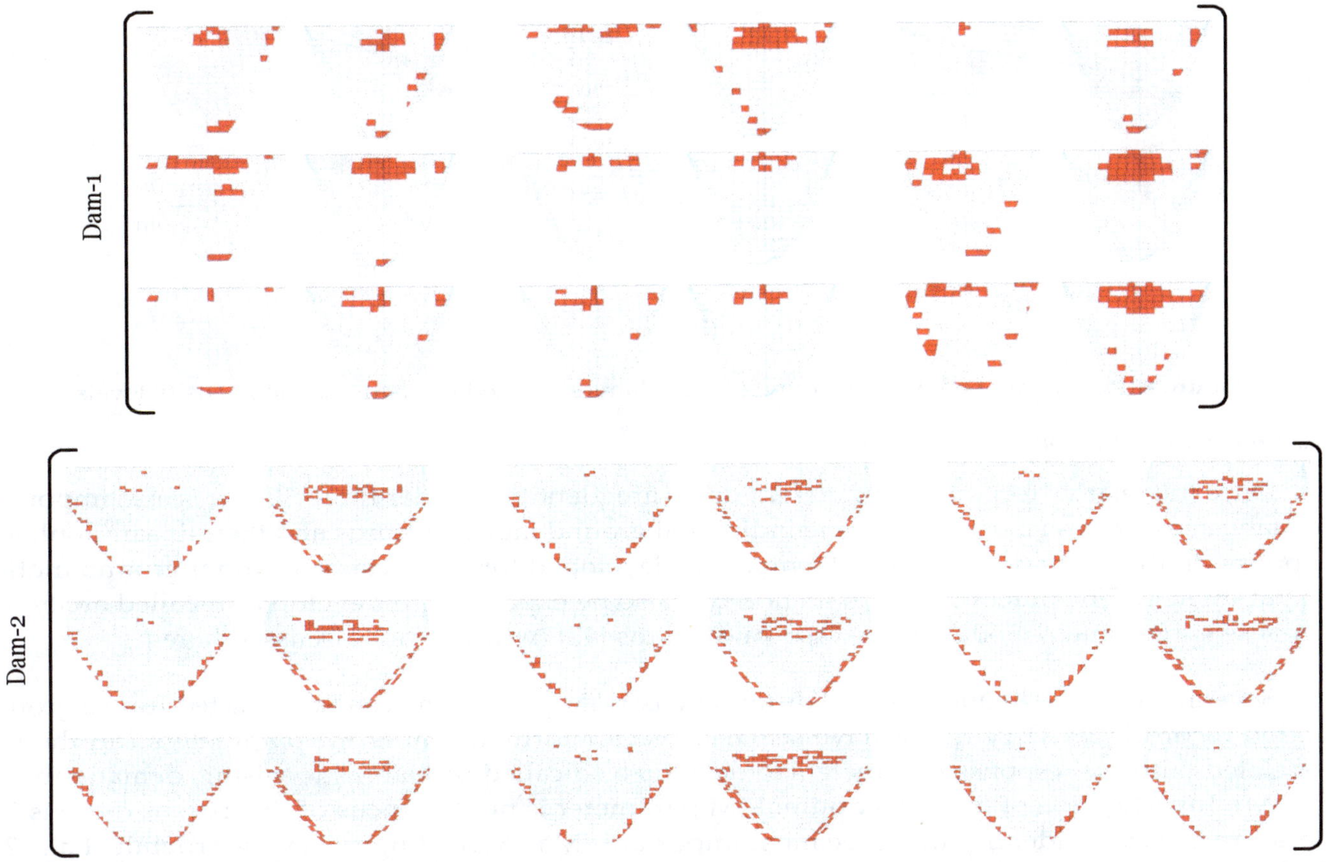

Figure 7. Crack profile of both dams from nonlinear analysis subjected to seismic intensity level-3 (SIL-3) ground motions.

Finally, Figure 8 presents the joint opening and sliding in terms of crack opening displacement (COD) and crack sliding displacement (CSD). Major observations can be summarized as follows:

- In general, the mean CSD is twice and three times the COD for Dam-1 and Dam-2, respectively.
- In general, joint opening is more critical than joint sliding (since the water can penetrate inside the joint and increases the pressure on the inner walls). The COD appears to be well-controlled in both dams.
- The CODs in both dams are very similar: limited to no more than 1 mm for the first 150 m of height. Moreover, there is practically no difference between three SILs up to height 150 m. The major joint opening occurs at the top 50 m (i.e., upper quarter of dam).
- On the other hand, the joint sliding has a more smoothed behavior along the height (especially for Dam-2). Again, the most critical zone is the upper quarter of dam height.

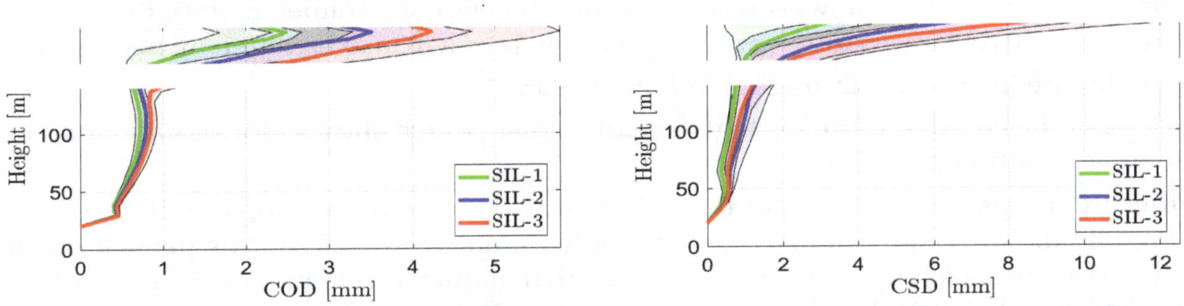

(**a**) Dam-1

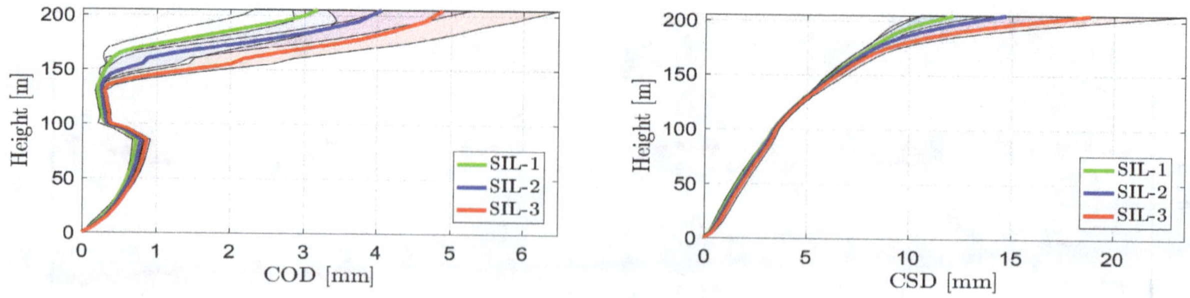

(**b**) Dam-2

Figure 8. (**a**,**b**) Joint opening and sliding of arch dams subjected to three seismic intensity levels.

5.2. Probabilistic Response Correlation

Once the relative responses of different dams are identified under three SILs, it is also important to establish a direct relation between the individual ground motion records and their linear/nonlinear response. In this section, a probabilistic model is developed for dam response under ground motion IM parameters. The outcome of this section can also be expanded to develop a so-called analytical response surface meta-model for the case study dams [42] (which is not discussed herein).

Several intensity-, frequency-, and duration-based IM parameters can be extracted from a ground motion record [42]. They might have scalar or vector format. These IM parameters can then be correlated with the response parameters to develop a so-called probabilistic seismic demand model (PSDM). Investigation of the most optimal IM parameter is not the focus of this paper; the existing literature is used to identify the three most important IMs. According to Hariri-Ardebili et al. [24], Hariri-Ardebili and Boodagh [42], peak ground acceleration (PGA), acceleration spectrum intensity (ASI), and the first-mode spectral accelerations ($S_a(T_1)$) are among top IM choices.

On the other hand, since three-component ground motions are used in this paper, the scalar IM parameters for different components need to be combined with an appropriate method. The square-root-of-sum-of-squares technique is used to combine them as:

$$IM^{combo} = \sqrt{IM_{H_1}^2 + IM_{H_2}^2 + IM_V^2} \tag{2}$$

where H_1, H_2, and V are two in-plane directions and one vertical direction.

Figure 9 provides PSDM for maximum crest displacement as a function of different IM parameters including the goodness-of-fit in terms of root mean square error (RMSE). Major observations can be summarized as follows:

- PGA and ASI are structure-independent IM parameters; therefore, both curves in Figure 9a,b have identical IM range. $S_a(T_1)$, however, is a structure-dependent parameter. Since the fundamental period of these two dams is different, the spectral values will also be different, See Figure 9c.
- The confidence interval for Dam-2 is larger than Dam-1.
- The slope of the curve in Dam-2 is higher than Dam-1, which shows more correlation between the input and output parameters.
- In order to increase the accuracy of the PSDM, it is possible to develop a multiple IM model, See Figure 9d. Although one may apply a polynomial with different degrees, a planer one is selected in this paper. It is already found that higher order models over-fit the results Hariri-Ardebili et al. [24]. In this plot the upper plane belong to Dam-2.

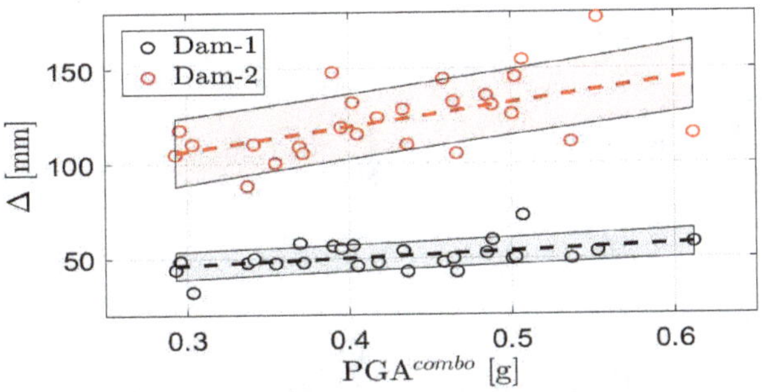

(a) Peak ground acceleration; $RMSE^1 = 6.86$; $RMSE^2 = 16.73$

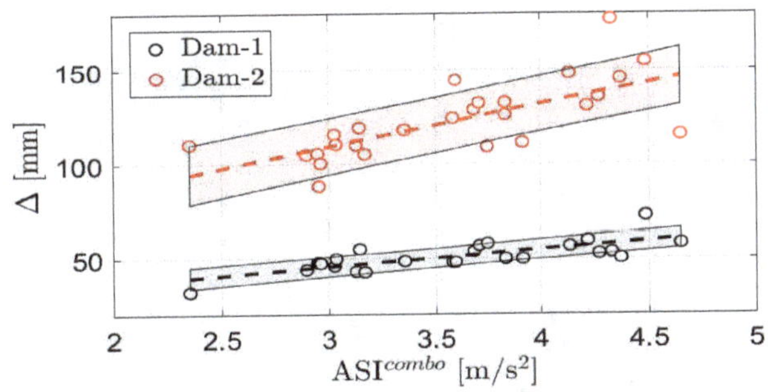

(b) Acceleration spectrum intensity; $RMSE^1 = 5.06$; $RMSE^2 = 14.39$

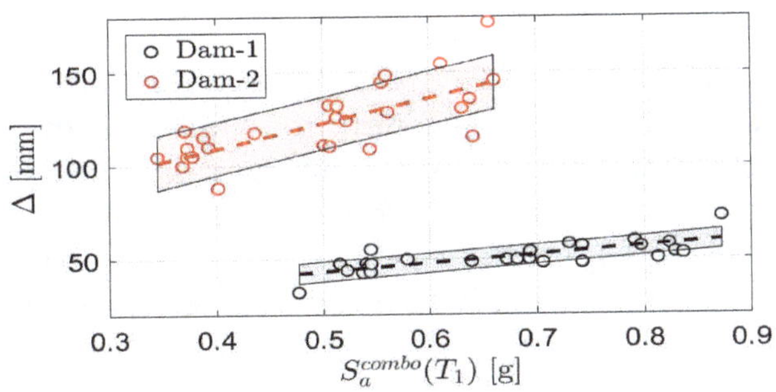

(c) First-mode spectral acceleration; $RMSE^1 = 5.05$; $RMSE^2 = 13.77$

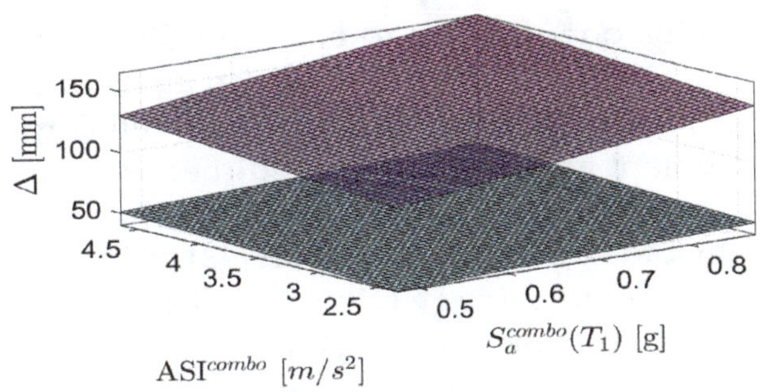

(d) Multiple intensity measures

Figure 9. (**a–d**) Correlation between the dam displacement and the seismic intensity measures.

A similar procedure can be applied in nonlinear analyses for the joint response. Figure 10 provides PSDMs for the maximum COD and CSD. Again, three IM parameters are contrasted. The most important finding is that the mean curves and the confidence intervals (from two dams) are nearly parallel (specially for ASI). This proves that there is a high correlation in nonlinear response of these two dams from a probabilistic point of view. Furthermore, this shows that ASI might be a good candidate to compare the joint capacity of different arch dams.

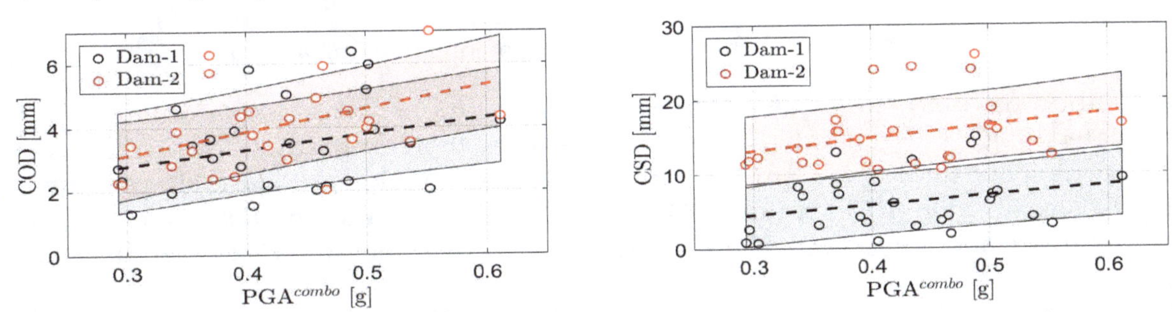

(a) Peak ground acceleration; $RMSE^1_{COD} = 1.34$; $RMSE^2_{COD} = 1.29$; $RMSE^1_{CSD} = 3.95$; $RMSE^2_{CSD} = 4.44$

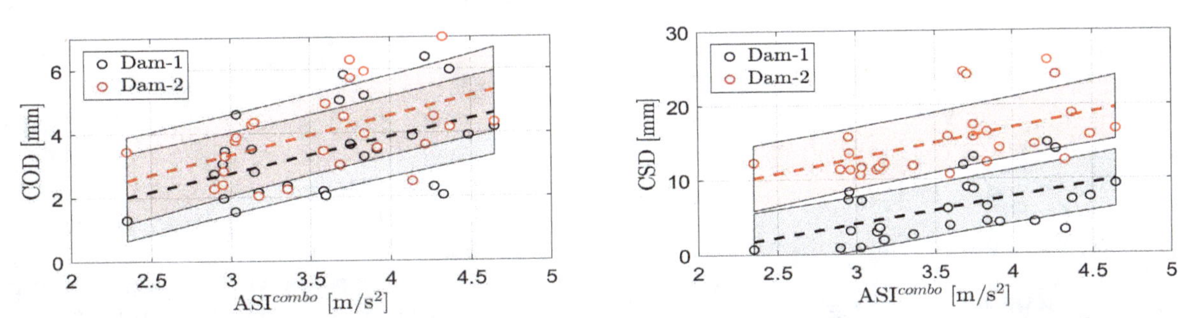

(b) Acceleration spectrum intensity; $RMSE^1_{COD} = 1.23$; $RMSE^2_{COD} = 1.23$; $RMSE^1_{CSD} = 3.49$; $RMSE^2_{CSD} = 3.97$

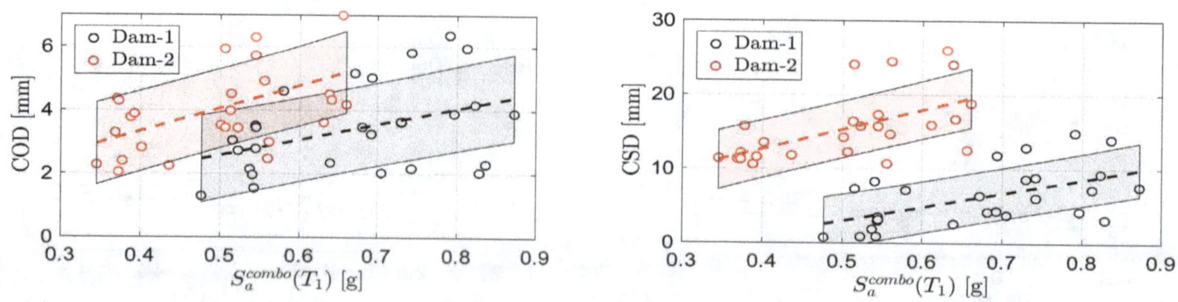

(c) First-mode spectral acceleration; $\text{RMSE}^1_{COD} = 1.27$; $\text{RMSE}^2_{COD} = 1.21$; $\text{RMSE}^1_{CSD} = 3.39$; $\text{RMSE}^2_{CSD} = 3.75$

Figure 10. (**a–c**) Correlation between the joint opening and sliding with seismic intensity measures.

Finally, one may try to correlate different dam responses together (and not in the context of seismic intensity measures). Such an attempt is shown in Figure 11. In order to develop a smooth curve or joint probability model, a relatively large number of simulations is required. Since the initial finite element analyses are limited (because they are computationally demanding), a procedure is used to expand the number of engineering demand parameters (EDPs). This method, which was originally developed by Yang et al. [43], is briefly explained in Algorithm 1. The input is the matrix of analytically determined EDPs (e.g., displacements, joint opening/sliding), **X**, and the output is the matrix of statistically determined EDPs, **W**.

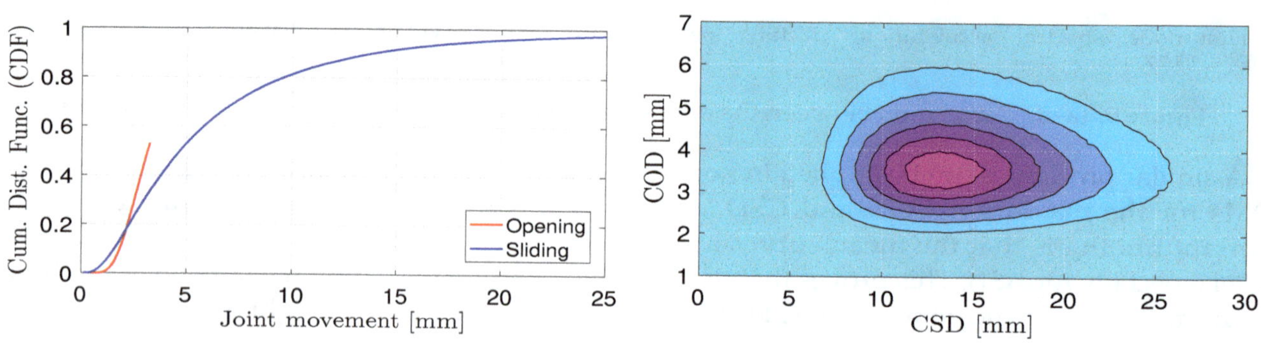

(**a**) Dam-1 (**b**) Dam-2

Figure 11. (**a**,**b**) Uncertainty quantification of joint opening and sliding.

This method is used to expand the initial simulations, and, subsequently, a total of 10,000 correlated EDPs are generated. Figure 11a shows the cumulative density function (CDF) of the joint opening and sliding in Dam-1 separately. On the other hand, Figure 11b presents the joint probability density function (PDF) of the COD and CSD for Dam-2. Both of these plots are very useful for identifying the relation of various EDPs and developing an integrated risk-based decision model.

Algorithm 1 Generating correlated EDPs from initial finite number of simulations

Inputs: $\mathbf{X}_{n\times k}$ ▷ EDP matrix, where k is the number of different EDPs.

Output: **W**

procedure
- Compute $\mathbf{Y}_{n\times k} = \ln \mathbf{X}_{n\times k}$ ▷ It has a joint normal distribution
- Compute the $\mathbf{M}_\mathbf{Y} = (\text{mean}(\mathbf{Y}))^t$ ▷ Mean vector
- $\mathbf{D}_\mathbf{Y} = \text{diag}(\text{std}(\mathbf{Y}))$ ▷ Diagonal standard deviation matrix
- $\mathbf{R}_\mathbf{YY} = \text{corrcoef}(\mathbf{Y})$ ▷ Correlation coefficient matrix
- $\Sigma_\mathbf{YY} = \mathbf{D}_\mathbf{Y}\mathbf{R}_\mathbf{YY}\mathbf{D}_\mathbf{Y}$ ▷ Covariance matrix
- $\mathbf{L}_\mathbf{Y} = (\text{chol}(\mathbf{R}_\mathbf{YY}))^t$ ▷ Lower-triangular decomposition
- Generate **U** with $\mu = 0$, $STD = 1$ ▷ A vector of uncorrelated standard normal RVs

$\mathbf{Z} = \mathbf{D_Y L_Y U} + \mathbf{M_Y}$ ▷ A linear transformation and translation from $\mathbf{U}$ to $\mathbf{Z}$
$\mathbf{W} = \exp(\mathbf{Z})$ ▷ Transfer back the generated joint normal logarithmic EDPs
end procedure

6. Conclusions

Risk-based management of a large portfolio of dams requires information about the relative performance of all the assets and particularly dams. Often, several dams are co-located or within a short distance from each other; thus, similar seismic scenarios might be applied to all of them. This paper presents the finite element results of two arch dams assuming they are subjected to similar seismic hazard scenarios.

Both dams are over 200 m in height; however, one of them has a larger crest length. In addition, one of them was constructed with Pulvino technology, while the other one is in direct contact with the foundation and has two massive concrete blocks on its left and right sides. The finite element model of the dam-foundation-reservoir coupled system is developed with both linear and nonlinear assumptions. For the nonlinear models, the concrete damage is modeled based on a rotating smeared crack approach, while the contraction (and peripheral) joints are simulated with tension-free contact elements. Thermal loads are also taken into account before dynamic analyses. Three seismic intensity levels are studied based on PSHA, and nine ground motion records are scaled in each one; subsequently, a total of over 160 transient analyses are performed for dams. Results are extracted and compared in terms of displacements, DCR, DSDR, CID, joint openings and sliding, and crack pattern.

The geometry of the dam (especially its crest length) found to be directly related with the deformation (See Figure 4). Moreover, it has a direct relation with the spatial distribution of the potentially damaged area on both the upstream and downstream faces of the dam (See Figure 6). It is not, however, related to the localized damage at the most critical point of the dam (See Figure 5). In fact, it has been shown that the localized potential damage between two dams can be well-correlated with strain-based criteria on CID (See Figure 5, right column).

Next, the crack pattern from nonlinear analyses is compared with potential overstressed/ overstrained area from linear simulations. It is found that the real crack pattern is more discrete compared to the continuous DSDR. Finally, the joint opening/sliding displacement from both dams are compared. Although the opening displacements were in the same range, the sliding displacement from the wider dam was twice that of the other dam (See Figure 8).

Last but not least, thanks to the multiple dynamic analyses, a PSDM is proposed for both dams. Surprisingly, the COD-based PSDMs on ASI were nearly identical. Moreover, PSDMs from CSD had parallel lines. This shows that in spite of all differences in geometry, the joint opening capacity of the dams is very close. Finally, a mathematical model is used to expand the initial engineering demand parameters and generate a large set of correlated EDPs. Using this new dataset (which has similar statistical parameters to the initial set), one can easily develop any joint PDF and CDF between the outputs.

Future research can be directed on developing a generalized PSDM for arch dams which is function of not only IM and limit state (LS) but also the geometry variables [44] in the form of $EDP = f(IM, LS, Geo)$.

Author Contributions: Conceptualization, M.A.H.-A.; Methodology, M.A.H.-A. and M.H.; Software, M.H.; Validation, M.A.H.-A., M.H., P.B. and J.W.S.; Formal Analysis, M.A.H.-A. and M.H.; Investigation, M.A.H.-A. and M.H.; Resources, M.A.H.-A.; Data Curation, M.A.H.-A., M.H. and P.B.; Writing—Original Draft Preparation, M.A.H.-A.; Writing—Review & Editing, M.A.H.-A., P.B. and J.W.S.; Visualization, M.A.H.-A. and P.B.; Supervision, M.A.H.-A. and J.W.S.; Project Administration, M.A.H.-A. and J.W.S.; Funding Acquisition, M.A.H.-A.

Acknowledgments: The first author would like to express his sincere appreciation to his former advisor (and current mentor), Victor E. Saouma, at the University of Colorado Boulder for his enthusiastic guidance and advice throughout this research. The authors would like to thank the reviewers for their helpful and constructive comments that greatly contributed to improving the final version of the paper.

Appendix A. Detailed DSDR Plots

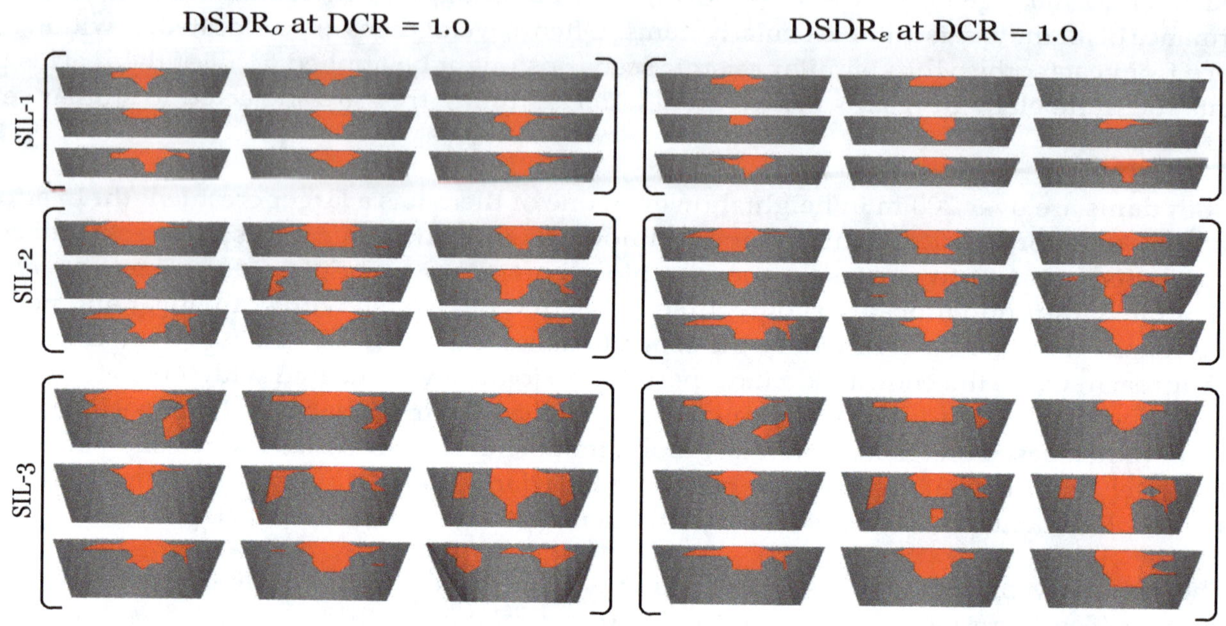

Figure A1. Matrix of over-stress and over-strain area on Dam-1 upstream face; Note: Each matrix includes 9 plots associated with 9 ground motions. In each matrix, top-left to bottom-right are #1 to #9.

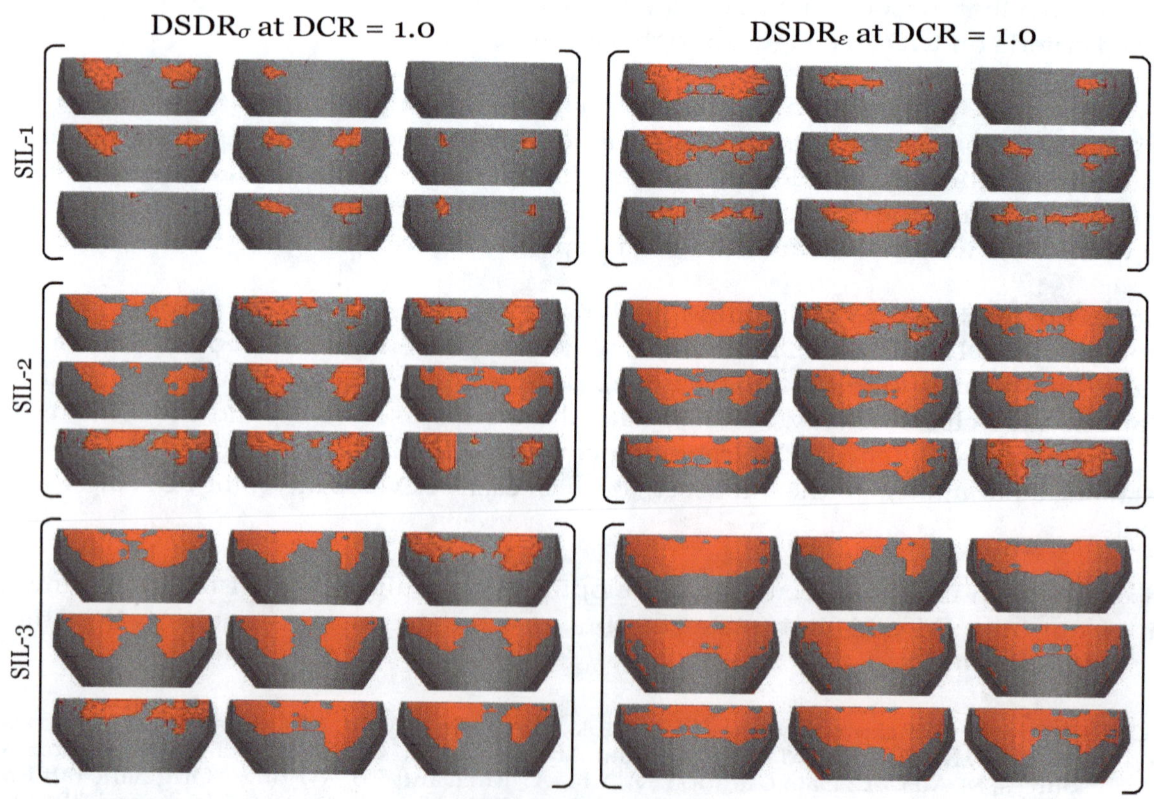

Figure A2. Matrix of over-stress and over-strain area on Dam-2 downstream face; Note: Each matrix includes 9 plots associated with 9 ground motions. In each matrix, top-left to bottom-right are #1 to #9.

References

1. Bowles, D.S.; Anderson, L.R.; Glover, T.F. A role for risk assessment in dam safety management. In Proceedings of the 3rd International Conference HYDROPOWER, Trondheim, Norway, 30 June–2 July 1997; Volume 97.
2. Hartford, D.; Baecher, G. *Risk and Uncertainty in Dam Safety*; Technical Report; Thomas Telford, Ltd: London, UK, 2004.
3. Hariri-Ardebili, M.A.; Nuss, L.K. Seismic risk prioritization of a large portfolio of dams: Revisited. *Adv. Mech. Eng.* **2018**, *10*. [CrossRef]
4. Bretas, E.; Batista, A.; Lemos, J.; Léger, P. Seismic analysis of gravity dams: A comparative study using a progressive methodology. In Proceedings of the EURODYN 2014-9th International Conference on Structural Dynamics, Porto, Portugal, 30 June–2 July 2014.
5. USACE-USBR. *Best Practices in Dam And Levee Safety Risk Analysis*; Technical Report; A Joint Publication by U.S. Department of the Interior, Bureau of Reclamation, and U.S. Army Corps of Engineers: Denver, CO, USA, 2015.
6. Canadian Dam Association (CDA). *Dam Safety Guidelines*; Technical Report; Canadian Dam Association: Edmonton, AB, Canada, 2007.
7. ICOLD. *Bulletin on Risk Assessment in Dam Safety Management*; Technical Report; International Commission on Large Dams: Paris, France, 2005.
8. NSW. *Risk Management Policy Framework for Dam Safety*; Technical Report; New South Wales Government Dams Safety Committee: New South Wales, Australia, 2006.
9. Italian Technical Code for dams. *Guidelines for Seismic Safety Reassessment of Existing Dams in Italy*; Technical Report; Dipartimento per I Servizi Tecnici Nazionali: Roma, Italy, 2001.
10. ASDSO. *State and Federal Oversight of Dam Safety Must be Improved*; Magazine of Association of State Dam Safety Officials (ASDSO): Lexington, KY, USA, 2011.
11. McCann, M.W.; Jr. Seismic Risk of a Co-located Portfolio of Dams–Effects of Correlation and Uncertainty. In Proceedings of the 3rd International Week on Risk Analysis, Dam Safety, Dam Security, and Critical Infrastructure Management, Polytechnic University of Valencia, Valencia, Spain, 17–18 October 2011; pp. 1–6.
12. Malm, R.; Gasch, T.; Eriksson, D.; Hassanzadeh, M. *Probabilistic Analyses of Crack Propagation in Concrete Dams: Part 1*; Elforsk: Stockholm, Sweden, 2013.
13. Cornell, C. Engineering seismic risk analysis. *Bull. Seismol. Soc. Am.* **1968**, *58*, 1583–1606.
14. Hariri-Ardebili, M.A. Concrete Dams: From Failure Modes to Seismic Fragility. In *Encyclopedia of Earthquake Engineering*; Beer, M., Kougioumtzoglou, I.A., Patelli, E., Au, I.S.K., Eds.; Springer: Berlin/Heidelberg, Germany, 2016; pp. 1–26.
15. ICOLD. *Selecting Seismic Parameters for Large Dams, Guidelines, Bulletin 148 (Revision of Bulletin 72)*; Technical Report; International Commission on Large Dams: Paris, France, 2010.
16. Wieland, M. Features of seismic hazard in large dam projects and strong motion monitoring of large dams. *Front. Archit. Civ. Eng. China* **2010**, *4*, 56–64. [CrossRef]
17. Wieland, M. Seismic hazard and seismic design and safety aspects of large dam projects. In *Perspectives on European Earthquake Engineering and Seismology*; Springer: Berlin/Heidelberg, Germany, 2014; pp. 627–650.
18. Wieland, M. ICOLD's revised seismic design and performance criteria for large storage dams. In Proceedings of AFRICA 2013—Water Storage and Hydropower Development for Africa, Addis Ababa, Ethiopia, 16–18 April 2013. .
19. Applied Technology Council. *Seismic Performance Assessment of Buildings: Methodology*; Technical Report ATC-58-1; Federal Emergency Management Agency: Redwood City, CA, USA, 2012; Volume 1.
20. PEER. Ground Motion Database. 2017. Available online: http://ngawest2.berkeley.edu/ (accessed on 15 June 2017);
21. Saouma, V.; Porter, K.; Nuss, L.; Hariri-Ardebili, M. *Performance Based Engineering Design Guidelines for Concrete Dams*; Technical Report; Enerjisa: Istanbul, Turkey, 2012.
22. Ghanaat, Y. Failure modes approach to safety evaluation of dams. In Proceedings of the 13th World Conference on Earthquake Engineering, Vancouver, BC, Canada, 1–6 August 2004.

23. USACE. *Earthquake Design and Evaluation of Concrete Hydraulic Structures*; Technical Report EM 1110-2-6053; Department of the Army, U.S. Army Corps of Engineers: Washington, DC, USA, 2007.
24. Hariri-Ardebili, M.A.; Saouma, V.; Porter, K.A. Quantification of seismic potential failure modes in concrete dams. *Earthq. Eng. Struct. Dyn.* **2016**, *45*, 979–997. [CrossRef]
25. Bayraktar, A.; Türker, T.; Akköse, M.; Ateş, Ş. The effect of reservoir length on seismic performance of gravity dams to near-and far-fault ground motions. *Nat. Hazards* **2010**, *52*, 257–275. [CrossRef]
26. Valamanesh, V.; Estekanchi, H.; Vafai, A.; Ghaemian, M. Application of the endurance time method in seismic analysis of concrete gravity dams. *Sci. Iran.* **2011**, *18*, 326–337. [CrossRef]
27. Wang, G.; Zhang, S.; Wang, C.; Yu, M. Seismic performance evaluation of dam-reservoir-foundation systems to near-fault ground motions. *Nat. Hazards* **2014**, *72*, 651–674. [CrossRef]
28. Wang, G.; Wang, Y.; Zhou, W.; Zhou, C. Integrated duration effects on seismic performance of concrete gravity dams using linear and nonlinear evaluation methods. *Soil Dyn. Earthq. Eng.* **2015**, *79*, 223–236. [CrossRef]
29. Yilmazturk, S.M.; Arici, Y.; Binici, B. Seismic assessment of a monolithic RCC gravity dam including three dimensional dam–foundation–reservoir interaction. *Eng. Struct.* **2015**, *100*, 137–148. [CrossRef]
30. Bezzi, A.M.A.; Fontana, C.; Fiorani, A.; Masciotta, S.; Pietrangeli, S.; Visconti, V.E. *Evaluation of Dam Performance Under Seismic Loads with Linear Time History Analysis, Case Study Grand Ethiopian Renaissance Rcc Main Dam*; HYDRO: London, UK, 2017. .
31. Zacchei, E.; Molina, J.L. Artificial accelerograms to estimate damage of dams by using failure criteria. *Sci. Iran.* **2018**. [CrossRef]
32. Bybordiani, M.; Arıcı, Y. Effectiveness of motion scaling procedures for the seismic assessment of concrete gravity dams for near field motions. *Struct. Infrastruct. Eng.* **2018**, *14*, 1339–1354. [CrossRef]
33. Malm, R. *Guideline for FE Analyses Of Concrete Dams*; Technical Report, Energiforsk: Malmo, Sweden , 2016.
34. Zhang, Y.; Chen, G.; Zheng, L.; Li, Y.; Zhuang, X. Effects of geometries on three-dimensional slope stability. *Can. Geotech. J.* **2013**, *50*, 233–249. [CrossRef]
35. Hariri-Ardebili, M.A.; Saouma, V.E. Quantitative failure metric for gravity dams. *Earthq. Eng. Struct. Dyn.* **2015**, *44*, 461–480. [CrossRef]
36. Ansari, M.I.; Agarwal, P. Categorization of Damage Index of Concrete Gravity Dam for the Health Monitoring after Earthquake. *J. Earthq. Eng.* **2016**, *20*, 1222–1238. [CrossRef]
37. Hariri-Ardebili, M.A.; Furgani, L.; Meghella, M.; Saouma, V.E. A new class of seismic damage and performance indices for arch dams via ETA method. *Eng. Struct.* **2016**, *110*, 145–160. [CrossRef]
38. Hariri-Ardebili, M.A.; Kianoush, M.R. Integrative seismic safety evaluation of a high concrete arch dam. *Soil Dyn. Earthq. Eng.* **2014**, *67*, 85–101. [CrossRef]
39. Mirzabozorg, H.; Hariri-Ardebili, M.; Heshmati, M.; Seyed-Kolbadi, S. Structural safety evaluation of Karun III Dam and calibration of its finite element model using instrumentation and site observation. *Case Stud. Struct. Eng.* **2014**, *1*, 6–12. [CrossRef]
40. ANSYS. *ANSYS Software Reference Manuals, Release Notes, Mechanical Apdl, Elements Reference, Commands Reference and Theory Reference, Version Release 11*; ANSYS: Berkeley, CA, USA, 2007.
41. Hariri-Ardebili, M.A.; Seyed-Kolbadi, S.M. Seismic cracking and instability of concrete dams: Smeared crack approach. *Eng. Fail. Anal.* **2015**, *52*, 45–60. [CrossRef]
42. Hariri-Ardebili, M.; Boodagh, P. Taguchi design-based seismic reliability analysis of geostructures. *Georisk Assess. Manag. Risk Eng. Syst. Geohazards* **2018**, *13*, 34–52. [CrossRef]
43. Yang, T.; Moehle, J.; Stojadinovic, B.; DerKiureghian, A. Seismic Performance Evaluation of Facilities: Methodology and Implementation. *Asce Struct. Eng.* **2009**, *135*, 1146–1154. [CrossRef]
44. Vicente, D.; San Mauro, J.; Salazar, F.; Baena, C. An Interactive Tool for Automatic Predimensioning and Numerical Modeling of Arch Dams. *Math. Probl. Eng.* **2017**, *2017*, 9856938. [CrossRef]

8

From Theory to Field Evidence: Observations on the Evolution of the Settlements of an Earthfill Dam, over Long Time Scales

Stella Pytharouli [1,*], Panagiotis Michalis [1,2] and Spyridon Raftopoulos [3]

1 Department of Civil and Environmental Engineering, University of Strathclyde, Glasgow G1 1XJ, UK; panagiotis.michalis@glasgow.ac.uk
2 School of Engineering, University of Glasgow, Glasgow G12 8LT, UK
3 Power Public Corporation, Athens 104 36, Greece; s.raftopoulos@dei.gr
* Correspondence: stella.pytharouli@strath.ac.uk.

Abstract: Unprecedented flooding events put dams and downstream communities at risk, as evidenced by the recent cases of the Oroville and Whaley bridge dams. Empirical models may describe expected 'normal' dam behaviour, but they do not account for changes due to recurring extreme weather events. Numerical modelling provides insights into this, but results are affected by the chosen material properties. Long-term field monitoring data can help with understanding the mechanical behaviour of earthfill dams and how this is affected by the environment over decades. We analyse the recorded settlements for one of the largest earthfill dams in Europe. We compare the evolution of these settlements to the reservoir level, rainfall, and the occurrence of earthquakes for a period of 31 years after first impoundment. We find that the clay core responds to the reservoir fluctuations with an increasing (from 0 to 6 months) time delay. This is the first time that a change in the behaviour of a central clay core dam, as observed from field data, is reported in the international literature. Seepage rates, as recorded within the drainage galleries, are directly affected by cumulative rainfall depths exceeding 67 mm per fortnight.

Keywords: earthfill dam; central clay core; downstream shoulder; settlements; long-term behaviour; reservoir level fluctuations; rainfall; seepage; geodetic monitoring

---

1. Introduction

Recent cases of dam failures or near failures, e.g., the Oroville dam (2017) and Whaley bridge dam (2019), highlighted the fact that disruptive dam incidents, even when the structures remain safe, are the cause of cascading effects to other types of critical infrastructure [1]. This leads to substantial direct and indirect costs for asset owners, insurers, and maintainers, and significant disruption to civil society. Research on dams is crucial and timely, the main reasons for which are listed below:

- Increasing population of ageing dams: According to the International Commission on Large Dams (ICOLD), worldwide, there are more than 58,000 registered large (with heights greater than 15 m) dams. However, there also exist thousands of smaller dams that are currently unregistered, which are usually not subject to proper maintenance and monitoring regimes. The European Environmental Agency (EEA) currently estimates that there exist over 7000 large dams in Europe, with thousands more planned to be constructed during the next decade. According to the National Inventory of Dams (NID) in the US, there are also in total 90,580 dams. The 2017 infrastructure report card published by the American Society of Civil Engineers (ASCE) estimates that the average age of existing dams in the US is 56 years old, and that by 2025, 70% of the existing dam infrastructure will exceed their 'useful' design lifespan. The number of high hazard dams is also on the rise, and currently accounts for almost 17%, while more than 2000 dams are listed as structurally deficient, requiring a repair investment of nearly $45 billion [2]. The vast majority of dam infrastructure worldwide was constructed during the period 1950–1970; therefore, higher failure rates are anticipated in these dams due to the lower quality design and construction methods. Investigating the time evolution of the response of ageing dams informs the early evaluation of risk and can help direct any necessary interventions.

- Extreme climatic conditions: Recent climate change projections [3,4] indicate that the frequency of shifting weather events will substantially increase, and severe flooding incidents are anticipated to double in Europe by 2050 [5]. Extreme variations in seasonal rainfall patterns are expected to affect dam infrastructure [6], with an increasing magnitude of degradation processes related to the depth of rainfall infiltration inside the dam body, the increase in the fill's saturation level, and changes in the shear strength and pore water pressure within saturated fill layers. For embankment dams, the hydrostatic pressure induced by flooding and the high permeability of the soil caused by desiccation can result in uncontrolled internal seepage [7]. Modern climatic conditions are expected to significantly affect the structural integrity of hydro-infrastructure due to the high potential of extreme drought and flood events, as evidenced by the incidents of Oroville dam (US) during February 2017 and the Whaley bridge dam (UK) in July 2019. Another main issue arises from the fact that past and existing methods that incorporate the risk of climatic hazards into dam design and assessment methods are based on historical records. This indicates that they do not capture both the current and projected significant climatic variability; therefore, ageing dam infrastructure is expected to be put under stress. Studying the effect that any deviations from historical precipitation records have on existing dams is crucial for the transition towards resilient infrastructure.
- Gap in the knowledge about the long-term behaviour of dams: Relationships and models describing the short and long-term behaviour of dams are based on empirical equations. Such models are limited at times, as dams of similar size and type can behave differently, albeit still within safety limits. These empirical models, developed as far back as 30 years ago, are not incorporating the changing environmental conditions. As a result, investigating the behaviour of dams over long time scales can assist mitigate potential failures due to ageing and their (in) direct impact.

With the advances in computational power, numerical modelling has been used to assess and even predict the mechanical behaviour of dams. In the case of earthfill and rockfill dams, numerical modeling is particularly challenging. The success of a numerical model depends on the chosen material properties for the dam, which might or might not be correct. In situ material properties, such as hydraulic conductivity, which greatly influence numerical results, e.g., for pore pressures within the dam, can sometimes be significantly different to those obtained from lab tests. As a result, a numerical model that uses lab values could yield, for example, different results for the pore pressure inside the dam to those in reality.

Monitoring is an important tool in the assessment of the ongoing performance of earthfill/rockfill dams, which are currently considered the most well-instrumented infrastructures. It can provide information on the geometry changes of the dam body (geodetic monitoring) as well as changes within the dam body (geotechnical monitoring). The challenge is accessing these monitoring records, which in many cases are considered confidential. This is one of the reasons that studies based on real, long-term field dam monitoring data are sparse, e.g., [8], and predictions on the long-term behaviour of these structures are based either on empirical models, or in most cases, on numerical modelling.

In concrete dams, displacements are generally better described and occur at a definite rate, usually originated by degradation factors, such as temperature and water level variations [9,10]. Hydrostatic–Season–Time models are commonly used to describe the deformations of concrete dams [11]. These models have recently been modified and applied to earthfill dams [12–14]. The deformation rates are significantly higher for earthfill and rockfill dams and reach maximum values during or immediately after construction and first impoundment of the reservoir [8,15]. This makes the probability of a dam failure significantly higher during the first years of its operation [16]. The failure risk is estimated to be substantially decreased in the long-term for mature earthfill dam structures [9]; however, this does not take into account dams subjected to extreme flooding, for example, in an annual or biannual cycle, which is a situation that has become more and more frequent lately [4].

The deformation rates of earthfill and rockfill dams are affected by many factors, including internal erosion, slope stability, secondary consolidation of the core, the creep of shoulder fill, and stress changes due to varying reservoir water levels and seismic activity [17].

The relationship between the reservoir level fluctuations and the deformation patterns of earthfill dams has been the focus of a number of studies, e.g., [17–20]. A decrease of the reservoir level is followed by a decrease of vertical stresses and higher settlement rates, and as a result, a rapid drawdown could potentially result in slope instabilities due to the heave of the upstream slope. A decrease of lateral pressure due to the lower hydraulic head results in consolidation, horizontal movement, and time-dependent settlement. In the case of re-filling, the lateral pressure imposed by the water causes the heave of the core as it is deformed towards the downstream fill [17].

Empirical models have been employed to evaluate the deformation patterns of earthfill dams. Ref. [21] suggested the settlement index (S_I) as presented below in Equation (1):

$$S_I = \frac{s}{1000 \times H \times \log\left(\frac{t_2}{t_1}\right)} \quad (1)$$

where s (mm) is the crest settlement between two different measurement periods t_1 and t_2 at each control point located at a height H (m) from the foundation level. In case the value of the dimensionless parameter S_I is greater than 0.02, the crest settlement is attributed to mechanisms other than creep, requiring further investigation to be conducted [17].

The study by [19] indicated the relationship between environmental parameters and the deformation rates of the earthfill, with central core Kremasta dam in Greece. The authors concluded that the critical values of the settlement index are exceeded when the reservoir level is above a specific threshold value, and both the rate of change for the reservoir level and the precipitation exceeded 1 m/month and 130 mm/month, respectively.

Ref. [22] examined the behaviour of 15 rockfill dams and suggested the annual rate of settlement (Sa) as a percentage of the dam height to assess the deformations of the crest, as presented in Equation (2):

$$S_a = \left(\frac{S_{ii} - S_i}{H}\right) \times 100 \quad (2)$$

where Sii and Si are consecutive yearly settlement measurements, and H is the height from the foundation level at each crest control point. The study concluded that when the (Sa) is equal or less than 0.02% of the height of the dam, the deformations were regarded as normal, and the dam was considered stabilised, with the stabilisation period varying from 24 to 30 months. However, other studies e.g., [8,23], indicated that the dam structural integrity is not necessarily at risk if the stabilisation period is not within the period suggested by [22], as dams of a similar size and type can behave differently and still operate within safety limits.

In this paper, we use a unique monitoring record consisting of the vertical movements of 32 control stations established on the crest and downstream shoulder of a central clay core earthfill dam. Our preliminary work on the same dam [8] has shown that there are no concerns regarding its structural integrity. This made the dam a favourable case study in our effort to better understand the response of earthfill dams with central clay cores to environmental factors such as reservoir level, rainfall, and earthquakes. Our methodology is based solely on field observations (geodetic measurements, reservoir level, rainfall depth, hydraulic head within the dam, seepage rates), and we focus on the time period starting at the end of the first filling and spanning for 31 years. The period between the end of construction and the first filling is not examined here, as it is outside the scope of this paper, which focusses on how the dam behaviour evolves in the long term.

2. Materials and Methods

2.1. The Pournari I Dam

Pournari I (Figure 1) is an earthfill dam with a central clay core and sand gravel shoulders. The clay core is enclosed by transition filter zones, and the upper part of the upstream face of the shell is protected by rip-rap (Figure 2). The width at the base of the dam is 453 m and its maximum height is 103 m (superstructure from the foundation level), while the crest length and width is 580 m and more than 10 m, respectively [24]. The dam is owned by the Public Power Corporation of Greece (PPC S.A.). It is located in the regional unit of Arta (Western Greece) on the Arachthos River.

Figure 1. Downstream shoulder of the Pournari I dam with the concrete spillway (left) and part of the hydroelectric station (right).

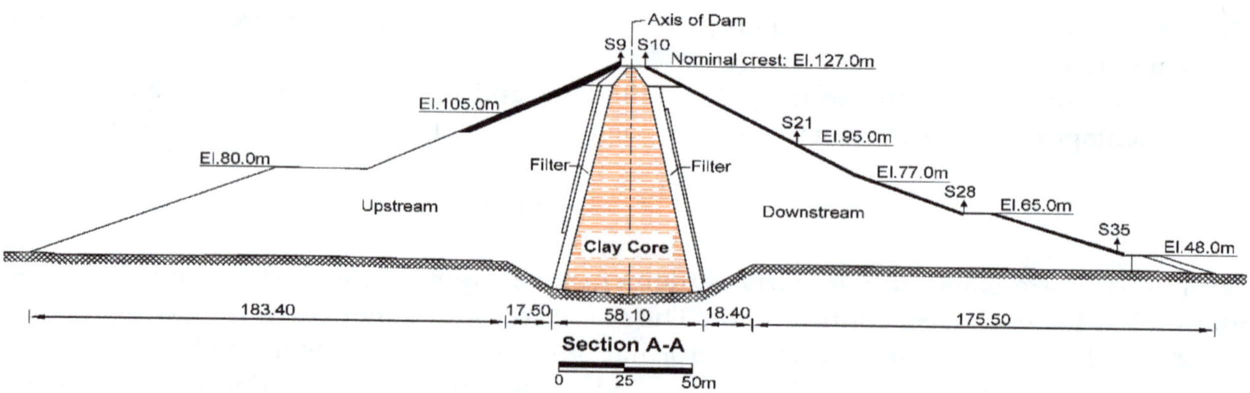

Figure 2. Maximum cross-section A-A of the Pournari I dam. For location, see Figure 3 (after [8]).

The Pournari I dam is the fifth largest dam in Greece (material volume of 9×10^6 m^3) and one of the largest dams in Europe. The construction of the dam was completed in 1981 and the first impoundment of the reservoir started immediately after, forming the artificial Arachthos reservoir with a maximum capacity of 865×10^6 m^3 covering an area of 20.6 km^2. The maximum and minimum water levels for hydroelectric power production are at elevations of 120 m and 100 m above mean sea level (AMSL), respectively, with the critical water level for the safety of the dam set at 126 m AMSL (nominal crest elevation: 127 m AMSL [24]). The main purpose of the dam is hydropower production and irrigation control. It is part of the Arachthos Hydroelectric Complex, which consists of four dams, the Pournari I dam being the largest both in size and electricity production capability.

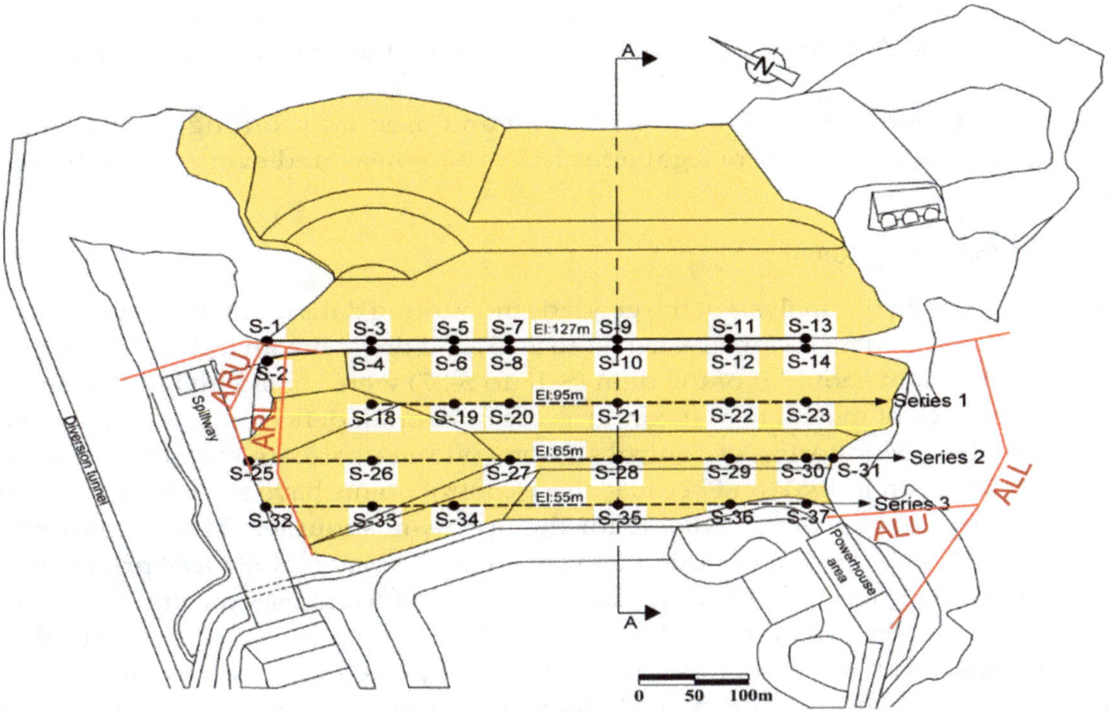

Figure 3. Part of the geodetic monitoring system at the Pournari I dam showing the control stations established on the crest and the downstream shoulder of the dam. In red are the four drainage galleries (i.e., Adit Right Abutment Upper - ARU, Adit Right Abutment Lower - ARL, Adit Left Abutment Upper - ALU, and Adit Left Abutment Lower - ALL) constructed at the dam to collect seepage. The location of the maximum cross-section A-A is also shown (modified after [8]).

2.1.1. Geology and Seismicity at the Dam Site

The dam has been founded on stratified sedimentary rocks (moderate to thickly bedded sandstones capping silty conglomerates above a sequence of thinly-bedded sandstone–siltstone) of flysch formations. At the dam site, beds strike N 30°W and dip NE at about 25°–45°, with no major faults. Most bedrock outcrops were found to be moderately to highly weathered. The unconfined compressive strength of the bedrock has values that range between 67 MPa and 118 MPa for the sandstones, and between 9.8 MPa and 49 MPa for the silty conglomerates and the sandstone/siltstone sequence. The modulus of elasticity was between 9.8 GPa and 19.6 GPa for all sequences. These values were considered adequate for the expected load on the foundations [25]. The area is also characterised by moderate to high seismicity. An increase of shallow seismicity (hypocentral depth ≤5 km) in the vicinity of the reservoir up to a 10-km distance that potentially triggered and was associated with the reservoir level fluctuations has been reported in [26].

2.1.2. Monitoring at Pournari I dam

The Pournari I dam has been monitored since its construction. The monitoring systems installed at the dam site mainly consist of geodetic and geotechnical instruments and five accelerographs.

The geotechnical monitoring system comprises various instruments (e.g., inclinometers, piezometers, pressure cells) installed inside the body of the dam at different locations. Over the years, a number of geotechnical instruments were damaged, and as a result, not all are currently operational. In addition, the reference points for some of the inclinometers and lateral movement instruments had to change after a few years of operation.

The geodetic monitoring system consists of 79 control points in total located on the crest, the upstream shoulder, and the downstream shoulder of the dam as well as the left and right abutment and the power house area. Vertical (from a reference datum) and horizontal movements (deflections from a straight line) of the control points are measured using reference stations that are located on stable ground. From the 79 control stations, 35 are located on the crest and downstream shoulder, as shown in Figure 3.

Four drainage galleries (see Figure 3) were built and used for grouting during construction of the dam. Seepage rates within the four galleries have been measured every fortnight since the end of construction.

2.1.3. Available Monitoring Record

This study focuses on the analysis of the geodetic monitoring dataset. Only the vertical movements of control points established on the upstream (S-1 to S-13) and downstream (S-2 to S-14) side of the crest as well at the downstream shoulder of the dam (S-18 to S-37) were considered in this analysis due to the uncertainty and lack of metadata as it concerns the horizontal deflections. No data were available for control station S-4. Monitoring of control stations on the upstream shoulder stopped after the completion of construction in 1981, and because these control points have since been submerged under the water. Therefore, no data were available for the upstream shoulder. The monitoring data were provided by PPC S.A. and cover a period of 34 years, from February 1981 to April 2015. This study focusses solely on the long-term mechanical behaviour of the dam and uses data from 1984 to 2015, thus avoiding the first three cycles of filling of the reservoir. The sampling rate for the vertical movements during the period under investigation was one per year. The procedure followed for the measurements is the same as that described in [27]. A typical accuracy for precise levelling is $a = 1$ mm/km [28]. The maximum distance of the levelling surveys at the Pournari dam was approximately $S = 1$ km. This means that the available measurements were significant within $\sigma = a^*\sqrt{(S)} = 1$ mm. Following the law of error propagation, we find that the accuracy of the settlement values is $\sigma_S = \sigma^*\sqrt{2} = 1.4$ mm. This value is an upper threshold as it refers to a single measurement. For repeated measurements, systematic errors are cancelled, as well as any gross errors that are identified, and the values are removed from the record. In this study, we adopt a very conservative threshold for any changes between consecutive measurements that are regarded as significant. This threshold takes into account a factor of safety equal to 2, i.e., ± 2.8 mm. This should suffice for any random errors such as misplacement of the levelling stuff during levelling, etc.

In addition, the record includes daily values of reservoir level and daily rainfall depths at the dam site for the period 1981–2015. Data that were also available were seepage rates, every fortnight, within the drainage galleries between 1987–2010 and monthly readings of the hydraulic head in four piezometric tubes installed within the upstream shoulder, the clay core, and the downstream shoulder of the dam between 1981 and 2010.

2.2. *Methodology*

Our methodology consists of the following steps:

Step 1: Time evolution of vertical movements for the control stations on the crest and downstream shoulder of the dam.

As a first step, the time evolution of the recorded vertical displacements is examined, aiming to identify the control points with maximum displacement on the crest and downstream shoulder, as well as compare the amplitude of settlements between the two sides of the crest.

In order to evaluate the amplitude of crest settlements, we calculate the settlement index (Equation (1)) and the annual rate of settlement (Equation (2)). We identify any time periods during which the settlement index and the annual rate of settlement exceed their threshold values equal to 0.02 [21] and 0.02% [22], respectively. We further examine these periods at a later stage of the analysis, as described in Step 2 of the methodology.

For the vertical movements of the downstream shoulder, we determine time periods that exhibit significant (above ± 2.8 mm as explained in Section 2.2) unexpected amplitude change, i.e., uplift.

We compare these time periods with the records of the reservoir level and rainfall at the next step of the analysis.

Step 2: Effect of environmental factors on the observed settlements.

In this step, we examine the settlement values to identify any potential relationship with environmental factors, i.e., the reservoir level fluctuations, rainfall, and occurrence of seismicity. We choose the observed settlements for control stations S-9, S-10, S-21, S-28, and S-35 that are located along the maximum cross-section A-A (Figure 3) for further analysis.

Since the sampling rate for the observed settlements is one per year, we use average monthly values for the reservoir level and monthly rainfall depths, instead of the daily values for our comparisons. The monthly values are calculated using the available daily data.

In order to investigate the effect of any climatic changes on the settlements, we calculate the deviation of the calculated monthly rainfall depth from corresponding historical rainfall depths for the same month that are available from the National Meteorological Service (EMY) of Greece (www.hnms.gr/emy/en/climatology/climatology_month; accessed on 08/07/2019). These historical monthly rainfall depths refer to the greater Arta region where the Pournari dam is located. Using the historical monthly values, we calculate the cumulative historical rainfall depths for the periods between consecutive measurements. The deviation (%) of the calculated rainfall depths between consecutive rainfall from the corresponding historical values are estimated as:

$$\text{deviation of rainfall } (\%) = \frac{\text{rainfall depth}_{(i)} - \text{historical depth}_{(i)}}{\text{historical depth}_{(i)}} \times 100 \quad (3)$$

where rainfall $\text{depth}_{(i)}$ is the cumulative rainfall depth for the time period between measurement epoch i and i-1, as calculated from the available daily rainfall depth values. Historical $\text{depth}_{(i)}$ is the cumulative rainfall depth that refers to the same months of the year as those included in the calculation for the rainfall $\text{depth}_{(i)}$, and is calculated from the available historical monthly rainfall depth values.

Since Greece is a highly seismically active region, we also studied the effect of earthquakes on the observed dam settlements solely in a qualitative form. More specifically, we examine whether any increase in the rate of settlements can be correlated with the occurrence of earthquakes within a 50-km radius from the dam site. We used the earthquake catalogue available from the National Observatory of Athens for earthquakes with magnitudes >3 within a radius of 50 km from the Pournari I dam site during the period 1984–2015. In structural engineering, earthquakes of magnitudes of M5 or more are considered as being significant to infrastructure ([29] and references therein); therefore, we also looked at the occurrence of earthquakes with magnitudes ≥M5.

3. Results

3.1. *Step 1: Time Evolution of Vertical Movements for the Control Stations on the Crest and Downstream Shoulder of the Dam*

In this section, the recorded settlements on the crest and downstream shoulder of the dam are discussed without any reference to external factors that could have an effect on them, e.g., reservoir level, rainfall, etc.

3.1.1. Crest Settlements

Figure 4a shows the time evolution of the crest settlements at the Pournari I dam. The solid lines with the filled symbols correspond to control stations at the upstream (U/S) side of the crest, while the dashed lines with the unfilled symbols correspond to control stations at the downstream (D/S) side of the crest. Symbols of the same shape and colour (filled and unfilled) correspond to points with locations facing each other, i.e., at the same distance along the crest but on different sides of the crest. From Figure 4a, it is observed that in general, settlement amplitudes are smaller closer to the abutments and increase as we move towards the middle of the crest. The maximum displacement up to 2015, equal to 623 mm, has been observed for control station S-9 on the U/S side of the crest. This corresponds to 0.6% of the dam height (from the foundation level).

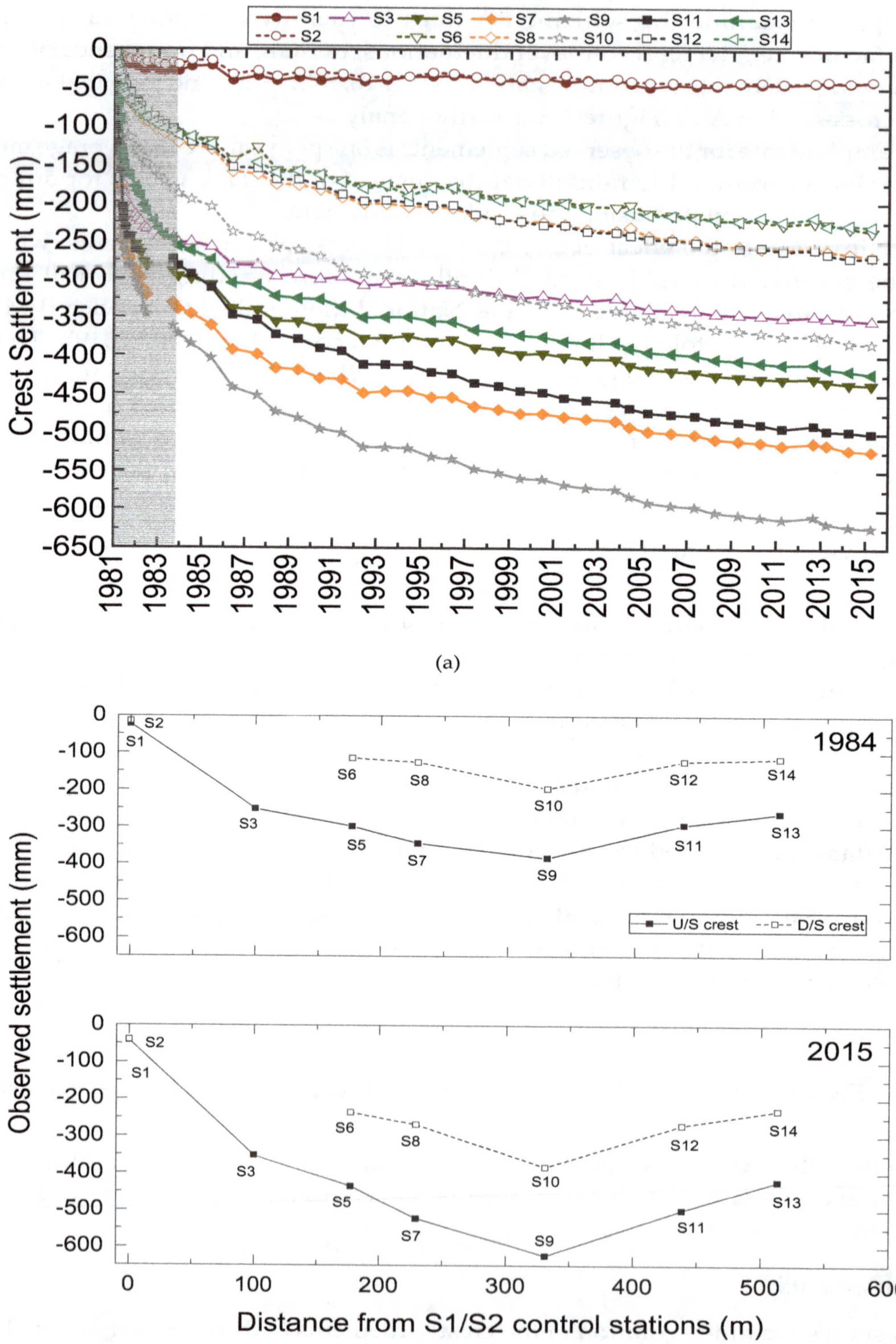

(a)

(b)

Figure 4. (**a**) Time evolution of the crest settlements of the Pournari I dam. The shadowed area corresponds to the first three filling cycles, and is not taken into account in this study. (**b**) Observed settlement for the control stations along the upstream (U/S) (solid line) and the downstream (D/S) (dashed line) side of the crest on 1 July 1984 (top) and on 1 April 2015 (bottom).

Differential settlement is observed on the crest. In particular, the U/S side appears to consistently have larger settlements by approximately 200 mm over the whole time period examined (Figure 4b), compared to the D/S side, for all stations but S-1, next to the left abutment. The difference is slightly bigger, 250 mm, between S-9 and S-10 at the location of the maximum cross-section. The U/S side of the crest closer to the left abutment has slightly larger settlements (control stations S-5 and S-7), by approximately 20 mm on average, compared to the settlements of the same side of the crest closer to the right abutment (control stations S-11 and S-13, respectively).

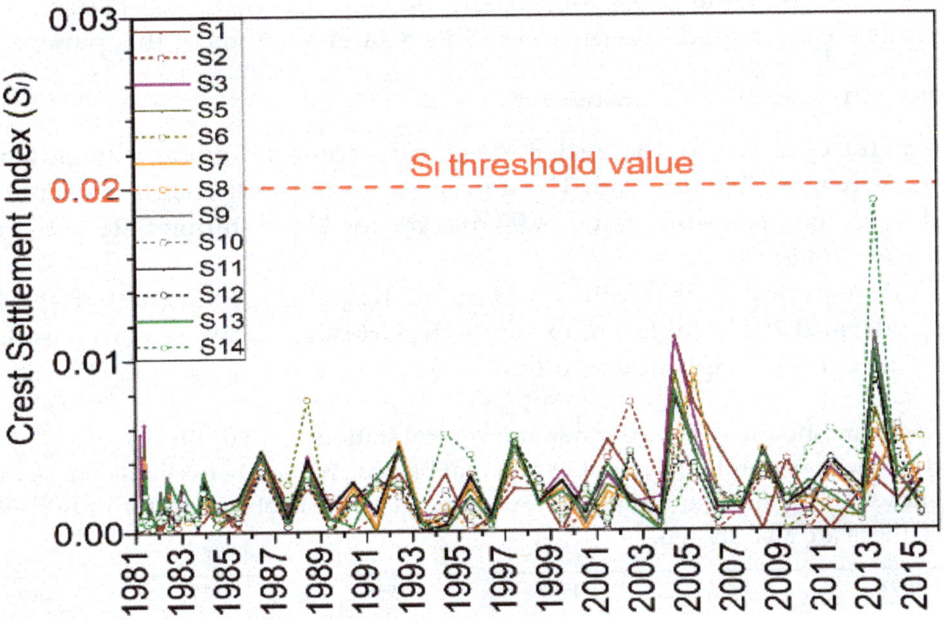

(a)

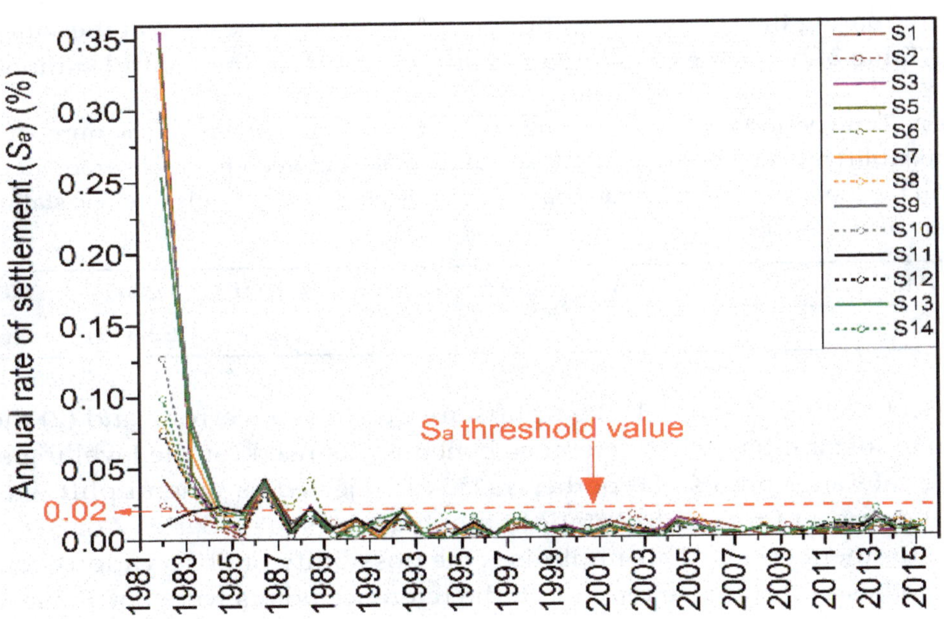

(b)

Figure 5. Crest settlements: (**a**) Settlement Index S_I and (**b**) Annual rate of settlement.

We calculated the crest settlement index (Figure 5a, Equation (1)) and the annual rate of settlements for the points on the crest (Figure 5b, Equation (2)) following the methodology by [22]. From Figure 4a, it can be shown that up until 2015, the settlement index has not exceeded the threshold value of 0.02 suggested by [22]. The annual rate of settlements seems to have stabilized for all control stations on the crest after 1984 (within three years since construction, consistent with the time period suggested by [22]) with the exception of measurement epochs 1986, 1988, and 1992, where values exceeded the 0.02% threshold for all or some of the control points. More specifically, all control stations on the crest appear to have an annual settlement rate higher than 0.02% of the dam height in 1986. In 1988 and 1992, only the control stations that were located on the U/S side of the crest surpassed the threshold value. These periods are examined in more detail in the later sections of this paper.

3.1.2. Settlements of the Downstream Shoulder

Figure 6 shows the evolution of the settlements for the control points on the downstream shoulder. No measurements were available for 1986. The amplitude of the settlements is smaller than that of the crest, as expected, with the maximum value of 95 mm recorded for station S-21 of Series 1 (for location of control points refer to Figure 3).

A number of time periods with uplift movements have been identified: 1989, 1992, 1994, 1996, 1997, 1999, 2001, 2003, and 2012. Table 1 shows which part of the downstream shoulder shows uplift and the corresponding uplift amplitude in mm.

Table 1. Downstream shoulder: Time periods and control stations with uplift movements. The numbers in brackets show the max uplift movement in mm that has taken place between the measurement epoch on the year shown in the table and the previous year. If uplift is observed for only specific stations along a series, these are also shown.

1989	1992	1996	1997	1999	2001	2003	2012
Series 1 (6)					Series 1 (10)	Series 1 (8)	
		Series 2 (6, S27-31)	Series 2 (13, S26-28)	Series 2 (13)	Series 2 (8)	Series 2 (6)	Series 2 (7)
	Series 3 (10, S34-37)		Series 3 (13)		Series 3 (7)	Series 3 (7)	Series 3 (7)

It is worth comparing the behaviour of the control stations of the crest at the same time periods as those in Table 1. Table 2 shows the calculated maximum uplift for the control stations at the crest.

Table 2. Crest: Time periods and control stations with uplift movements. The numbers in brackets show the max uplift movement in mm that has taken place between the measurement epoch on the year shown in the table and the previous year. If uplift is observed for only specific stations along the crest, these are also shown.

1989	1992	1996	1997	1999	2001	2003	2012
S1-2 (5)	-	-	-	-	S2 (4)	S2-3 (3)	(6)

We examined two hypotheses: (1) the uplifts are due to gross errors, and (2) the uplifts are not real, but are due to instabilities of the reference benchmarks used for the levelling surveys, i.e., the settlement of the reference points. Hypothesis (1) was rejected as when uplift was observed, this was for either all or most of the control stations along a series (e.g., Series 2 or 3) or multiple series at the same measurement epoch. This minimises the possibility of observations being attributed to gross errors. Hypothesis (2) was also rejected. If the reference points were not stable, this would have been identified during the surveys of the control points that take place regularly. In addition, if this hypothesis was true, the effect on the recorded measurements would have been continuous, not on isolated time periods only.

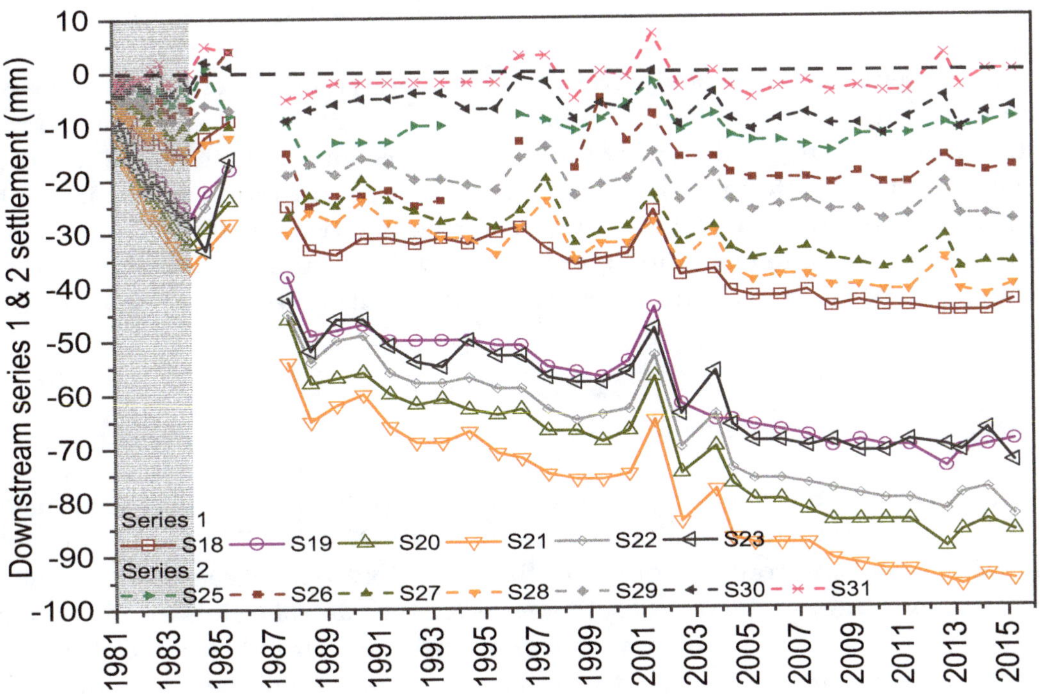

(a)

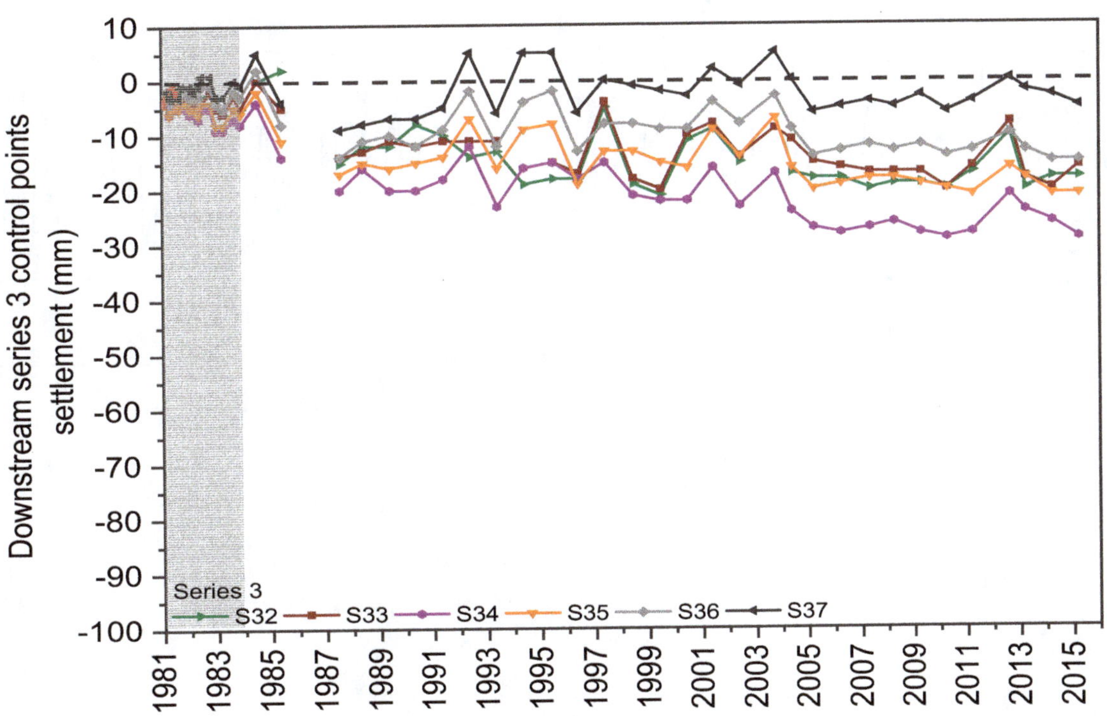

(b)

Figure 6. Time evolution of settlements for the D/S shoulder: **(a)** Series 1 control stations S-18 to S-23, and Series 2 control stations S-25 to S-31; **(b)** Series 3 control stations S-32 to S-37. Both graphs have the same scale on the x and y-axis for comparison purposes. Points with the same symbols, filled or unfilled, lie along the same cross-section, perpendicular to the dam axis. The shadowed area corresponds to the first three filling cycles (1981–1984), and is not taken into account in this study.

3.2. *Step 2: Effect of Environmental Factors on the Observed Settlements*

In this step, the settlements are analysed in combination with the monthly average reservoir level values, the monthly cumulative rainfall depths, and the occurrence of seismicity. From Figure 4a, Figure 6a,b, it is evident that the settlements of control stations that lie along the same line and elevation (e.g., crest, Series 1, etc.), are following very similar patterns, although with different amplitudes. Hence, the settlements of only one station per monitoring line (U/S crest, D/S crest, Series 1-3), those along the maximum cross section (A-A in Figure 3; S-9, 10, 21, 28 and 35), are examined further in this paper. These were the points that exhibit maximum settlements amongst the control points along the same elevation.

3.2.1. Settlements and Reservoir Level

Figure 7 shows the monthly average reservoir level and the observed settlements of the control stations along the maximum cross-section A-A (S-9, S-10, S-21, S28, S35). A first observation is that the rate of settlements of the control points on the crest does not increase following the reservoir drawdown in 1987. A change in the rate of settlements for the control points on the crest (S-9 and S-10) in 1986, 1992, 1997, and 2002 follows the lower level in the reservoir in December 1985, February 1992, November 1996, and February 2002, respectively (Figure 7). This could be a possibility if a time lag of 4 to 6 months between the two parameters, i.e., water level and settlements, exists.

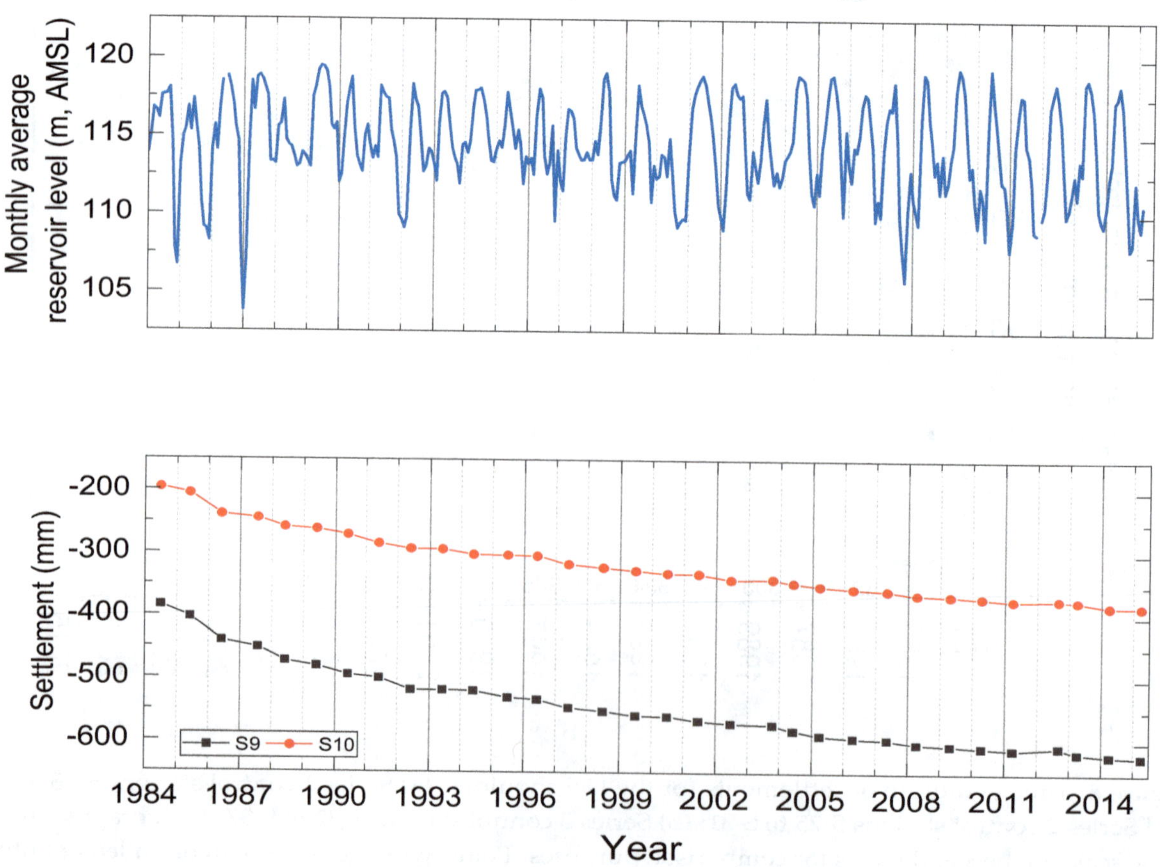

Figure 7. *Cont.*

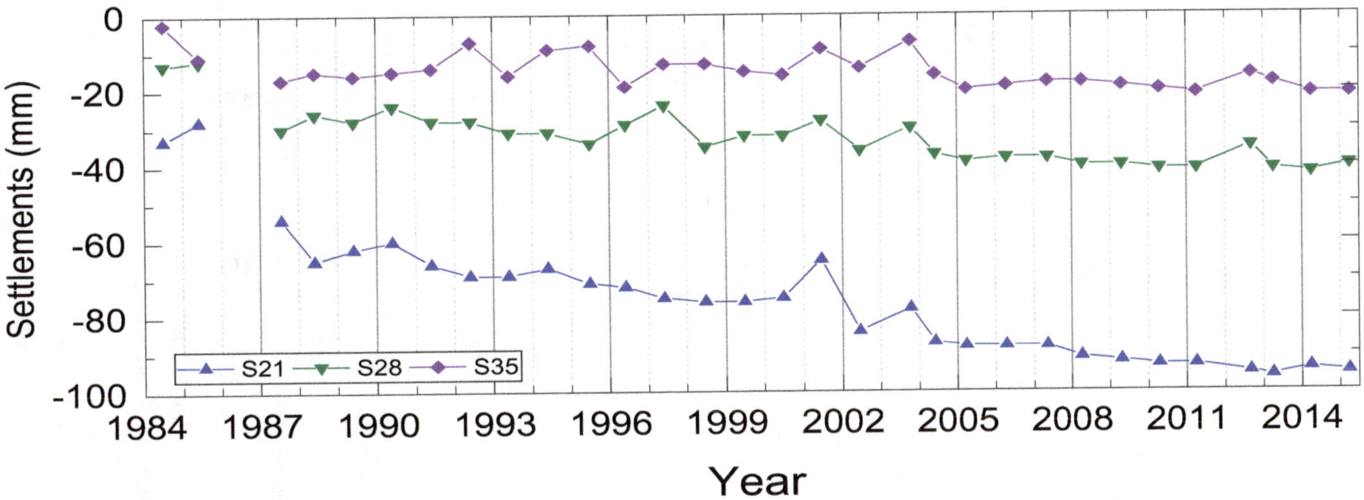

Figure 7. Monthly averages of reservoir level and observed settlements of control stations along cross-section A-A (see Figure 3 for location). The settlements of the control stations on the downstream shoulder (bottom plot) are plotted at a different y-axis scale for easier comparisons.

For the D/S shoulder, the two distinct periods of recorded uplift in 2001 and 2003 visible in the records of all series could not be linked to the reservoir level from the analysis so far.

3.2.2. Settlements and Rainfall

Next, we investigate whether a relationship exists between settlements and rainfall at the dam site. Figure 8a shows the time history of the settlements and the annual rainfall depth. Here, the 'annual' rainfall depth corresponds to the amount of rainfall between consecutive measurement epochs (which is on average one year), not within a calendar year. In order to examine whether wet or dry years, i.e., changing climatic conditions, played any role in the evolution of settlements, we compare the observed settlements with the percentage of deviation of the rainfall depth between consecutive measurement epochs, to the corresponding historical depths.

One observation that can be made from Figure 8b is that the crest settlements do not have a direct relationship with the recorded rainfall events. On the other hand, there is some evidence that the D/S shoulder does. More specifically, uplift is observed when the amount of cumulative rainfall changes significantly either way (positive or negative) compared to the amount of rainfall that fell within the previous measurement epoch. In 1989, 1992, 2001, and 2012, the rainfall depth is substantially lower than in the 1988, 1991, 2000, and 2011 measurement epochs, respectively. The opposite stands for 1996, 1997, 1999, and 2003, with the rainfall depths being higher than in 1995, 1998, and 2002. These years correspond to times when uplift is observed at the D/S shoulder.

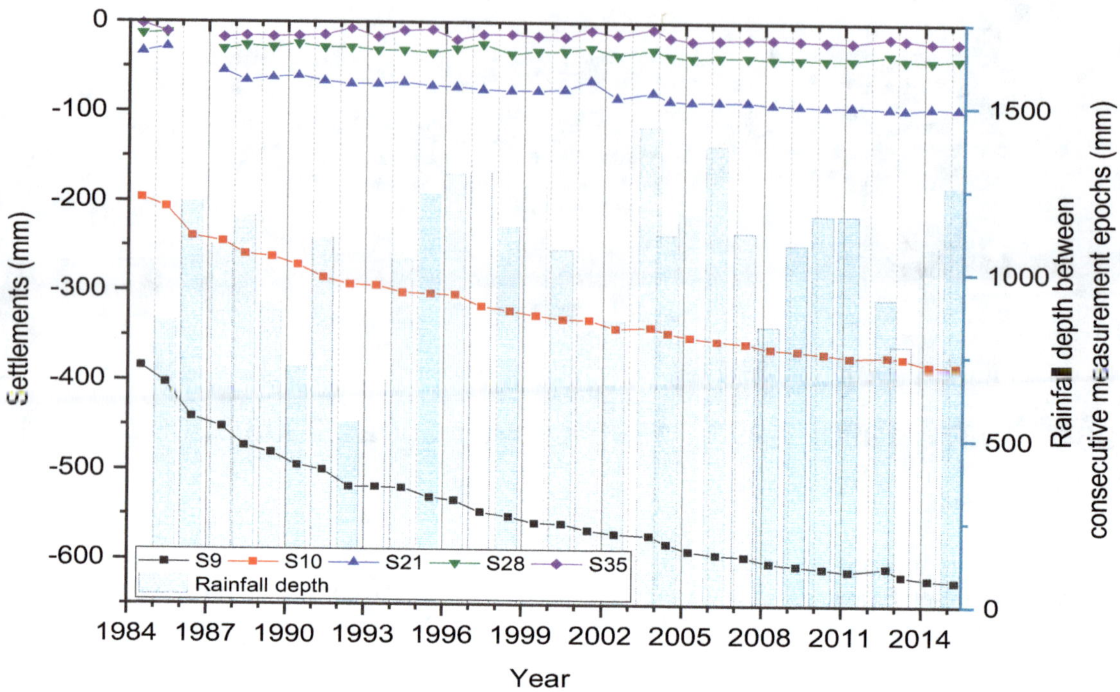

(a)

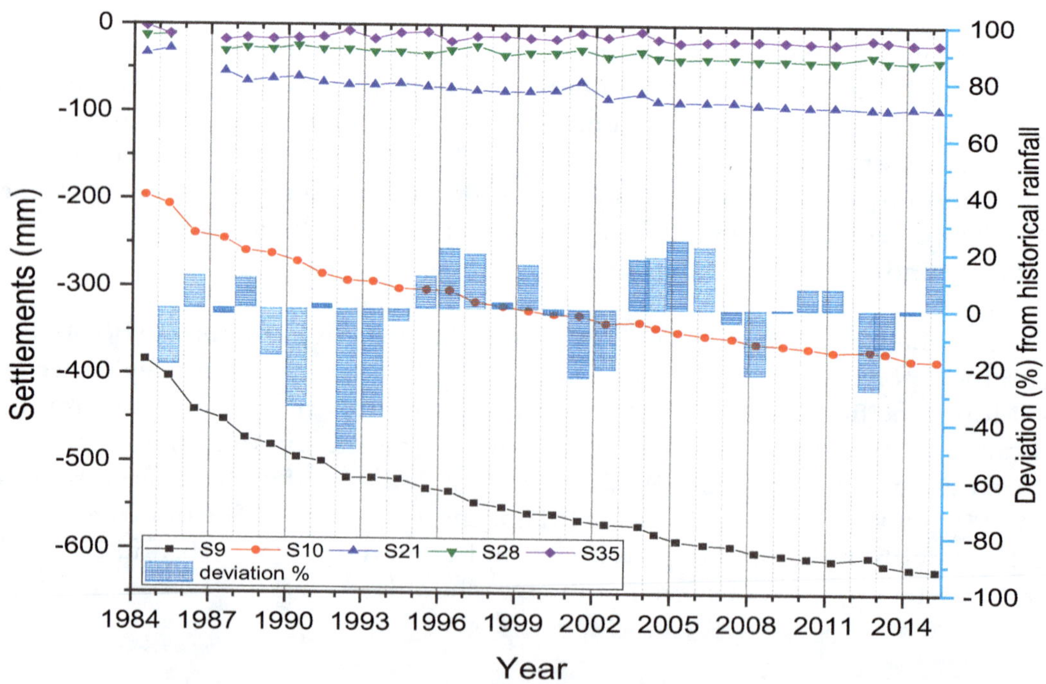

(b)

Figure 8. Settlements across A-A section and (**a**) rainfall depth between consecutive measurement epochs, and (**b**) deviation (%) of rainfall depths between consecutive measurement epochs from historical rainfall depths.

3.2.3. Settlements and Seismicity

We investigated, in a qualitative way, how/if the occurrence of seismicity affects the dam and whether the observed settlements at the times when higher rates are observed (1986, 1988, and 1992) could be earthquake-induced. Our study does not make use of any accelerometer data, as these were not available. Earthquakes are known to induce permanent settlements in earthfill dams, even failure, e.g., the Fujinuma dam [30]. Regarding the possible uplift, this could also be a response to earthquakes, which are known to cause an increase in water levels within wells [31] and/or an increase in the seepage rates [32], which could result in an increase in the pore pressures within the dam.

Figure 9 shows the settlements and the occurrence of earthquakes with magnitudes 3 or above, within 50 km from the dam between 1984 and 2015. The number of earthquakes for a specific measurement epoch corresponds to the number of earthquakes that occurred in the time period between that specific measurement epoch and the one preceding it. Earthquakes can have an impact on the geometry changes of the dam, especially on the settlements recorded at the crest [33]. No such significant effect can be seen from Figure 9 for the Pournari I dam. According to the earthquake catalogue from the Institute of Geodynamics, National Observatory of Athens, there have been three earthquakes with magnitudes of M_L 5.4, M_L 5.3, and M_L 5.2 in 1993, 2000, and 2014, respectively (shown as stars in Figure 9) within a 50-km radius from the dam. The 13 June 1993 M_L 5.4 earthquake appears under the 1994 measurement epoch in Figure 9. This is because it occurred after the 1993 measurement epoch (1 June 1993) had taken place. No earthquakes with magnitudes larger than the three reported occurred during the examined time period and within 50 km from the dam site. The notable increase of earthquakes at lower magnitudes over time could be due to the detection capability of the Greek National Seismic Network having improved since 2000. However, this should have had no impact on the earthquakes with numbers of magnitude of 5 or above.

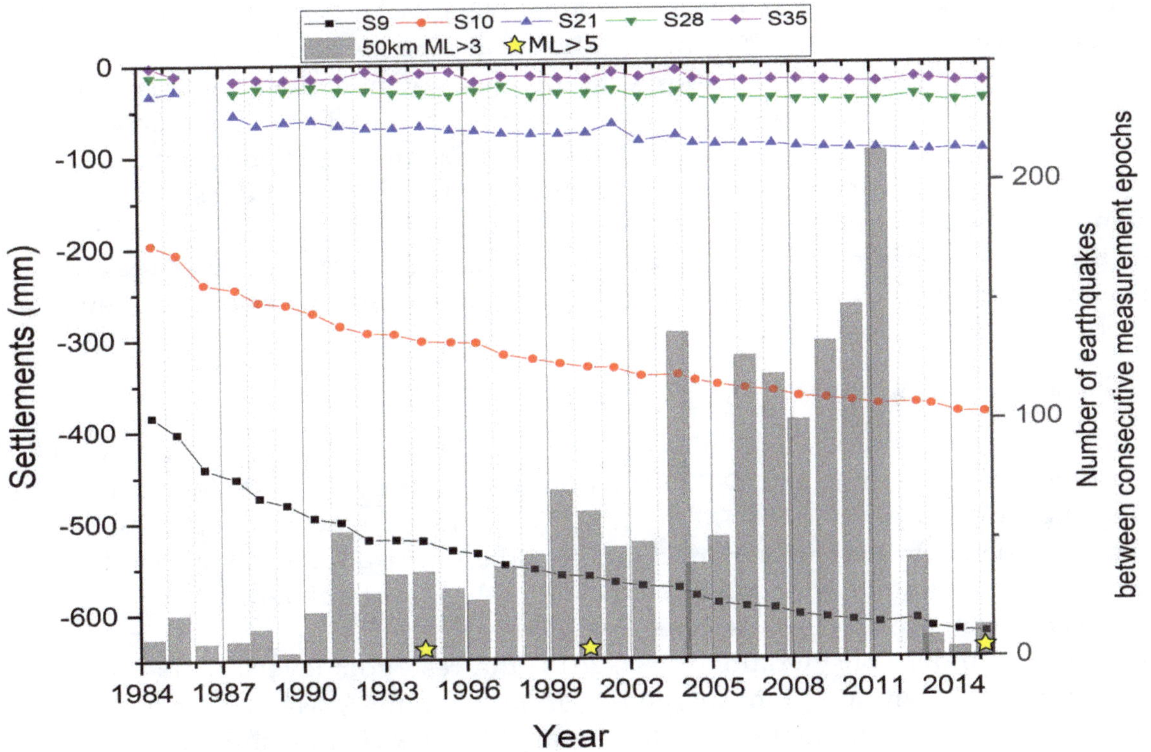

Figure 9. Settlements and occurrence of earthquakes (M_L >3) within a radius of 50 km from the dam site between consecutive measurement epochs. Stars indicate the largest earthquakes during the examined period (M_L 5.4 in 1993, M_L 5.3 in 2000, and M_L 5.2 in 2014 at distances of 36.8 km, 15.4 km, and 16.3 km from the dam, respectively). Earthquake data were sourced from the National Observatory of Athens.

The absence of evidence on the effect of earthquakes on Pournari I dam settlements is consistent with what is expected from the international literature. It has been observed that in general, earthfill and rockfill dams are quite resilient to earthquakes of magnitudes less than 6.5 [32]. All reported cases of dams having suffered severe damage in the form of longitudinal cracks or settlements following an earthquake refer to larger earthquakes than those that occurred close to the Pournari I dam area during the examined time period. Base sliding is another earthquake-induced failure, but it affects mainly concrete dams [34] rather than earthfill dams, and therefore is not examined in this paper. Liquefaction can occur for earthfill dams founded on alluvium. This is not the case for the dam under investigation. It can be concluded that the Pournari I dam has been unaffected by the seismic activity of the region (magnitudes up to ML 5.3), but this does not necessarily mean that the settlement rate won't be affected in the case of earthquakes with larger magnitudes in the future.

4. Discussion

The analysis of the recorded crest settlements of the Pournari I dam following the methodology of [21] showed that the crest settlements are mainly due to the creep/secondary consolidation of the clay core. The annual rate of crest settlement as a percentage of the dam height dropped for the first time below the threshold of 0.02% [22] in 1985 and permanently after 1986, i.e., within 3 years (36 months) and 4 years (48 months) since the completion of the construction, respectively. Dascal [22] suggests that the annual rate of settlements of the crest should drop below 0.02% within 24–30 months since the completion of construction. The Pournari I dam appears to take longer to stabilise, but it is still within the maximum period of 8–10 years suggested by Lawton and Lester [35].

The maximum settlement observed at control station S9 at the U/S side of the crest corresponds to 0.6% of the dam height from the foundation. A number of studies have suggested threshold values below which the settlement, as a percentage of the dam height, is regarded expected/normal: [36] suggest the range 0.25–1% of the dam height. Dascal [22] gives a 0.35% threshold value, while [37] gives a range between 0.1% and 0.4%. The threshold values suggested by [22,37] were found to be conservative based on the evaluation of the crest settlements of more than 40 dams with a central clay core by [20]. More than 50% of these dams had a maximum recorded crest settlement higher that 0.4% of the dam height without these dams being at risk. The maximum settlement of the Pournari I dam, recorded 34 years since construction, is well below the 1% threshold, and does not pose any concerns for the dam's structural integrity.

The differential settlement between the U/S and the D/S side of the crest is of the order of 250 mm, and can be attributed to the effect of the impoundment of the reservoir and the wetting of the U/S shoulder of the dam.

4.1. *Observed Settlements and Reservoir Level Fluctuations*

Our analysis provided no evidence of a direct relationship between the reservoir level fluctuations and the settlements of the crest and D/S of the Pournari I dam. The reservoir level fluctuations is expected to affect primarily the U/S shoulder [38]. The lack of direct relationship between reservoir level and crest settlements has been reported for other dams as well, e.g., [39]. Therefore, the absence of significant settlements at the crest and D/S shoulder at the Pournari I dam as a result of the reservoir fluctuations is not unexpected.

In the period under consideration, 1984–2015, the reservoir level fluctuates annually. The fluctuations are mostly within ± 5 m. Only once, in January 1987, was the reservoir level lowered significantly to 100.67 m AMSL (daily value), compared to levels before and after that date, reaching almost the minimum reservoir level for the production of electricity (100 m AMSL). This drawdown in July 1986 corresponded to 16% of the original reservoir depth (120 m AMSL). No significant settlements or settlement rates were observed for the crest and D/S shoulder immediately following this drawdown (Figure 7).

Tedd et al. [40] describe the effect of reservoir drawdown on the settlements of the U/S shoulder, the dam core, and the dam crest. They studied the crest settlements of the Ramsden dam, and concluded that crest settlements continued during drawdown, and while the reservoir was empty, this was reversed at some degree with refilling. For what they regarded as a minor reservoir drawdown (30% of the reservoir depth), they calculated a settlement of 12 mm for the U/S shoulder, which was equal to the measured value, while 50% of that settlement was reversed during refilling. The measured settlement of the crest was reported to be almost 1.5 times bigger than that of the U/S shoulder. They suggest that significant crest and U/S shoulder settlements take place every time the reservoir is lowered.

The fact that no significant settlements were recorded for the Pournari I dam following the 1987 reservoir drawdown does not necessarily mean that such settlements did not take place. A possibility could be that any settlements due to drawdown were partially reversed during refilling, and this is not evident in the settlement record because of the very low sampling rate (annual). By the time of measurement in June 1987, the reservoir level had already returned to 119 m elevation. The recorded settlements along the maximum cross-section could well reflect a net settlement. To investigate whether this hypothesis could be viable, we followed the approach and Equation (9) described in [40] to approximately calculate the expected settlement of the U/S shoulder at the Pournari I dam due to the drawdown in January 1987. We found it equal to 15 mm. Assuming that at least 50% of that settlement could have been recovered by June 1987 when measurements took place, a net settlement of 7–8 mm could be expected. We have no available observations for the U/S shoulder to validate this. According to [40], the crest settlements should have been bigger than our estimated 15 mm. Unfortunately, the approach of [40] cannot be used to estimate settlements on the crest due to reservoir drawdown. The measured settlement at control station S-9 on the U/S side of the crest in June 1987 was 11 mm. Assuming that a 50% recovery had taken place by that time, this would give a 22-mm settlement due to drawdown. This value is 1.5 times higher than the calculated settlement value for the U/S shoulder (15 mm) at the Pournari I dam, and is also consistent with the observations by [40] at Ramsden dam.

4.2. *Observed Settlements and Rainfall*

Prolonged extreme rainfall events have been identified in [41] as a potential factor affecting the crest and D/S shoulder through surface erosion. As extreme rainfall, the authors of [41] use examples of storms with rainfall depths from 76 mm to 210 mm/day. Figure 10 shows the recorded daily rainfall depths at the Pournari I dam during the time period 1984–2015. There are 19 occasions when this rainfall exceeded the lower threshold of what is characterised as extreme rainfall in [41]. None of these occasions could be directly related to the times when uplift was observed.

There are very few studies in the international literature that investigate the effect of rainfall on earthfill and rockfill dams. Most of the existing studies focus on the effect of rainfall or extreme rainfall events on the occurrence of flooding events and the risk of overtopping. Ref. [42] simulated the effect of rainfall on rockfill and suggested that long-term deformations of the rockfill are controlled by the wetting history of the dam shoulders. They showed that the settlement rate increases at times of extreme rainfall events with an indicative value of 300 mm/month (Figure 3 in [42]). Although there were occasions when the rainfall at the Pournari dam exceeded 300 mm/month, no increased settlement rates were observed at those times. This could be due to the material used for the Pournati I dam shell (sand and gravels), which is finer than the rockfill material of the dam in the study of [42], and hence the infiltration rate is lower. A higher infiltration rate and consequently increased pore pressure results in the decrease of the effective stress, i.e., weakening of the dam. This does not appear to be the case for the Pournari I dam.

Ref. [43] link rainfall and leakage as a possible mechanism for internal erosion with leakage being dominated by rainfall. Following the methodology on threshold correlation by [19], we compare the rainfall depths to leakage reported within four drainage galleries at the Pournari I dam (for location see Figure 3). We find that the volume recorded in the drainage galleries is strongly correlated to rainfall for cumulative rainfall depths over the same fortnightly period above 67 mm (horizontal line in Figure 11). This correlation is consistent with published literature [38], and taking into account that the recorded leakage rate in the drainage galleries is low, we do not expect the Pournari I dam to be at risk under the conditions examined in the present study.

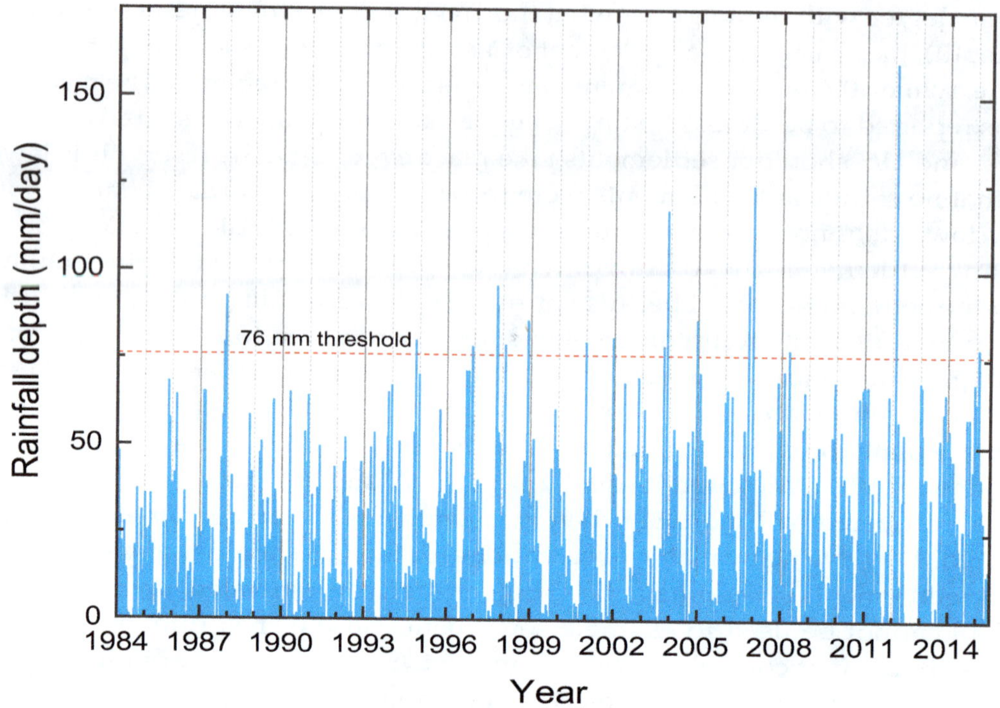

Figure 10. Daily rainfall at the Pournari I dam site between 1984 and 2015. The dashed line corresponds to a low threshold of 76 mm, above which the rainfall is regarded as a storm in [41].

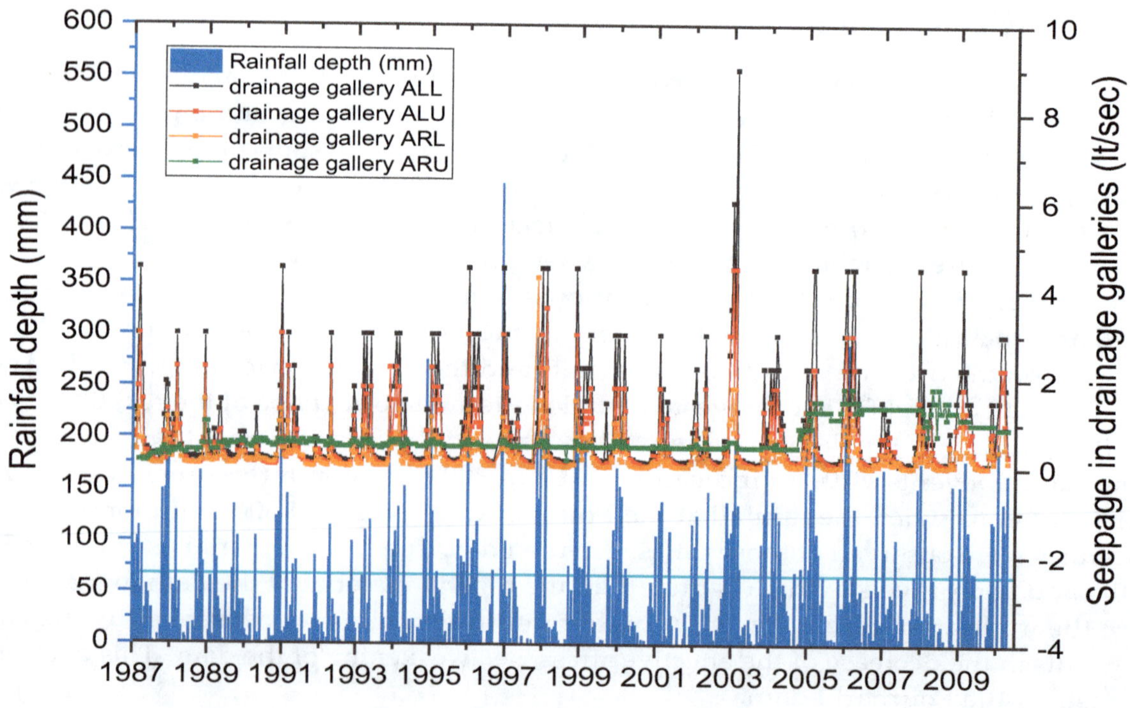

Figure 11. Left y-axis: rainfall depth between consecutive drainage measurement epochs. The light blue horizontal line corresponds to the threshold of 67 mm, above which a direct correlation between rainfall depth and seepage volume is observed. Right y-axis: Seepage recorded in the drainage galleries of the Pournari I dam.

The D/S shoulder can also be affected by inflow from seepage from springs, mainly at areas prone to karstification. At the Pournari I dam site, there are no springs (it is founded on flysch); however, the existence of small lenses of conglomerates within the sandstone strata exhibiting high weathering has been reported [25]. Our results on the correlation between rainfall depths and seepage are consistent with observations during the investigation of the foundation bedrock conditions that preceded grouting. It had been found that the ground water profile fluctuated over several meters annually, with some water inflow recorded in the drainage galleries after rainstorms. This was attributed to the presence of open joints and bedding planes [25].

4.3. Time Lag between Recorded Settlements and Reservoir Level Fluctuations

In our analysis, we did not consider a time lag between the settlements and the reservoir level. A potential time lag could be due to either the hydraulic conductivity values of the material of the upstream shoulder and the core (in the case of steady state) or the storage capacity of the material [44]. Due to the different hydraulic conductivities of the dam materials, it should be expected that any potential changes in the geometry of the dam observed at the crest and D/S shoulder as a result of the reservoir level fluctuations occur with some delay. In the absence of information on the hydraulic conductivity of the U/S and D/S shoulder as well as the filter and the clay core, precise estimation of such a time lag is not possible. The investigation of the role of the storage capacity would require numerical modelling and unsaturated conditions, and this is outside the scope of this paper. Instead, we used available elevations of the water (hydraulic head) in four piezometers installed within the dam (for location, see Figure 12).

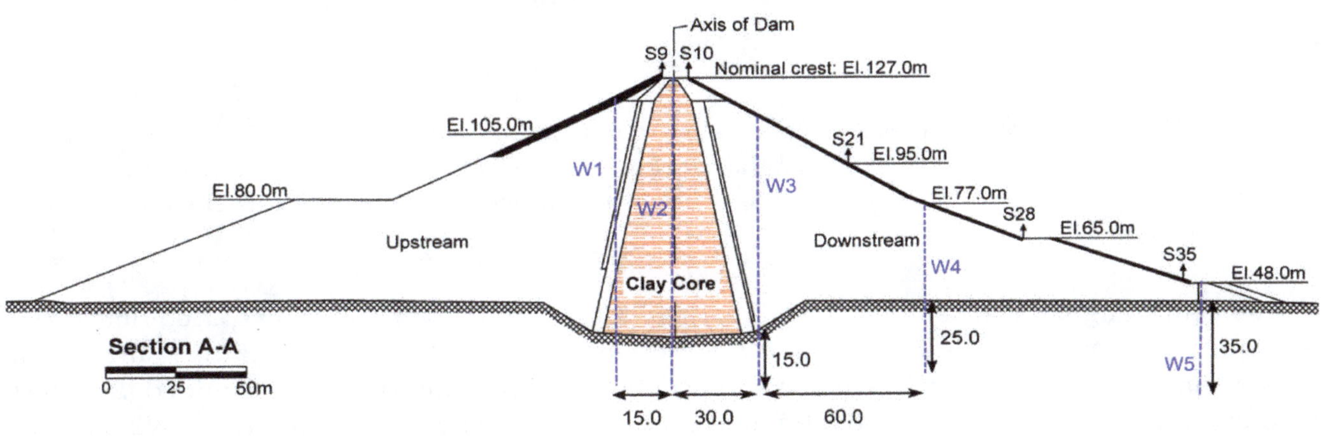

Figure 12. Location of piezometers W1–W5 along cross-section A-A.

The time history between 1982 and 2010 of the hydraulic head in W1–W4 is shown in Figure 13. From Figure 13, it is evident that the upstream shoulder (W1) responds immediately to changes in the water level in the reservoir. The core (W2) was following the changes up to around late 1989 (see further for details), after which there appears to exist a time lag between the water elevation in W2 and the reservoir level fluctuations. For example, the local minimum in 1990 in W1 values appears slightly earlier than the 1990 local minima in W2 values (Figures 13 and 14). This time lag appears to be getting bigger over time if one compares the W1 peaks with the W2 peaks and attempts to align them. Precise determination of this time lag has not been possible for the period 1982–2010 from the available data mainly because the two time series are not evenly spaced, and the traditional cross-correlation analysis [45] cannot be applied. Interpolation could be used to overcome this, but such an approach would increase the possibility of biased results.

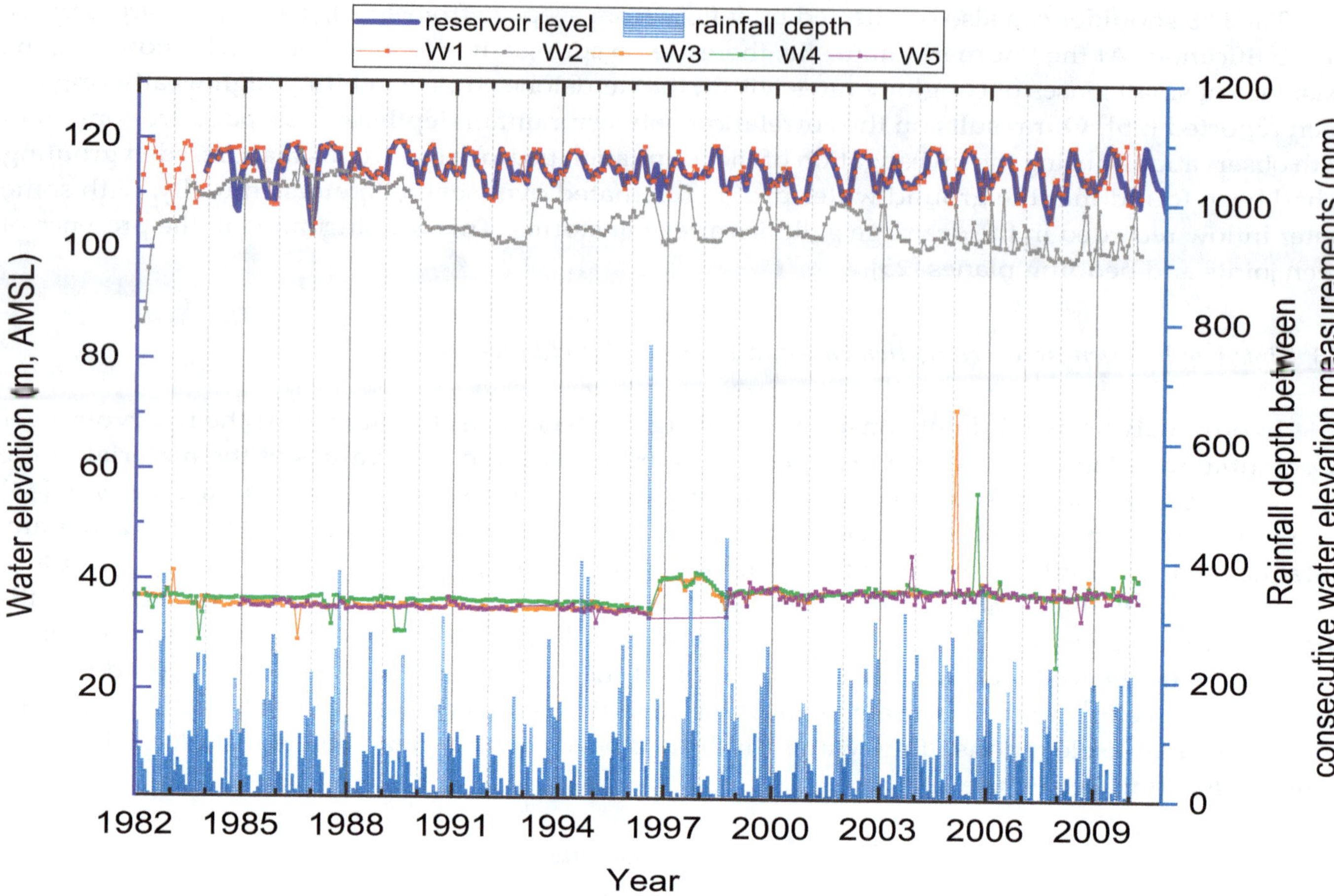

Figure 13. Variation of water level elevation along cross-section A-A. Changes of the reservoir level appear immediately at the D/S shoulder, the core (at centre axis) appears to respond with a varying time lag. The phreatic surface in the D/S shoulder is unaffected by the reservoir-level variations in the reservoir.

Ref. [46] along with [47] suggest a methodology for the analysis of unevenly spaced time series in the frequency domain. This could have been an alternative approach in the case of the Pournari I dam, for a constant time lag between W1 and W2. Since the time lag varies over the examined time period, the different time lags could only be estimated graphically: We manually shifted the W2 time series over the W1 time series. By doing so, we were able to identify three time periods with different time lag characteristics, as described below. For clarity purposes and to help with the following discussion, we present again the evolution of the hydraulic head in W1 and W2 in Figure 14. In this figure, W1 record is used to compare W2 with the reservoir level fluctuations, as both W1 and W2 records share the exact same sampling times. We documented the almost perfect match between the reservoir-level fluctuations and the hydraulic head in W1 (see also Figure 13) earlier; therefore, the use of W1 instead of the reservoir level record is valid.

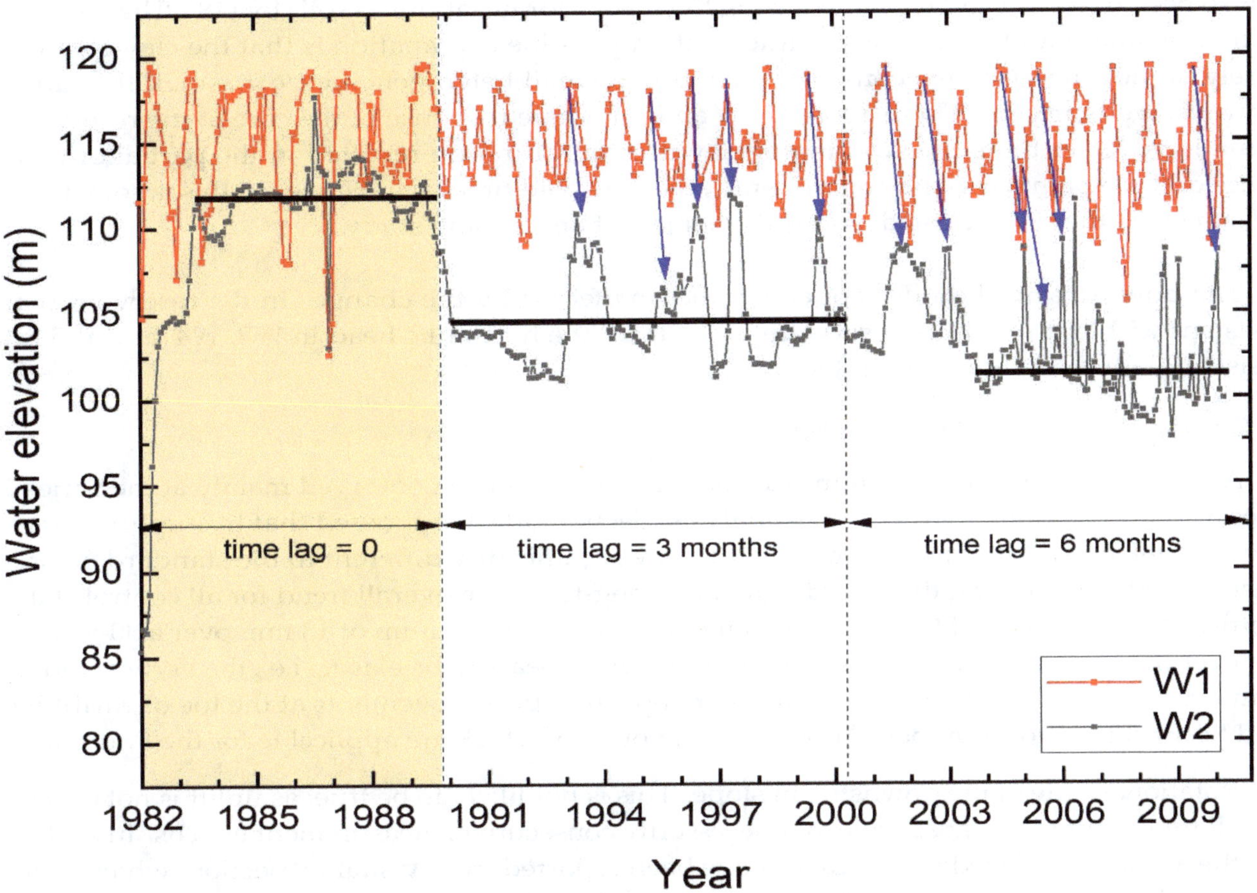

Figure 14. Hydraulic head in W1 and W2 between 1982 and 2010. Horizontal thick black lines represent the mean value of the hydraulic head in W2 during periods 1983–1989, 1990–2001, and 2003–2010. The dashed lines indicate the approximate time periods when the time lag between W1 and W2 changes. The blue arrows are connecting corresponding peaks between W1 and W2. These peaks, when aligned, result in maximum correlation between sections of data. The time difference between corresponding peaks is equal to the time lag between W1 and W2 for that section of data.

- 1982 to December 1989: During this time period, the hydraulic head in the core (W2 location in Figure 12) does not appear to have a time lag with the reservoir-level fluctuations despite the fact that it does not exactly follow the fluctuation pattern. More specifically, both W1 and W2 appear to have a drop in the hydraulic head in January 1987, with the last simultaneous drop in July 1989 (first dashed line from the left in Figure 14). The shape of the W2 curve between 1982 and late 1989 (highlighted area in Figure 14) is characteristic for the period during which the clay core gets saturated: the hydraulic head increases, and then reaches a plateau, at which point the core is regarded as saturated. The latter has been observed in early studies on unsaturated clays, e.g., [39].
- Early 1990 to mid 2000: During this period, the mean value (104.46 m) of the hydraulic head (second horizontal black line in Figure 14) is decreased compared to the previous time period (111.83 m). The observed peaks in the W2 values appear to have a time lag with the corresponding peaks in W1 (corresponding peaks shown with blue arrows in Figure 14). This time lag is 3 months.
- Mid 2000 to end of record in 2010: During this time period, the mean of the hydraulic head values has dropped further (now at 101.84 m). The observed time lag has increased to 6 months.

The decreasing mean value of the hydraulic head over time has been reported in the literature (see [39]), and is expected. A time lag of a few days up to a few months, consistent with the time lag estimated for the Pournari I dam, between reservoir-level fluctuations and the downstream shoulder has also been observed and explained using the theory for unsaturated soils by [48]. The varying time lag, on the other hand, is not easy to interpret. A possible explanation is that the clay core was not entirely homogeneous when constructed, and that small heterogeneities, e.g., small fractures/fluid pathways, were present. These allowed for an immediate response of the core to the reservoir-level fluctuations. Over time, the swelling of clay during saturation resulted in the partial or complete closure of these pathways, and consequently, the observed time lags. However, this is something that needs to be further investigated, and is the subject of an ongoing study.

The downstream shoulder appears to be unaffected by the changes in the reservoir, which is consistent with our previous observations. Instead, the hydraulic head in W3, W4, and W5 is more consistent with the rainfall depth (Figure 13)

4.4. Observed Uplift at the Toe of the Dam

A few millimetres (up to 13 mm, see Table 1) uplift has been observed mainly at the Series 2 and Series 3 control stations. Earlier in the analysis (Section 3.1), we showed that this uplift is not likely to be an artefact, and that the recorded values are significantly different to the standard error of the measurements. Examining the recordings in the long-term, the overall trend for all control stations is to settle. The magnitude of the uplift movement is small: a maximum of 13 mm over a 30-year period. Whatever the causing effect, the response of the dam appears to be elastic, i.e., the vertical movement became a settlement by the next measurement epoch. Uplift movements at the toe of earthfill dams could be attributed to a number of possibilities, not all of which are applicable for the Pournari I dam:

1. Rotational slide of the downstream slope. This is not likely to be true, as uplift is not observed at all times, but rather at specific, not necessarily consecutive, measurement epochs. In addition, to the authors' knowledge, no cracks have been reported from visual inspection, which is another sign that indicates a potential slide.
2. Effect of groundwater. This could be a likely factor; however, the absence of relevant data, e.g., pore pressures at the toe of the dam, do not allow any analysis to be made regarding its effect to the observed vertical movements.
3. Swelling rocks beneath the dam. Swelling of rocks resulting in deformations has been reported for tunnels. The main contributing factor is the presence of clays or clayey rocks [49]. Ref. [25] mentions the presence of some clay material at the dam foundation level. This could be a possible cause of the observed uplifts, although the fact that this uplift is reversed by the next measurement epoch makes this scenario unlikely.
4. Internal erosion that could result in uplift at the toe of the dam. From the available data used in this study, there is no evidence to support this.

4.5. Time-Dependent Settlements

Figure 5a shows that the settlement index for the control points located at the dam crest has remained under the 0.02 threshold value since the settlement records begun. This indicates that the recorded settlements are due to the secondary consolidation/creep of the clay core [17]. At preliminary stages of the analysis presented in this paper, we attempted to remove the time-dependent component of the recorded settlements, aiming to continue the analysis with the residuals only. Initially, we attempted to fit an exponential model to the time series of settlements or even a linear model to the logarithmic time series of the settlements without success. As a last resort, we fitted, based on least-squares, a third-degree polynomial which resulted in satisfactory fits (correlation coefficient >0.93 for the vast majority of control points), although without a physical meaning for the fitted parameters [50]. However, the derived residuals were significantly affected by numerical effects; hence, any further analysis based on these values could result in biased outcomes.

4.6. Future Research Directions

A limitation of the available monitoring record is that settlement measurements only took place once per year. Annual epoch measurements for surface control stations have been a common long-term monitoring strategy for dams that do not show any evidence that their structural integrity is at risk, as is the case of the Pournari I dam. A higher sampling rate for the dam settlements, at least monthly, is required for more detailed studies. This can be an expensive exercise for the dam owner, but with new automatic monitoring technology that is now routinely used for dam monitoring (e.g., [11]). This should not be an onerous task, and in the long-term is cost-effective. The geotechnical instrumentation in old dams can be easily damaged within the first 5–10 years. Hence, geodetic measurements together with geophysical monitoring [7] could be, in many cases, the only source of information for continuous monitoring. Remote sensing (INSAR) can also be used; however, it is more expensive, the computational expense is high, and it requires photogrammetric expertise. In addition, nowadays, the accuracy of INSAR images is within a few mm. This could be sufficient for concrete dams, but is relatively low for the purposes of examining the response of earthfill dams to environmental factors.

Due to the limitations posed by the annual sampling of the settlements, and in the absence of any geotechnical observations, a causative relationship could not be quantified between the settlements and the reservoir level and rainfall. However, backward (data-driven) numerical modelling could assist interpret the geodetic observations further, and is the focus of an ongoing study.

5. Conclusions

Despite the limitations of the low sampling rate of the recorded settlements (one per year), our results provide evidence of the role of the reservoir-level fluctuations and the rainfall on an earthfill dam over 31 years of operation after its first filling. For the case of the Pournari I dam, the reservoir level seems to affect the recorded settlements on the crest up to a degree. A more interesting outcome is the increasing trend in the time lag between reservoir-level changes and the response of the dam core. This is the first time that observed changes in the long-term response of a central clay core dam to reservoir-level fluctuations, based on field data, are reported in the international literature. This time lag appears later in the life of the dam, after 8 years since construction, and has the tendency to increase with time. By 2010, this time lag was 6 months. This can be an indication of homogenisation of the dam core, i.e., existing small fractures that could have been present after construction have closed following saturation and swelling of the clay core. Such a behaviour would be difficult to predict solely based on numerical modelling, but the observations of this study could be used to inform future constitutive models for the engineering design of earthfill dams. The downstream shoulder is mainly affected by rainfall via effects on seepage. No evidence was found in the available data of the dam being affected by seismicity at the magnitudes (up to M_L5.3) recorded within 50 km from the dam during the examined time period.

Overall, based on the available data and from the current analysis, without any numerical modelling, the Pournari I dam appears to respond satisfactorily to environmental changes. More frequent monitoring on dams, especially as they age and in view of the changing climatic conditions would allow for detailed studies of their long-term behaviour based on field evidence.

Author Contributions: All authors contributed to various aspects of this paper: conceptualisation, S.P.; methodology, S.P.; formal analysis, P.M. and S.P.; resources, P.M. and S.R.; data curation, P.M.; writing—original draft preparation, P.M. and S.P.; writing—final draft preparation, S.P.; visualisation, S.P. and P.M.; proof-reading, all.

Acknowledgments: The authors would like to thank the Public Power Corporation (P.P.C. S.A.) of Greece for providing access to the data used on this study, and the management and personnel of the Hydroelectric Station of the Pournari I dam. Stathis Stiros, Alessandro Tarantino, and Alessia Amabile are thanked for useful discussions on the interpretation of results. This paper has benefitted by the comments and suggestions of three anonymous reviewers.

References

1. Michalis, P.; Konstantinidis, F.; Valyrakis, M. The road towards Civil Infrastructure 4.0 for proactive asset management of critical infrastructure systems. In Proceedings of the 2nd International Conference on Natural Hazards & Infrastructure (ICONHIC), Chania, Greece, 23–26 June 2019.
2. American Society of Civil Engineers 2017 Infrastructure Report Card. Available online: https://www.infrastructurereportcard.org/cat-item/dam (accessed on 31 July 2019).
3. Rojas, R.; Feyen, L.; Bianchi, A.; Dosio, A. Assessment of future flood hazard in Europe using a large ensemble of bias corrected regional climate simulations. *J. Geophys. Res. Atmos.* **2012**, *117*, D17109.
4. Forzieri, G.; Bianchi, A.; e Silva, F.B.; Herrera, M.A.M.; Leblois, A.; Lavalle, C.; Aerts, J.C.J.H.; Feyen, L. Escalating impacts of climate extremes on critical infrastructures in Europe. *Glob. Environ. Chang.* **2018**, *48*, 97–107. [CrossRef]
5. Jongman, B.; Hochrainer-Stigler, S.; Feyen, L.; Aerts, J.C.J.H.; Mechler, R.; Botzen, W.J.W.; Bouwer, L.M.; Pflug, G.; Rojas, R.; Ward, P.J. Increasing stress on disaster risk finance due to large floods. *Nat. Clim. Chang.* **2014**, *4*, 264–268. [CrossRef]
6. Preziosi, M.-C.; Micic, T. An adaptive methodology for risk classification of small homogeneous earthfill embankment dams integrating climate change projections. *Civ. Eng. Environ. Syst.* **2014**, *31*, 111–124. [CrossRef]
7. Michalis, P.; Sentenac, P.; Macbrayne, D. Geophysical assessment of dam infrastructure: The mugdock reservoir dam case study. In Proceedings of the 3rd Joint International Symposium on Deformation Monitoring (JISDM), Vienna, Austria, 30 March–1 April 2016.
8. Michalis, P.; Pytharouli, S.; Raftopoulos, S. Long-term deformation patterns of earth-fill dams based on geodetic monitoring data: The Pournari I Dam case study. In Proceedings of the 3rd Joint International Symposium on Deformation Monitoring (JISDM), Vienna, Austria, 30 March–1 April 2016.
9. He, X.Y.; Wang, Z.Y.; Huang, J.C. Temporal and spatial distribution of dam failure events in China. *Int. J. Sediment Res.* **2008**, *23*, 398–405. [CrossRef]
10. Yigit, C.O.; Alcay, S.; Ceylan, A. Displacement response of a concrete arch dam to seasonal temperature fluctuations and reservoir level rise during the first filling period: Evidence from geodetic data. *Geomat. Nat. Hazards Risk* **2016**, *7*, 1489–1505. [CrossRef]
11. Boudon, R.; Blin, S.; Pons, E.; Ajzenberg, A. Automatic follow-up of the tri-directional displacements of the Sainte-Croix arch dam (Verdon-France) by motorized total station. In Proceedings of the 4th Joint International Symposium on Deformation Monitoring (JISDM), Athens, Greece, 15–17 May 2019.
12. Bonelli, S.; Royet, P. Delayed response analysis of dam monitoring data. In Proceedings of the ICOLD European Symposium on Dams in a European Context, Geiranger, Norway, 25–27 June 2001.
13. Gamse, S.; Oberguggenberger, M. Assessment of long-term coordinate time series using hydrostatic-season-time model for rock-fill embankment dam. *Struct. Control Health Monit.* **2017**, *24*, e1859. [CrossRef]
14. Gamse, S.; Zhou, W.-H. Adaptive parametric identification in dam monitoring by Kalman filtering. In Proceedings of the 4th Joint International Symposium on Deformation Monitoring (JISDM), Athens, Greece, 15–17 May 2019.
15. Gikas, V.; Sakellariou, M. Settlement analysis of the Mornos earth dam (Greece): Evidence from numerical modeling and geodetic monitoring. *Eng. Struct.* **2008**, *30*, 3074–3081. [CrossRef]
16. Baecher, G.B.; Christian, J.T. The practice of risk analysis and the safety of dams. In Proceedings of the Conference of Geotechnical Engineering, Cairo, Egypt, January 2000.
17. Tedd, P.; Charles, J.A.; Holton, I.R.; Robertshaw, A.C. The effect of reservoir drawdown and long-term consolidation on the deformation of old embankment dams. *Geotechnique* **1997**, *47*, 33–48. [CrossRef]
18. Guler, G.; Kilic, H.; Hosbas, G.; Ozaydin, K. Evaluation of the movements of the dam embankments by means of geodetic and geotechnical methods. *J. Surv. Eng.* **2006**, *132*, 31–39. [CrossRef]
19. Pytharouli, S.I.; Stiros, S.C. Investigation of the parameters controlling crest settlement of a major earthfill dam based on the threshold correlation analysis. *J. Appl. Geod.* **2009**, *3*, 55–62. [CrossRef]
20. Pytharouli, S. Study of the Long-Term Behaviour of Kremasta Dam Based on the Analysis of Geodetic Data and Reservoir Level Fluctuations. Ph.D. Thesis, University of Patras, Patras, Greek, August 2007. (In Greek).

21. Charles, J.A. The significance of problems and remedial works at British earth dams. In Proceedings of the BNCOLD-IWES Conference on Reservoirs 1986, Edinburgh, UK, 3–6 September 1986; pp. 123–141.
22. Dascal, O. Post-construction deformation of rockfill dams. *J. Geotech. Eng.* **1987**, *113*, 46–59. [CrossRef]
23. Pytharouli, S.I.; Stiros, S.C. Estimation of a safety threshold for the crest settlements of embankment dams. In Proceedings of the 2nd National ICOLD Conference on Dams and Reservoirs, Athens, Greece, 7 November 2013.
24. Public Power Corporation (P.P.C.) S.A. Brochure: Hydroelectric project Pournari. Available online: http://users.itia.ntua.gr/nikos/arx_int/CDfrag/scanarismena/POURNARI/Untitled.pdf (accessed on 1 August 2019).
25. Papageorgiou, S.A. Engineering Geology of Dam Foundations in North-Western Greece. Ph.D. Thesis, Durham University, Durham, UK, 1983.
26. Pavlou, K.; Drakatos, G.; Kouskouna, V.; Makropoulos, K.; Kranis, H. Seismicity study in Pournari reservoir area (W. Greece) 1981–2010. *J. Seismol.* **2016**, *20*, 701–710. [CrossRef]
27. Pytharouli, S.; Stiros, S. Reservoir level fluctuations and deformation of Ladhon Dam. *Int. J. Hydropower Dams* **2004**, *11*, 82–84.
28. Bannister, A.; Raymond, S.; Baker, R. *Surveying*, 7th ed.; Pearson Education Limited: London, UK, 1998; 512p.
29. Bommer, J.J.; Crowley, H. The purpose and definition of the minimum magnitude limit in PSHA calculations. *Seismol. Res. Lett.* **2017**, *88*, 1097–1106. [CrossRef]
30. Harder, L.F.; Kelson, K.I.; Kishida, T.; Kayen, R. Preliminary Observations of the Fujinuma Dam Failure Following the March 11, 2011 Tohoku Offshore Earthquake, Japan. Geotechnical Extreme Events Reconnaissance Report No. GEER-25e, 2011. Available online: http://www.geerassociation.org (accessed on 18 September 2019).
31. Nespoli, M.; Todesco, M.; Serpelloni, E.; Belardinelli, M.E.; Bonafede, M.; Marcaccio, M.; Rinaldi, A.P.; Anderlini, L.; Gualandi, A. Modeling earthquake effects ongroundwater levels: Evidences from the 2012 Emilia earthquake (Italy). *Geofluids* **2016**, *16*, 452–463. [CrossRef]
32. United States Society on Dams. Observed Performance of Dams During Earthquakes, Volume III. 2014. Available online: http://www.ussdams.org/wp-content/uploads/2016/05/EQPerfo2_v3.pdf (accessed on 18 September 2019).
33. Swaisgood, J.R. Embankment Dam Deformations Caused by Earthquakes. In Proceedings of the 7 Pacific Conference on Earthquake Engineering, Christchurch, New Zealand, 13–15 February 2003; Available online: https://www.nzsee.org.nz/db/2003/View/Paper014s.pdf (accessed on 1 October 2019).
34. Fiorentino, G.; Furgani, L.; Nuti, C.; Sabetta, F. Probabilistic evaluation of dams base sliding. In Proceedings of the COMPDYN 2015 5th ECCOMAS Thematic Conference on Computational Methods in Structural Dynamics and Earthquake Engineering, Crete Island, Greece, 25–27 May 2015; Papadrakakis, M., Papadopoulos, V., Plevris, V., Eds.; National Technical University of Athens: Athens, Greece, 2015.
35. Lawton, F.L.; Lester, M.D. Settlement of rockfill dams. In Proceedings of the 8th International Congress on Large Dams, Edinburgh, UK, 4–8 May 1964; Volume III, pp. 599–613.
36. Sowers, G.F.; Williams, R.C.; Wallace, T.S. Compressibility of Brocken rock and the settlement of rockfills. In Proceedings of the 6th International Conference on Soil Mechanics and Foundation Engineering, Toronto, ON, Canada, 8–15 September 1965; Volume 2, pp. 561–565.
37. Fell, R.; McGregor, P.; Stapledon, D.; Graeme, B. *Geotechnical Engineering of Dams*; Taylor & Francis: London, UK, 2005; p. 900.
38. Twort, C.A.; Ratnayaka, D.D.; Malcolm, J.B. 5-Dams, impounding reservoirs and river intakes. In *Water Supply*, 5th ed.; Twort, A.C., Ratnayaka, D.D., Brandt, M.J., Eds.; Eliane Wigzell: London, UK, 2000; pp. 152–195.
39. Ventrella, C.; Pelekanos, L.; Skarlatos, D.; Pantazis, G. Finite element analysis of earth dam settlements due to seasonal reservoir level changes. In Proceedings of the XVII ECSMGE-2019 Geotechnical Engineering Foundation of the Future, Reykjavik, Iceland, 1–7 September 2019; ISBN 978-9935-9436-1-3.
40. Tedd, P.; Charles, J.A.; Holton, J.R.; Robertshaw, A.C. Deformations of embankment dams due to changes in reservoir level. In Proceedings of the 13th International Conference on Soil Mechanics and Foundations Engineering, New Delhi, India, 5–10 January 1994; pp. 951–954.
41. British Dams, Climate Change Impacts on the Safety of British Reservoirs. Available online: https://britishdams.org/assets/documents/defra-reports/200201Climate%20change%20impacts%20on%20the%20safety%20of%20British%20reservoirs.pdf (accessed on 1 August 2019).

42. Alonso, E.E.; Olivella, S.; Pinyol, N.M. A review of Beliche dam. *Geotechnique* **2005**, *55*, 267–285. [CrossRef]
43. Almog, E.; Kelham, P.; King, R. Delivering benefits through evidence: Models of dam failure and monitoring and measuring techniques. In *Flood and Coastal Erosion Risk Management Programme*; Project: SC080048/R1; Environment Agency: Bristol, UK, 2011.
44. Chang, C.S.; Duncan, J.M. Analysis of Consolidation of Earth and Rockfill Dams, Volume I, Main Text and Appendices, A. and B.; U.S. Army Engineer Waterways Experiment Station, Contract Report S-77-4, 1977. Available online: https://apps.dtic.mil/dtic/tr/fulltext/u2/a045332.pdf (accessed on 2 October 2019).
45. Davis, J.C. *Statistics and Data Analysis in Geology*, 3rd ed.; John Wiley & Sons: Hoboken, NJ, USA, 2002; p. 656.
46. Schultz, M.; Stattegger, K. SPECTRUM: Spectral analysis of unevenly spaced paleoclimatic time series. *Comput. Geosci.* **1997**, *23*, 929–945. [CrossRef]
47. Olafsdottir, K.B.; Schulz, M.; Mudelsee, M. REDFIT-X: Cross-spectral analysis of unevenly spaced paleoclimate time series. *Comput. Geosci.* **2016**, *91*, 11–18. [CrossRef]
48. Bonelli, S.; Radzicki, K. Impulse response function analysis of pore pressures in earthdams. *Eur. J. Environ. Civ. Eng.* **2008**, *12*, 243–262. [CrossRef]
49. Kontogianni, V.; Tzortzis, A.; Stiros, S.C. Deformation and failure of the Tymfristos Tunnel, Greece. *J. Geotech. Geoenvironmental Eng.* **2004**, *130*, 1004–1013. [CrossRef]
50. Michalis, P. Pournari I dam (Greece): Analysis of Its Structural Integrity Based on Long-Term Monitoring Data. Master's Thesis, University of Strathclyde, Glasgow, UK, August 2010.

9

Hydro-Thermo-Mechanical Analysis of an Existing Gravity Dam Undergoing Alkali–Silica Reaction

Martina Colombo and Claudia Comi *

Department of Civil and Environmental Engineering, Politecnico di Milano, Piazza Leonardo da Vinci 32, 20133 Milan, Italy

* Correspondence: claudia.comi@polimi.it

Abstract: The alkali–silica reaction is a chemical phenomenon that, by inducing expansion and the formation of cracks in concrete, can have a severe impact on the safety and functioning of existing concrete dams. Starting from a phenomenological two-phase isotropic damage model describing the degradation of concrete, the effects of alkali-silica reaction in an existing concrete gravity dam are evaluated and compared with real monitoring data. Considering the real temperature and humidity variations, the influence of both temperature and humidity are considered through two uncoupled diffusion analyses: a heat diffusion analysis and a moisture diffusion analysis. The numerical analyses performed with the two-phase damage model allow for prediction of the structural behaviour, both in terms of reaction extent and increase of crest displacements. The crest displacements are compared with the real monitoring data, where reasonably good agreement is obtained.

Keywords: alkali-silica reaction; damage; existing concrete dam; finite element analysis; temperature; saturation degree

---

1. Introduction

The alkali–silica reaction (ASR) is a chemical phenomenon that may occur over time in concrete between the alkaline cement paste and the reactive silica present in some kinds of aggregates. This reaction causes expansion of the altered aggregate by the formation of a viscous gel, which increases in volume when absorbing water. This gel exerts an expansive pressure inside the siliceous aggregate and/or at the interface between aggregates and the cement paste, causes spalling and loss of stiffness of the concrete and, finally, leads to the formation of cracks [1]. While ASR can be avoided in new concrete structures by the selection of non-reactive aggregates and/or by the chemical composition of cement (which determines the alkali ion concentration in the pore solution [2]) and additions [3–5], ASR can lead to serious cracking in concrete structures built several decades ago in very wet environments (e.g., dams), resulting in critical structural problems. It takes many years for the reaction to develop and the symptoms to become visible, in terms of an increase in displacements and of a pattern of diffuse localized micro- and macro-cracks [6]. Several examples of important concrete structures, such as dams or shielding in nuclear plants, subject to ASR have been reported in the literature (see, e.g., [7–9]).

The expansion of concrete as a result of the reaction between cement and aggregates was first studied in [1]. Since then, much research effort has been devoted to describing the reactions occurring between aggregates and cement paste, in order to experimentally characterize the consequent expansion (as in [10]) and to develop material models to describe the mechanical effects at different scales (see, e.g., [11,12]) but, very often, there is a lack of validation of the proposed models at a structural level, as has been pointed out in [13]. Significant examples of analyses conducted on dams affected by ASR can be found in [14–17].

The purpose of this paper is to further validate the bi-phase damage model proposed in [18] by simulating the experimental campaign reported in [19] and to present a case study concerning the analysis, using that model, of a real concrete dam affected by ASR. As the kinetics of ASR strongly depend on the evolution of the temperature and humidity, as described in [20,21], we use real monitoring data of temperature variations and air humidity to compute the temperature and degree of saturation fields through preliminary heat and moisture diffusion analyses. Then, following a weakly coupled strategy, we perform an elasto-damage analysis, which allows for computation of the displacement evolution and the damage patterns.

2. Bi-Phase Damage Model

In this section, a brief summary of the chemo-damage model is given (for details, see [18]). At the meso-scale, concrete affected by ASR is described as a two-phase heterogeneous material constituted by a solid skeleton s and a wet gel g produced by ASR. The total volume V is, hence, divided into V_s, the solid volume, and V_g, the volume occupied by the dry gel dg, liquid water w, and the vapor and air phase va (see Figure 1).

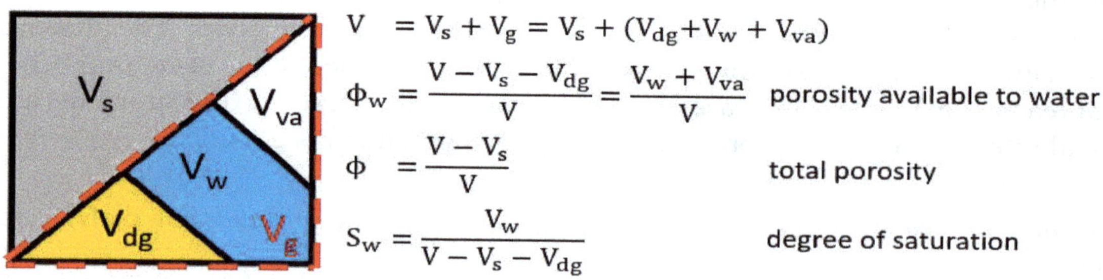

Figure 1. Schematic representation of the assumed bi-phase model.

Assuming small strains and quasi-static conditions, the compatibility and equilibrium equations for the bi-phase solid are

$$\boldsymbol{\varepsilon} = \frac{1}{2}\left(grad\mathbf{u} + grad^T\mathbf{u}\right), \tag{1}$$

$$div\boldsymbol{\sigma} + \rho\mathbf{b} = \mathbf{0}, \tag{2}$$

where $\boldsymbol{\varepsilon}$ is the small strain tensor of the skeleton; $\mathbf{u}$ is the skeleton displacement; $\boldsymbol{\sigma}$ is the Cauchy stress; $\rho = \rho_s(1-\phi) + \rho_g\phi$ is the mass density (ρ_s and ρ_g being the intrinsic solid and fluid mass densities, respectively); and $\rho\,\mathbf{b}$ is the body force of the combination of solid and fluid. In this model, the variable which accounts for the gel formation and expansion (considered simultaneously) is the volumetric fraction of the wet gel,

$$\zeta_g = \frac{V_g}{V}. \tag{3}$$

To take into account the influence of humidity and temperature, the degree of saturation $S_w = V_w/(V - V_s - V_{dg})$ and the temperature T are included as additional state variables and proper conservation and transport laws are necessary. In particular, by combining the mass conservation law for liquid water and the transport law for moisture in its liquid form, as proposed in [22], and the relation between the water pressure and the degree of saturation given by the experimental capillarity curve (see, e.g., [23]), one obtains the non-linear transport law for moisture (in its liquid form) as

$$div(D_w\, gradS_w) + \phi_w \frac{\partial S_w}{\partial t} = 0, \tag{4}$$

where D_w is the permeability of concrete, dependent on the degree of saturation S_w

$$D_w(S_w) = m_1\, m_2 \left(1 - \frac{1}{m_2}\right) k \frac{[1-(1-S_w^{1/m_3})^{m_3}]^2}{[\eta_w\, S_w^{(1/2+m_2)}(S_w^{-m_2}-1)^{1/m_2}]}, \tag{5}$$

where m_1, m_2, and m_3 are material parameters; k is the intrinsic permeability; and η_w is the dynamic viscosity.

Regarding the temperature, the heat conservation law, combined with linear and isotropic Fourier conduction laws, gives the following heat transport law

$$C\,\frac{\partial T}{\partial t} = div(D_T\, gradT), \tag{6}$$

where C is the volumetric heat capacity of concrete and D_T is the isotropic heat conductivity coefficient.

In a bi-phase model, the stress $\boldsymbol{\sigma}$ is the sum of the effective stress $\boldsymbol{\sigma}'$ (acting on the skeleton) and of the stress -$bp\mathbf{1}$ (acting on the fluid phase), where b is Biot's coefficient and p is the fluid pressure (see [24]). In the bi-phase model proposed in [18,25], the fluid phase is constituted of the wet gel and the material behaviour is assumed to be elastic with isotropic damage. The state equations, hence, read

$$\boldsymbol{\sigma} = (1-D)\,[2G\,\mathbf{e} + K\,(tr\varepsilon - \alpha\,\theta)\mathbf{1}] - b\,p\,\mathbf{1}, \tag{7}$$

$$p = -(1-D)\,M\,(b\,tr\varepsilon - \zeta_g - \alpha_g\,\theta), \tag{8}$$

where G and K are the shear and bulk moduli of the homogenized concrete skeleton, respectively; $\mathbf{e}$ is the deviatoric strain tensor; M is the Biot modulus; α and α_g are the volumetric coefficients of thermal expansion of the skeleton and of the gel, respectively; $\theta = T - T_0$ is the temperature variation, with respect to a reference temperature T_0; and D is the isotropic damage variable.

The constitutive model is completed by the definition of the evolution law for the gel volume content, assumed to be proportional to the reaction extent ξ, a phenomenological internal variable varying from 0 to 1,

$$\zeta_g = \frac{K + M\,b^2}{M\,b}\,\varepsilon^{\infty}_{ASR}\,\xi, \tag{9}$$

where ϵ^{∞}_{ASR} is the free asymptotic volumetric expansion due to ASR in the isothermal fully saturated case ($S_w = 1$).

The following form for the reaction rate is considered:

$$\dot{\xi} = \frac{\langle f_{Sw} - \xi \rangle}{\tilde{t}}, \tag{10}$$

with

$$f_{Sw} = \frac{1 + b_1\, exp(-b_2)}{1 + b_1\, exp(-b_2\, S_w)}, \tag{11}$$

where b_1 and b_2 are material parameters, which are calibrated with experimental data.

In the present work, the following law for the intrinsic time $\tilde{t}$ is assumed, as in [18]

$$\frac{1}{\tilde{t}} = \frac{\frac{\xi}{f_{Sw}} + exp(-\frac{\tau_{lat}}{\tau_{ch}})}{\tau_{ch}\,(1 + exp(-\frac{\tau_{lat}}{\tau_{ch}}))}, \tag{12}$$

where τ_{lat} and τ_{ch} are defined as (with $i = lat, ch$)

$$\tau_i(T, S_w) = \left[\tau_i(\bar{T},1) + \frac{\tau_i(\bar{T},0) - \tau_i(\bar{T},1)}{1 + c_{1i} exp\left[-\frac{c_{2i}(1-S_w)}{S_w(1-S_w)}\right]}\right] exp\left[U_i\left(\frac{1}{T} - \frac{1}{\bar{T}}\right)\right], \tag{13}$$

where U_{ch} and U_{lat} [K] denote the Arrhenius activation energies, representing the energy that must be reached for a chemical reaction to occur.

For simplicity, and in view of the specific application of this paper, we only consider the activation of damage for prevaling tensile stress state, instead of the more general bi-dissipative damage model [26]. The evolution of the tensile damage variable D is governed by the loading–unloading conditions, defined in terms of σ and p through the inelastic effective stress σ'' (with $0 < \beta \leq b$):

$$\sigma'' = \sigma + \beta\, p\, \mathbf{1}, \tag{14}$$

$$f = \frac{1}{2}\mathbf{s} : \mathbf{s} - a_t\,(tr\sigma'')^2 + b_t\,(tr\sigma'')\,h - k_t\,h^2, \tag{15}$$

$$f \leq 0 \quad \dot{D} \geq 0 \quad f\,\dot{D} = 0, \tag{16}$$

where f is the damage activation function; $\mathbf{s}$ is the deviatoric stress; a_t, b_t, and k_t are material parameters governing the shape and dimensions of the elastic domain (see [26] for details); and h is the softening function, defined as

$$h(D) = [1 - (D)^c]^{0.75}. \tag{17}$$

The parameter c in Equation (17) tunes the specific fracture energy of the model.

3. Model Validation

The model used in this work, already employed in [18,25] to simulate the experimental campaigns in [21,27], is here further validated by simulating the recent experimental campaign of [19,28]. In these works, the expansion of cubic specimens (of 254 mm side) of concrete affected by ASR and subject to multi-axial stresses was measured. The post-tensioning method of pre-stressing was chosen as the technique to apply and sustain the desired compressive stresses in the concrete specimens. The stress application in one direction required four high-strength bolts to be inserted into four rigid sheaths embedded in the concrete during casting. Figure 2a shows the experimental configuration for a three-axial stress-state application. Each bolt was stretched to a desired stress level by producing a reaction against a pair of bearing plates placed symmetrically on the opposite surfaces of a concrete specimen. Different dimensions of bolts allow for obtaining the average nominal compression stress in the concrete, in a range of 0–9.6 MPa.

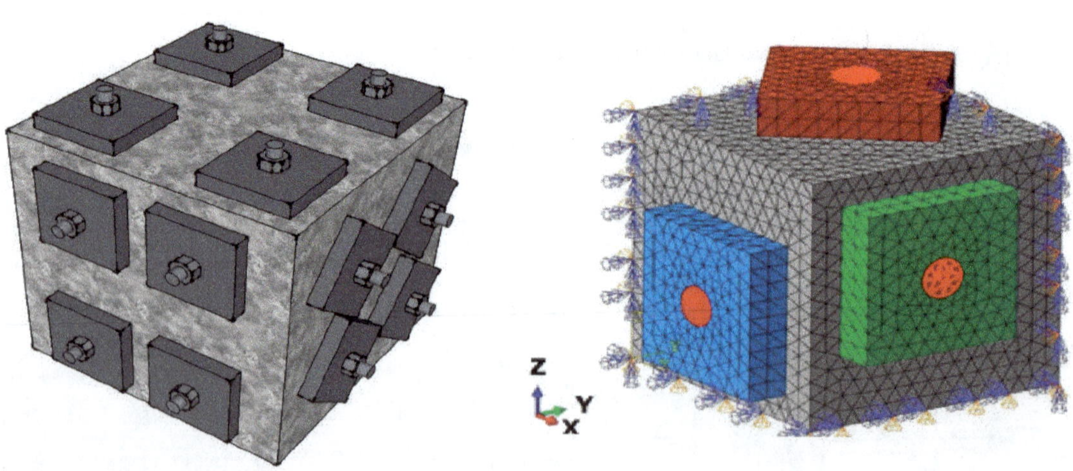

(a) (b)

Figure 2. (**a**) Experimental configuration of a concrete cubic specimen with plates and bolts in the three mutually perpendicular directions, adapt from [19]; and (**b**) adopted finite element (FE) mesh of one eighth of the specimen.

Four (three reactive and one non-reactive to serve as the control specimen) cubic specimens were cast for each of the different stress states. Expansion for reactive specimens due to ASR can be decoupled from the creep and shrinkage strains by taking the algebraic difference between the expansion of a reactive specimen and the corresponding expansion of the control specimen (non-reactive concrete) at an identical stress state. We will refer to these data as *ASR-only* results. The stress was applied for 52–56 days after casting at room temperature. After 180 days, the specimens were conditioned inside an acceleration chamber maintained at 50 °C and close to 100% relative humidity to promote and accelerate ASR. During these tests, the strains were monitored in the three directions as the average of the four measurements.

To simulate these experimental results, we assumed the mechanical parameters of concrete given in [19] and calibrated the material parameters governing the ASR evolution with the difference between the free expansion curves of reactive and non-reactive specimens (*ASR-only* data). Other material parameters were defined according to reference values, such as the porosity $\phi = 0.16$ (from which it is possible to evaluate Biot's coefficient $b = 0.41$) and Biot's modulus $M = 5000$ MPa. A comparison between experiments and the model result is shown in Figure 3, in terms of mean strain evolution. It should be noted that, in these tests, the temperature conditions changed while the degree of saturation was constant and equal to 1; therefore, according to Equation (13), the latency and characteristic times changed. The following values were identified at 23 °C: $\tau_{lat} = 478$ days and $\tau_{ch} = 274$ days while, at 50 °C, we have $\tau_{lat} = 34$ days and $\tau_{ch} = 60$ days. The identified asymptotic volumetric strain is $\varepsilon^{\infty}_{ASR} = 4.9 \times 10^{-3}$.

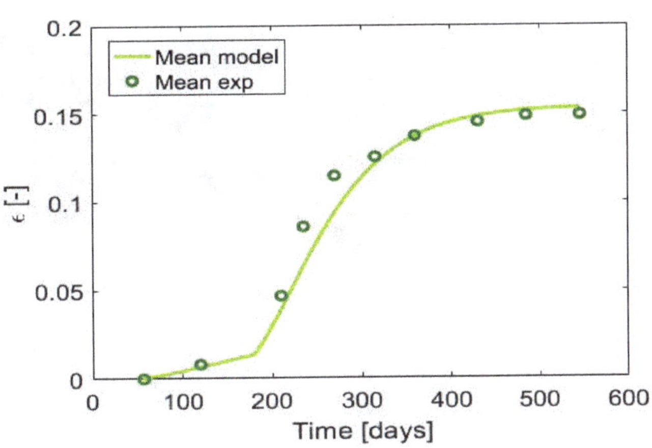

Figure 3. Mean strain evolution in the free expansion test: Comparison between experimental *ASR-only* data by Gautam, et al. [28] and model result.

In order to simulate the loaded tests in a realistic way, a finite element (FE) mechanical analysis was carried out. In fact, the stress state within the specimens induced by the pre-stressed bolts was non-homogeneous. In Figure 2b, the finite element model used for the simulation of the experimental tests is shown (as the problem is symmetric, only one eighth of the cubic specimen was modelled), in which the plates were modelled and the load was applied in the circular regions representing the bolts. With the above parameters calibrated with the free exansion test, we simulated the tests with uniaxial compression and biaxial compression. In Figure 4, comparisons between the experimental data and numerical results are represented in terms of strain evolutions in the three directions. Even though the material model was isotropic, the load-induced anisotropy of the expansion curves was well predicted.

The difference between the expansion in the X and Y directions in the bi-axial loaded case (Figure 4b) was due to the different locations of the loading plates on the two faces of the cube, which induced a stress state far from being equi-biaxial at each point. The numerical simulation was able to predict this complex stress state and the consequent non-uniform damage pattern (see Figure 5).

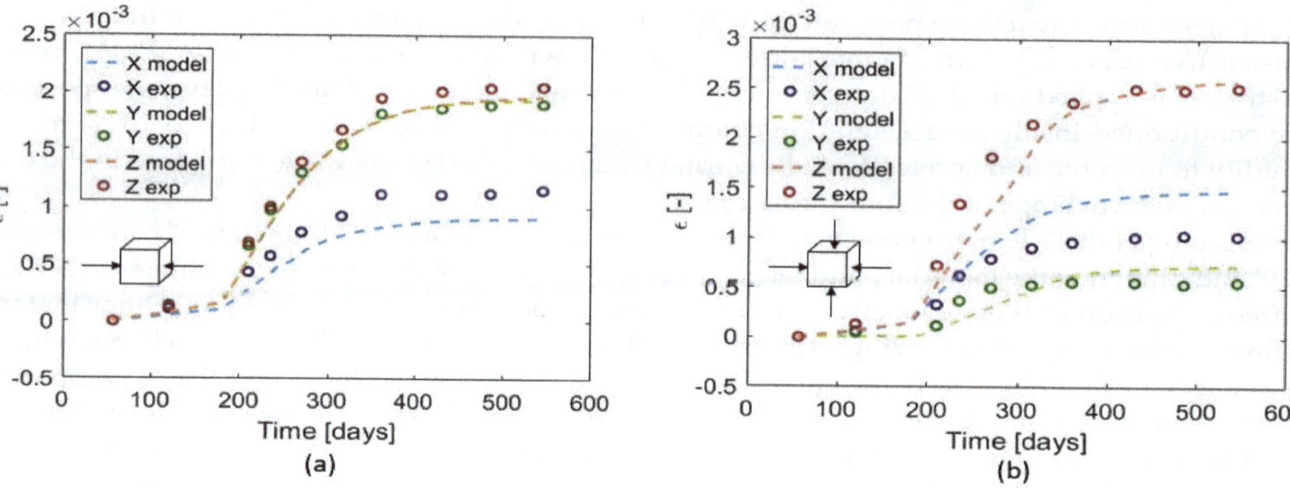

Figure 4. Comparison between experimental data of Gautam's campaign and model results obtained through 3D FE analysis, in terms of X, Y, and Z strains: (**a**) Uniaxial compression with $\sigma_x = -3.9$ MPa, and (**b**) biaxial case with $\sigma_x = \sigma_y = -3.9$ MPa.

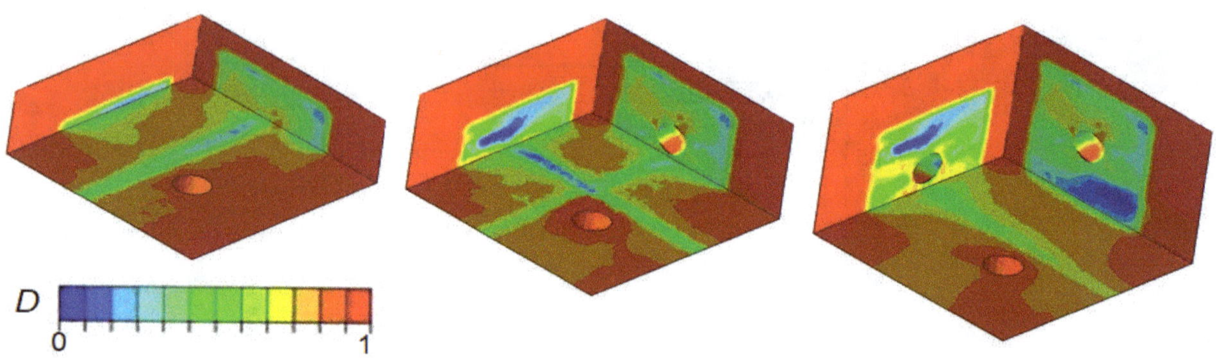

Figure 5. Contour plot of damage at the end of the biaxal expansion test in different sections of the specimen.

4. Results and Discussion. Structural Analysis of a Concrete Dam Affected by Alkali–Silica Reaction

4.1. Numerical Approach

The model presented above allows us to perform numerical simulations, using the finite element method, of the mechanical consequences of ASR on real concrete structures.

The approach we followed is weakly coupled. First, the histories of degree of saturation and temperature are computed by solving the moisture diffusion problem (Equation (4)) and the heat diffusion problem (Equation (6)) with corresponding boundary conditions, possibly varying in time. Then, the reaction extent evolution can be evaluated, through Equation (10). As this evolution depends on temperature and humidity, the development of the reaction is non-uniform inside the structure; in particular, the reaction is faster in warmer parts of the dam and expansion is higher in parts where the degree of saturation is higher.

The reaction extent, together with the external loads and temperature variations, is the input of the elasto-damage analysis, which allows us to compute the evolution of displacements and the damage patterns leading to the formation of cracks.

4.2. Case Study: Gravity Dam in Canada

Reference is made to the left wing of the gravity dam of the Beauharnois power plant (Québec, Canada), shown in Figure 6, courtesy of Hydro-Québec. The construction of the whole plant begun in 1932 and the gravity dam, considered here, was built between 1956 and 1960. It is 21 m high (h) and it is characterized by a base width of b = 15 m and a crest dimension of a = 3 m (Figure 6b). The mechanical properties of the concrete and foundation provided by Hydro-Québec are collected in Table 1. The fracture energy in tension typical of concrete characterized by a large aggregate size was also given: G_{ft} = 350 N/m.

Table 1. Mechanical and thermal parameters of concrete and rock for the Beauharnois dam.

Parameter	Value	Unit
$E_{concrete}$	26,000	MPa
$\nu_{concrete}$	0.21	-
$\rho_{concrete}$	2365	kg/m^3
f_{ct}	3	MPa
$\alpha_{concrete}$	1×10^{-5}	1/°C
E_{rock}	30,000	MPa
ν_{rock}	0.2	-
α_{rock}	1×10^{-5}	1/°C

Figure 6a shows the locations of the various instruments for monitoring. The monitoring system for the dam-foundation is represented by topographic measurements along dam crest through the reference targets shown by black dots in Figure 6a. Displacements in the three directions (X, Y, and Z) are measured at all reference targets. Due to ASR, an increase in the mean annual crest displacements has been observed since 1990.

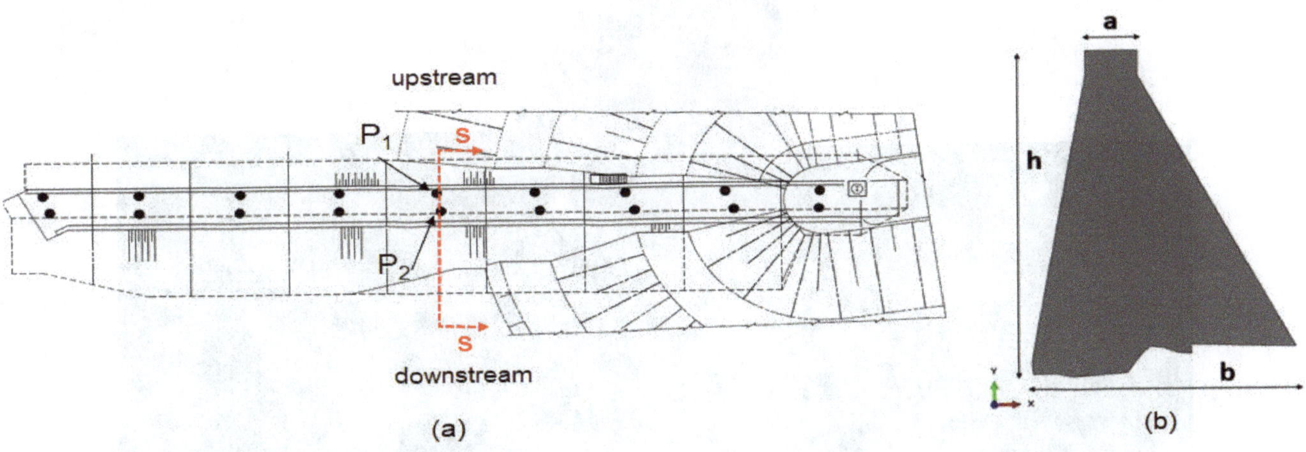

Figure 6. (**a**) Location of the monitoring instruments in the studied gravity dam and (**b**) the 2D section s–s considered for the analyses.

Furthermore, a set of thermometers determined the temperature of the concrete in contact with air. These values, recorded over six decades, have been interpolated by a periodic function with a period of one year. Figure 7a shows the annual profile thus obtained: $T(t) = \bar{T} + \Delta T g(t)$, where $\Delta T = T_{max} - T_{min}$ and $g(t)$ is the time function with period of one year.

The reservoir-level oscillations were very limited (with a maximum oscillation of 0.2 m, from a maximum value of 46.1 m to a minimum of 45.9 m). For this reason, a mean value of 46 m was considered in all analyses.

The degree of saturation inside the dam was, however, non-constant due to the variation of relative humidity of the air. The annual variation of relative humidity of the air is given in Figure 7b.

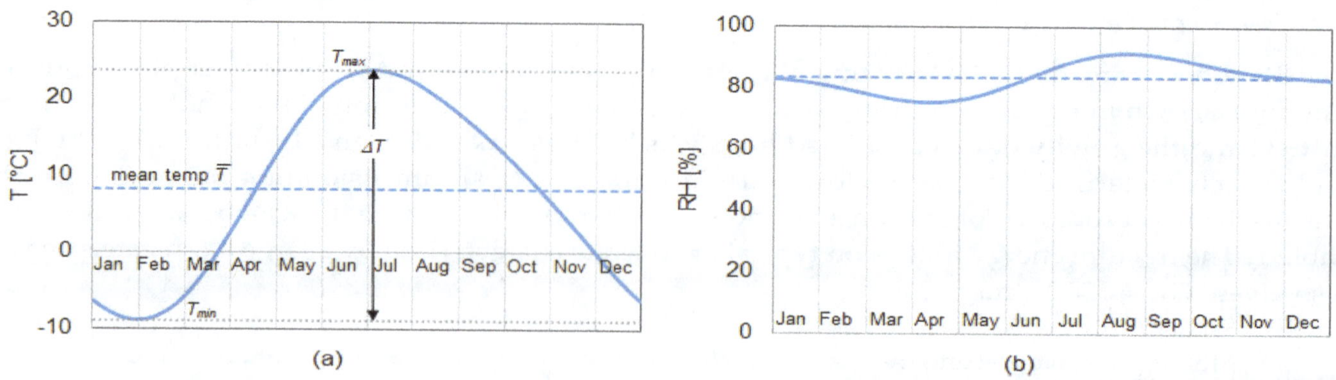

Figure 7. Annual variation of (**a**) temperature in °C and (**b**) relative humidity of the air (courtesy of Hydro-Quebec).

In view of the geometry, plane strain conditions were assumed and a 2D mesh with triangular (3 nodes) elements was employed, as depicted in Figure 8 (for the dam foundation system). Firstly, the temperature and humidity effects were evaluated using the diffusion analyses described previously. Then, the mechanical analysis of the studied dam was carried out, in order to compare the structural response obtained by the model with the real monitoring data.

Figure 8. Finite element model for the dam foundation system.

To determine the history of the temperature field within the dam, a heat diffusion analysis (by solving Equation (6) with proper initial conditions and varying boundary conditions) should be performed. The boundary conditions (BC) were set as follows (see Figure 9). The temperature of the dam surface in contact with air (the downstream face and the upstream face above water) was assumed equal to the air temperature, whose annual variation is known (as previously discussed) and shown in Figure 7a. The upstream surface below the reservoir level, assumed constant and equal to 46 m, was subject to a temperature linearly varying with the depth (as proposed in [29]), with an annual oscillation similar to that of the air, $T(t) = \bar{T} + \Delta T g(t)$, but with different maximum and minimum values (from 1 °C to 15 °C at the water surface and from 3 °C to 7 °C at the dam foot, as depicted in Figure 9). The rock temperature at the external boundary was assumed constant and equal to

4 °C. The specific heat capacity and thermal conductivity of concrete, given by Hydro-Québeq, were $C = 917$ J/kg°C and $D_T = 2.9$ W/m°C, respectively. These values were all within the quite large range of variability of concrete thermal properties [30,31].

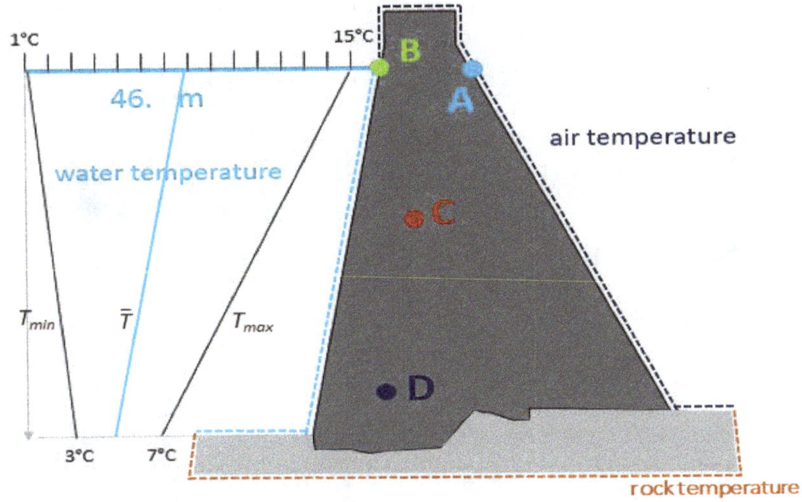

Figure 9. Boundary condition for the upstream face of the dam and locations of the nodes considered for the diffusion analyses.

The procedure followed to determine the temperature at each point of the dam, at each time of the year, and then periodically repeating, $T(\mathbf{x}, t)$, is schematically shown in Figure 10.

In the first step, fixing the temperature at different portions of the boundary of the dam as equal to the mean value over one year, $\bar{T}$, in air, water, and rock, a steady state analysis was used to evaluate the stabilized temperature in the internal nodes of the mesh $T(\mathbf{x}, 0)$, which was used as initial conditions for the following transient analysis.

In Figure 11a, the contour plot of the stabilized initial temperature $T(\mathbf{x}, 0)$ is shown.

The second step consisted of a thermal transient analysis, in which the real periodic annual variation of air and water temperature was applied at the buondary $T = \bar{T} + \Delta T g(t)$. The analysis was performed until a stabilized periodic pattern was obtained throughout the whole body of the dam. This stabilized periodic temperature field was used as input for the following chemo-mechanical analysis. In Figure 12a, the yearly histories of temperature at different points of the dam (A = in contact with air, B = close to the reservoir level, and C and D = internal nodes; see Figure 9) obtained through this second transient analysis are depicted.

The evaluation of the yearly evolution of the degree of saturation field followed the same procedure. As in the case of thermal analyses, a preliminary moisture diffusion analysis was performed before the transient moisture diffusion analysis.

For the first analysis, we assumed an initial uniform field for the degree of saturation, $S_w = 0.85$, and boundary conditions corresponding to a water level of 46 m, such that the submerged nodes were characterized by a complete saturation, while, for the nodes in contact with air, the mean of the sinusoidal annual variation (equal to 0.66 and corresponding to RH = 84%) was applied. The parameters m_1, m_2, and m_3 of the permeability were assumed, as in [22]; namely $m_1 = 18.62$ MPa, $m_2 = 2.27$, and $m_3 = 0.44$. The obtained pattern of saturation degree is shown in Figure 11b. Then, the transient moisture diffusion analysis was performed, considering the annual oscillation of Figure 7b. Figure 12b reports the obtained stabilized yearly variations of the saturation degree at different nodes. It should be noted that the moisture diffusive process affects the external boundary of the structure only to a depth of about 20 cm (i.e., the depth of one element) and the degree of saturation is constant elsewhere, equal to the imposed initial value of 0.85. Therefore, the effect of humidity on the ASR evolution is expected to have a marginal role in the structural response.

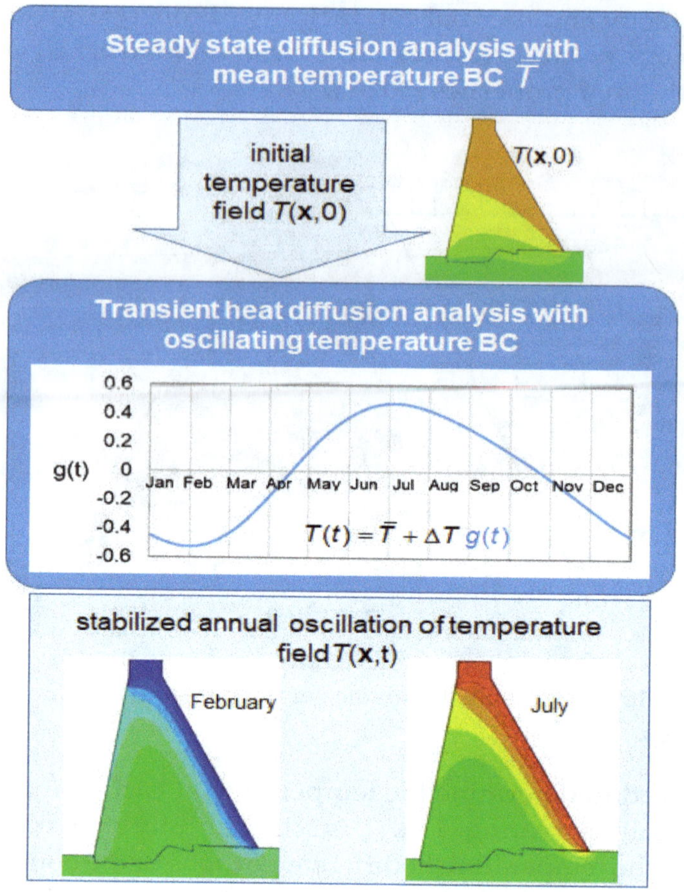

Figure 10. Flowchart of the heat diffusion analyses to determine the yearly evolution of the temperature field.

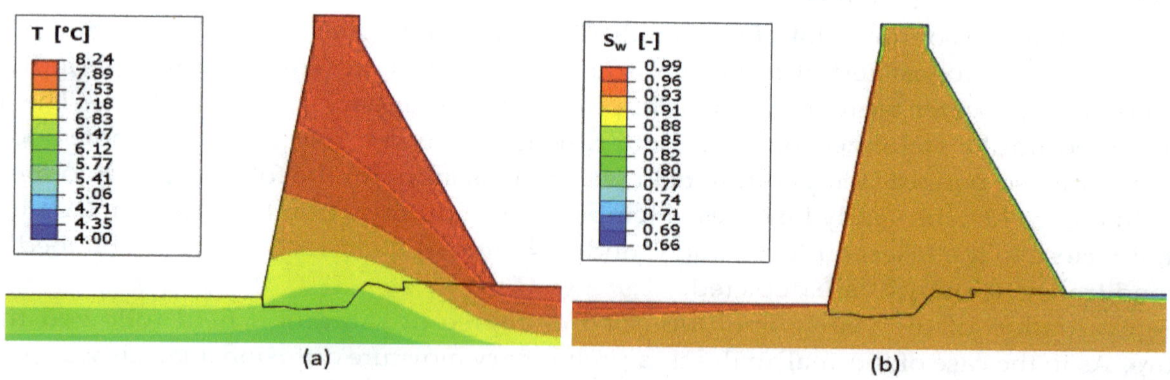

Figure 11. Contour plots: (**a**) Stabilized temperature (in °C) and (**b**) stabilized saturation degree.

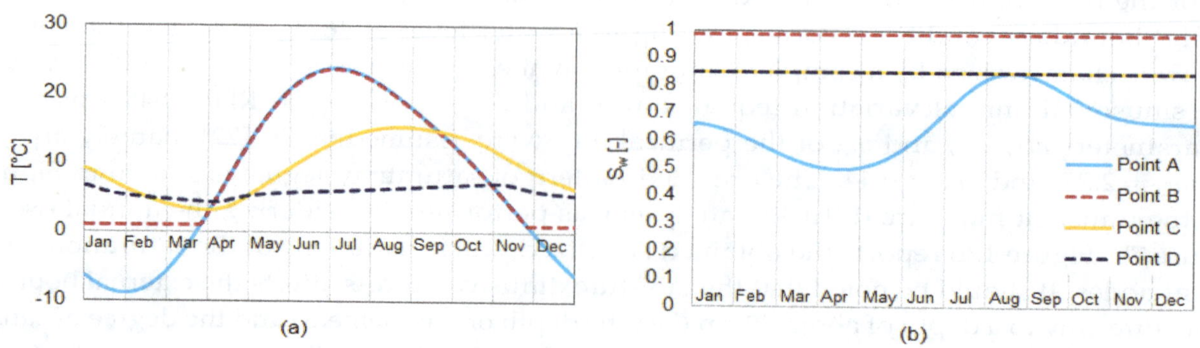

Figure 12. Histories of (**a**) temperature and (**b**) saturation degree for some nodes (A, B, C, and D) in the dam.

The reaction extent evolution was then computed by integrating Equation (10). No specific tests of ASR expansion on concrete specimens of this dam were available. Therefore, we conducted several analyses by changing the latency, the characteristic times, and the asymptotic ASR expansion within the ranges previously established (e.g., in [18]) and, finally, we arrived at the values given in Table 2.

Table 2. Parameters for the alkali–silica reaction (ASR) model used for the simulation of Beauharnois dam.

Parameter	Value	Unit
$\tau_{lat}(\bar{T},1)$	500	days
$\tau_{lat}(\bar{T},0)$	1100	days
$\tau_{ch}(\bar{T},1)$	350	days
$\tau_{ch}(\bar{T},0)$	700	days
ϵ^{∞}_{ASR}	0.01	-

Figure 13 shows the evolution of the reaction extent for the points A, B, C, and D of Figure 9. The reaction extent developed more quickly at nodes A and B, located in the upper part of the dam, where the temperature was higher. The influence of the different degrees of saturation between points A and B was very limited. The temperature and the reaction extent fields were then the input into the chemo-thermo-mechanical analysis of the dam, aimed at evaluating the response due to ASR.

The development of ASR produced an overall expansion of the dam, with a permanent variation of upstream–downstream displacements, which was superposed to the normal seasonal oscillation. Furthermore, ASR induced damage in the material. Figure 14 shows the damage at two different time steps. At the beginning, the damage appeared only on the external skin of the structure; the only part affected by the external humidity conditions. This was the first manifestation of ASR development, due to the different expansions at points with different degrees of saturation. The first macroscopic crack in the body of the dam and visible on the surface of the structure in contact with the rock developed 30 years after construction. One should observe that, as expected in a real massive structure at moderate environmental temperatures, the time scale involved in the damage process due to ASR was completely different from that of one typical in a standard test on mortar specimens accelerated by high temperature (see, e.g., [32]). Figure 15 shows a comparison between the monitoring data (mean values at points P_1 and P_2 of Figure 6a) and the numerical results, in terms of horizontal crest displacement, positive in the downstream–upstream direction. A good agreement between them was obtained.

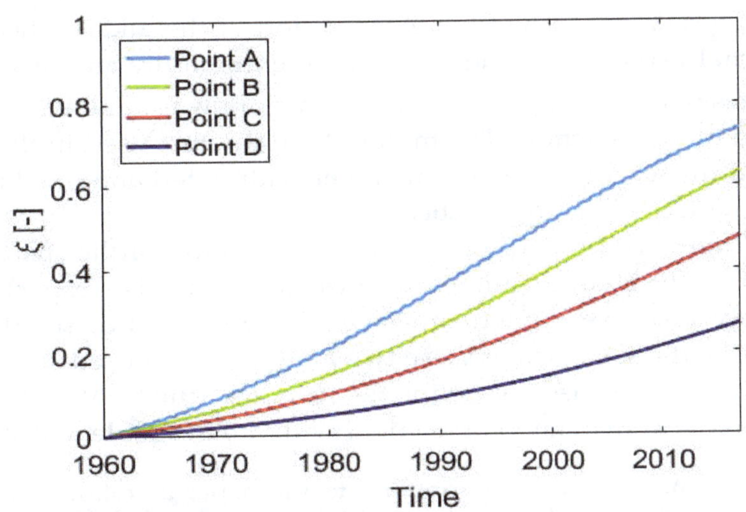

Figure 13. Reaction extent of point A (downstream face), B (upstream face, upper part), C (in the bulk), and D (close to the rock foundation) of the dam (see Figure 9 for point locations).

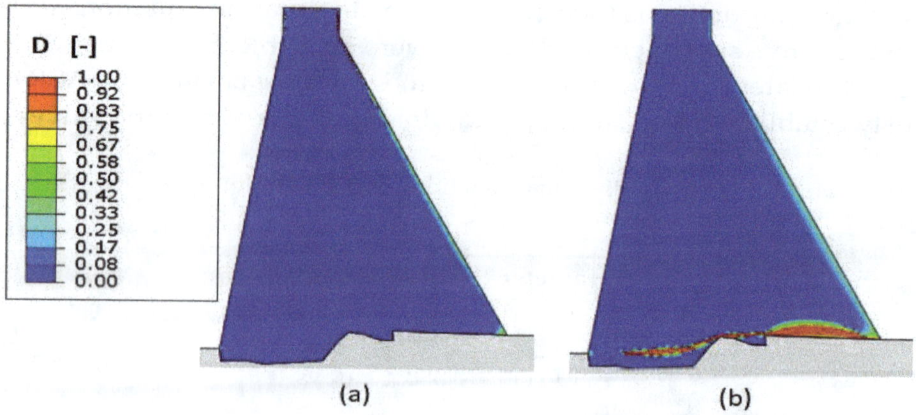

Figure 14. Patterns of damage after (**a**) 10 and (**b**) 30 years.

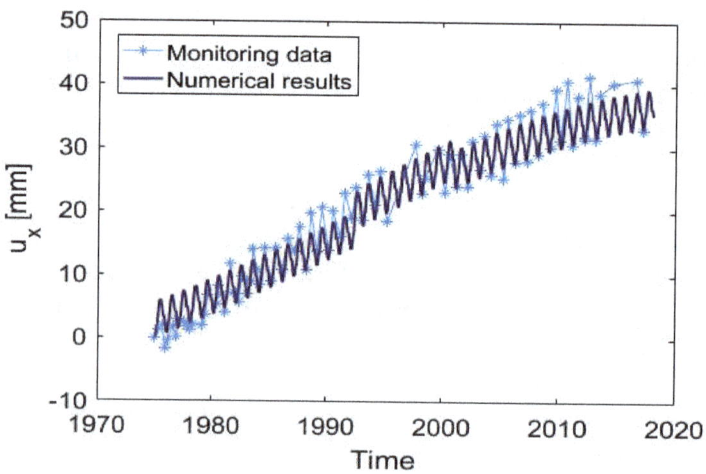

Figure 15. Comparison between monitoring data and numerical results, in terms of horizontal crest displacements.

5. Conclusions

In this work, a bi-phase damage model for concrete subject to the alkali-silica reaction present in the literature was used to simulate a recent experimental campaign, as well as to perform a mechanical analysis of an existing concrete gravity dam. In the adopted model, based on Biot's theory of multi-phase materials, the progressive expansion of the alkali–silica gel is described by a phenomenological internal variable, depending both on temperature and on degree of saturation.

The structural analysis of an existing concrete gravity dam was aimed at assessing the extent of structural damage caused by the reaction. The material parameters used in the model to define the mechanical behaviour and the reaction development were calibrated and fixed using the monitoring information available, provided by Hydro-Quebec.

Firstly, the effects of temperature and of humidity on the hydraulic structure were evaluated through a heat diffusion analysis and a moisture diffusion analysis, respectively. The moisture gradient was present only near the dowstream face of the dam and, so, its effect on the ASR evolution was limited. Then, the solutions of these two analyses were used as input for a consequent chemo-thermo-mechanical analysis, used to define the response due to ASR. The results, in terms of horizontal crest displacement, were in good agreement with the available real structural data.

Author Contributions: The authors confirm contribution to the paper as follows: conceptualization, M.C. and C.C.; methodology, C.C.; software, M.C.; validation, M.C.; analysis, M.C.; data curation, M.C. and C.C.; writing—original draft preparation, M.C.; writing—review and editing, M.C.and C.C.; supervision, C.C.

Acknowledgments: The authors acknowledge Simon-Nicholas Roth (Études de sécurité Direction Barrages et Infrastructures, Hydro-Quebèc) for providing the monitoring data of the studied gravity dam.

Abbreviations

The following abbreviations are used in this manuscript:

ASR Alkali–Silica Reaction

References

1. Stanton, T. Expansion of Concrete Through Reaction between Cement and Aggregate. *Proc. Am. Soc. Civ. Eng.* **1940**, *66*, 1781–1812.
2. Sanjuán, M.Á.; Estévez, E.; Argiz, C. Alkali Ion Concentration Estimations in Cement Paste Pore Solutions. *Appl. Sci.* **2019**, *9*, 992. [CrossRef]
3. Stanton, T. Studies of Use of Pozzolans for Counteracting Excessive Concrete Expansion Resulting from Reaction between Aggregates and the Alkalies in Cement. In *Symposium on Use of Pozzolanic Materials in Mortars and Concretes*; ASTM: West Conshohocken, PA, USA, 1950; p. 178.
4. Thomas, M. The effect of supplementary cementing materials on alkali-silica reaction: A review. *Cem. Concr. Res.* **2011**, *41*, 1224–1231. [CrossRef]
5. Sanjuán, M.A.; Argiz, C. Coal fly ash alkalis content characterization by means of a full factorial design. *Mater. Lett.* **2016**, *164*, 528–531. [CrossRef]
6. Alnaggar, M.; Di Luzio, G.; Cusatis, G. Modeling Time-Dependent Behavior of Concrete Affected by Alkali Silica Reaction in Variable Environmental Conditions. *Materials* **2017**, *10*, 471. [CrossRef]
7. Léger, P.; Coté, P.; Tinawi, R. Finite element analysis of concrete due to alkali-aggregate reactions. *Comput. Struct.* **1996**, *60*, 601–611. [CrossRef]
8. Ingraffea, A. Case studies of simulation of fracture in concrete dams. *Eng. Fract. Mech.* **1990**, *35*, 553–564. [CrossRef]
9. Józwiak-Niedzwiedzka, D.; Glinicki, M.A.; Gibas, K.; Baran, T. Alkali-Silica Reactivity of High Density Aggregates for Radiation Shielding Concrete. *Materials* **2018**, *11*, 2284. [CrossRef]
10. Grymin, W.; Gawin, D.; Koniorczyk, M. Experimental and numerical investigation of the alkali-silica reaction in the cement-based materials. *Arch. Civ. Mech. Eng.* **2018**, *18*, 1698–1714. [CrossRef]
11. Grymin, W.; Koniorczyk, M.; Pesavento, F. Macroscopic and mesoscopic approach to the alkali-silica reaction in concrete Macroscopic and Mesoscopic Approach to the Alkali-Silica Reaction in Concrete. *AIP Conf. Proc.* **1922**. [CrossRef]
12. Cuba Ramos, A. Multi-Scale Modeling of the Alkali-Silica Reaction in Concrete. Ph.D. Thesis, Ecole Polytechnique Federale de Lausanne, Lausanne, Switzerland, 2017.
13. Esposito, R.; Hendriks, M.A.N. Literature review of modelling approaches for ASR in concrete: A new perspective. *Eur. J. Environ. Civ. Eng.* **2017**, *8189*, 1–21. [CrossRef]
14. Fairbairn, E.M.R.; Ribeiro, F.L.B.; Lopes, L.E.; Toledo-filho, R.D.; Silvoso, M.M. Modelling the structural behaviour of a dam affected by alkali-silica reaction. *Commun. Numer. Methods Eng.* **2006**, *22*, 1–12. [CrossRef]
15. Saouma, V.; Perotti, L.; Shimpo, T. Stress Analysis of Concrete Structures Subjected to Alkali-Aggregate Reactions. *ACI Struct. J.* **2007**, *104*, 532–541.
16. Noret, C.; Molin, X. Theme A. Effect of concrete swelling on the equilibrium and displacements of an arch dam. In Proceedings of the XI ICOLD Benchmark Workshop on Numerical Analysis of Dams, Valencia, España, 20–21 October 2011.
17. Carrazedo, R.; Sanches, R.A.K.; de Lacerda, L.A.; Divino, P.L. Concrete Expansion Induced by Alkali-Silica Reaction in a Small Arch Dam. *Int. J. Civ. Eng.* **2017**. [CrossRef]
18. Comi, C.; Kirchmayr, B.; Pignatelli, R. Two-phase damage modeling of concrete affected by alkali-silica reaction under variable temperature and humidity conditions. *Int. J. Solids Struct.* **2012**, *49*. [CrossRef]
19. Gautam, B.P.; Panesar, D.K. A new method of applying long-term multiaxial stresses in concrete specimens undergoing ASR, and their triaxial expansions. *Mater. Struct. Constr.* **2016**, *49*, 3495–3508. [CrossRef]
20. Ulm, F.J.; Coussy, O.; Kefei, L.; Larive, C. Thermo-chemo-mechanics of ASR expansion in concrete structures. *J. Eng. Mech.* **2000**, *126*, 233–242. [CrossRef]

21. Multon, S.; Toutlemonde, F. Effect of moisture conditions and transfers on alkali silika reaction damaged structures. *Cem. Concr. Res.* **2010**, *40*, 924–934. [CrossRef]
22. Mainguy, M.; Coussy, O.; Baroghel-Bouny, V. Role of air pressure in drying of weakly permeable materials. *J. Eng. Mech.* **2001**, *127*, 582 – 592. [CrossRef]
23. Baroghel-bouny, V.; Mainguy, M.; Lassabatere, T.; Coussy, O. Characterization and identification of equilibrium and transfer moisture properties for ordinary and high-performance cementitious materials. *Cem. Concr. Res.* **1999**, *29*, 1225–1238. [CrossRef]
24. Coussy, O. *Poromechanics*; John Wiley & Sons: New York, NY, USA, 2004.
25. Comi, C.; Fedele, R.; Perego, U. A chemo-thermo-damage model for the analysis of concrete dams affected by alkali-silica reaction. *Mech. Mater.* **2009**, *41*, 210–230. [CrossRef]
26. Comi, C.; Perego, U. Fracture energy based bi-dissipative damage model for concrete. *Int. J. Solids Struct.* **2001**, *38*, 6427–6454. [CrossRef]
27. Multon, S.; Toutlemonde, F. Effect of applied stresses on alkali-silica reaction-induced expansions. *Cem. Concr. Res.* **2006**, *36*, 912–920. [CrossRef]
28. Gautam, B.P.; Panesar, D.K.; Sheikh, S.A.; Vecchio, F.J. Multiaxial expansion-stress relationship for alkali silica reaction-affected concrete. *ACI Mater. J.* **2017**, *114*, 171–184. [CrossRef]
29. Huang, M.; Pietruszczak, S. Modeling of thermomechanical effects of alkali-silica reaction. *J. Eng. Mech.* **1999**, *125*, 476–485. [CrossRef]
30. Asadi, I.; Sha, P.; Fitri, Z.; Abu, B. Thermal conductivity of concrete—A review. *J. Build. Eng.* **2018**, *20*, 81–93. [CrossRef]
31. Pomianowski, M.; Heiselberg, P.; Jensen, R.L.; Cheng, R.; Zhang, Y. A new experimental method to determine specific heat capacity of inhomogeneous concrete material with incorporated microencapsulated-PCM. *Cem. Concr. Res.* **2014**, *55*, 22–34. [CrossRef]
32. Shin, J.H.; Struble, L.J.; Kirkpatrick, R.J. Microstructural Changes Due to Alkali-Silica Reaction during Standard Mortar Test. *Materials* **2015**, *8*, 8292–8303. [CrossRef] [PubMed]

10

Shedding Light on the Effect of Uncertainties in the Seismic Fragility Analysis of Existing Concrete Dams

Giacomo Sevieri [1,*], Anna De Falco [2] and Giovanni Marmo [3]

1 Department of Civil, Environmental & Geomatic Engineering, University College London, London WC1E 6BT, UK
2 Department of Energy, Systems, Territory and Construction Engineering, University of Pisa, 56126 Pisa, Italy; a.defalco@ing.unipi.it
3 Directorate of Dams, Italian Ministry of Infrastructure and Transport, 00161 Rome, Italy; giovanni.marmo@mit.gov.it
* Correspondence: g.sevieri@ucl.ac.uk

Abstract: The seismic risk assessment of existing concrete gravity dams is of primary importance for our society because of the fundamental role of these infrastructures in the sustainability of a country. The seismic risk assessment of dams is a challenging task due to the lack of case histories, such as gravity dams' seismic collapses, which hinders the definition of limit states, thus making the application of any conventional safety assessment approach difficult. Numerical models are then fundamental to predict the seismic behaviour of the complex dam-soil-reservoir interacting system, even though uncertainties strongly affect the results. These uncertainties, mainly related to mechanical parameters and variability of the seismic motion, are among the reasons that, so far, prevented the performance-based earthquake engineering approach from being applied to concrete dams. This paper discusses the main issues behind the application of the performance-based earthquake engineering to existing concrete dams, with particular emphasis on the fragility analysis. After a critical review of the most relevant studies on this topic, the analysis of an Italian concrete gravity dam is presented to show the effect of epistemic uncertainties on the calculation of seismic fragility curves. Finally, practical conclusions are derived to guide professionals to the reduction of epistemic uncertainties, and to the definition of reliable numerical models.

Keywords: concrete gravity dams; seismic fragility analysis; uncertainty quantification; performance based earthquake engineering

---

1. Introduction

Existing concrete gravity dams are fundamental assets for all countries around the world because of their multiple connections with local communities for flood control, water supply and energy production [1]. Most of the existing large concrete gravity dams have been designed in the last century without considering any seismic design criteria [2]. However, concrete dams have behaved well when subjected to earthquakes. They have never collapsed after experiencing seismic events [3] and only few of them have exhibited damage.

Despite the good seismic performance shown by existing concrete gravity dams, their seismic safety must be quantified in order to allow decision-makers to prioritize Disaster Risk Reduction (DRR) strategies, thus increasing the resilience of communities. This justifies why professionals and scientists are still investing resources in this field [4].

Two different approaches are commonly adopted to assess the seismic safety of a structure: the Load-and-Resistance-Factor Design (LRFD) [5] and the Performance-Based Earthquake Engineering (PBEE) [6,7]. In the LRFD approach, the structural safety is primarily assessed in terms of failure of individual building components. The resistance of the component under investigation is compared to the effects of the actions affecting it in a particular scenario, the so-called Limit State (LS).

The LRFD approach is the basis of numerous modern codes [8] because it allows standardising design and safety assessment of a large number of new and existing structural typologies. Code parameters, such as combination factors, characteristic values, confidence levels and safety factors, are needed to make the LRFD approach applicable to real cases. These factors allow matching a target structural reliability in a particular scenario and accounting for all the uncertainties involved in the analysis. The calibration of those parameters can be based on: expert judgment, fitting methods, code optimization or a combination of these [9].

Expert judgment was the common approach until computational tools were developed to solve complex optimization problems. In that case, the parameters of the code were initially guessed based on expert knowledge and then verified over the years.

When expert judgment cannot be derived, for instance, because of lack of historical data, and the construction features of the analysed structural typology do not allow technicians to define clusters of similar buildings characterised by common failure modes, then the application of LRFD approach is vague, as in the case of concrete gravity dams.

In the PBEE approach, structural performances are assessed at system levels in terms of risk of collapse, fatalities, direct and indirect costs; this is usually known as "*dollars, deaths and downtime*" [10]. The outputs are then probabilistic performance metrics which are relevant to decision-makers for the definition of decision criteria. The PBEE aims to estimate the frequency of threshold exceedance by a particular performance metric for a given design at a given location.

The PBEE framework, in the version proposed by Pacific Earthquake Engineering Research Center (PEER) [10], can be thought as being composed of four steps: hazard analysis, structural analysis, damage analysis and loss analysis. The result of each step is then combined to obtain a probabilistic description of the Decision Variable (DV).

The hazard analysis makes use of the Probabilistic Seismic Hazard Analysis (PSHA) for the selection of ground motions to describe the annual frequency of exceedance of a seismic excitation, for a given geographic area. This step requires the choice of an Intensity Measure (IM) for the seismic excitation. The IM has to be a quantity that captures attributes of the ground motion hazard at a specific site. Simple scalar IMs which are readily available from the seismic hazard analysis, such as the Peak Ground Acceleration (PGA), may introduce a broad variability in structural analysis results, thus requiring a large number of nonlinear analyses in further steps. The choice of a proper IM should involve considerations related to both site characteristics and dynamic properties of the analysed structure [11]. The selection of the best IM for the seismic assessment of existing concrete gravity dams is still an important open issue [12].

During the structural analysis step, numerical models of the facilities are built to estimate the uncertain structural behaviour, which is synthetically expressed in terms of Engineering Demand Parameters (EDPs). EDPs must be selected based on the performance target and the type of system of interest in order to be relevant. In this context, there are fundamental differences in tracking damage or collapse LSs. Assuming that the computational model is refined enough to be able to properly describe the material degradation due to seismic effects, the collapse is implicitly considered within the simulations and collapse LSs become just a binary limit point. Whereas, pre-collapse damage accumulates gradually and requires further steps to relate EDPs to appropriate Damage Measures (DMs). Moreover, this aspect is an important open issue in dam engineering.

Unlike other structure typologies, dams are characterised by construction features which make the definition of dam-specific EDP-to-DM relationships particularly complex. In fact, the great number of different dam features and the variety of the ancillary works' arrangement prevent the generalization of the EDP-DM relationships.

The loss analysis, the last stage of the PBEE process, can be seen as the probabilistic estimation of structural performances parameterized through Decision Variables (DVs), such as direct economic losses, down-time and life-safety. When the analysed structure is characterised by a high collapse risk, DVs will reflect full replacement costs, rebuild time and risk of causalities associated with the collapse. The calculation of the relationship between collapse and direct losses is fairly direct, while the quantification of causalities requires further considerations related to the number of people who will suffer due to the collapse, the rescue resources, the importance of the assets, etc.

Assuming that $p\,[X|Y]$ is the probability density of X, conditioned to the event Y, and $g\,[X|Y]$ is the frequency of occurrence of X, given Y, the PBEE framework can be represented as in Figure 1.

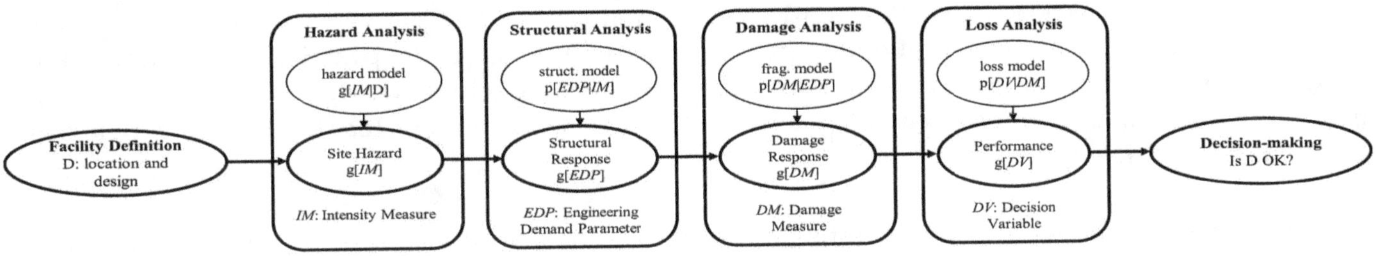

Figure 1. PBEE framework, adapted from [10].

Whereas, if D is the facility definition, the frequency of occurrence of DV for a given structure can be quantified as,

$$g\,[DV|D] = \int\int\int p\,[DV|DM,D]\,p\,[DM|EDP,D]\,p\,[EDP|IM,D]\,p\,[IM|D]\,dIMdEDPdDM \quad (1)$$

Equation (1) directly follows the total probability theorem, where uncertainties in each step of the process are described in terms of independent conditional probabilities.

DVs are usually controlled by less severe and more frequent earthquakes for which the collapse risk is low and the progression of damage is more gradual. In this case, the aforementioned problems related to the estimation of indirect costs still remain but also the direct loss quantification is more complicated because of the uncertainties involved in loss predictive models and in the definition of DM-to-DV relationships for a given facility. This is exactly the case of existing concrete dams, for which the definition of reliable DM-to-DV relationships is even more complicated by the specific use of the analysed dam. A closer collaboration among scientists, public agencies, dam owners and (re)insurance companies could help to solve these issues, thus making the first step towards the application of PBEE in dam field.

A more accurate estimation of the dam safety and a more comprehensive treatment of the involved uncertainties may be achieved by moving from the LRFD approach to the PBEE one. This is even more important considering that national and international codes relate concrete dam failure to unsatisfactory performances that could lead to uncontrolled water release [13]. The limited availability and poor quality, in terms of data aggregation, of historical damage data associated with several seismic-prone areas make the derivation of numerical fragility functions (i.e., based on computational models of the structures) an essential component of probabilistic seismic risk assessment for dams. These needs have been already highlighted by Hariri-Ardebili in their pioneering work [14], which paved the way for the application of PBEE to dam engineering, addressing some of the main issues related to the dam fragility analysis.

The main focus of this paper was to study how epistemic uncertainties affect the fragility analysis of existing concrete gravity dams. Starting from an extended literature review on the fragility curve derivation for concrete gravity dams, this study presents a classification of the main sources of uncertainties involved in the seismic analysis of such infrastructures. Finally, an illustrative case of study is presented to show how epistemic uncertainties affect the calculation of the seismic fragility, thus highlighting the need for considering them in the application of the PBEE to existing concrete gravity dams.

2. Source of Uncertainties Involved in the Seismic Analysis of Existing Concrete Gravity Dams

Uncertainties in the field of numerical modelling can be divided into two categories: epistemic and aleatory [15]. The term aleatory, which is derived from the Latin *alea*, means the rolling of the dice and it represents the intrinsic randomness of a physical phenomenon. The word epistemic derives from the Greek $\epsilon\pi\iota\sigma\tau\eta\mu\eta$ (*episteme*), which means knowledge. An epistemic uncertainty is then related to the lack of knowledge. Pragmatically, the main difference between them is that epistemic uncertainties can be reduced while aleatory uncertainties cannot. In practical cases, uncertainty reduction process involves the subjectivity of the analyst who decides which uncertainties of his/her model are random and which epistemic, based on the problem typology, his/her own experience and the available information.

In the context of seismic assessment of existing structures, aleatory uncertainties are related to the variability of ground motions (the so called record-to-record variability), while epistemic uncertainties are mostly related to the variability of the mechanical characteristics of materials [16].

Numerical models simulating the behaviour of complex systems are sources of uncertainty. In this context, the choice of the best modelling approach is known as *model class selection* [17]. The selection of the best modelling approach is an aspect that must be considered in the seismic assessment of existing concrete gravity dams [18]. De Falco et al. [19,20] address this topic highlighting how different modelling approaches involve the variation of analysis results. The authors mainly focus on three modelling aspects for gravity dams: geometrical modelling, soil–Structure Interaction (SSI) and fluid–Structure Interaction (FSI). The geometrical modelling approach is related to the choice of 2-dimensional (2D) or 3-dimensional (3D) numerical models. De Falco et al. [21] show that only 3-dimensional models are able to properly describe the seismic behaviour of the dam-basin-soil system. From the perspective of the dam body, 3D models allow the peculiarities of each dam to be described, thus catching failure modes which cannot be reproduced in 2D models. Three-dimensional modelling also affect the FSI and SSI results. Regarding the FSI approach, the authors show the main differences between the Westergaard added mass [22] and the acoustic Finite Element (FE) modelling [19]. The comparison between simple parameterized models representing prototype dams and complex models of real dams shows that the two approaches lead to the same results only around the fundamental period of the system, while they provide completely different results for higher frequency values.

Finally, the authors investigated the effects of the SSI modelling approach by comparing several commonly adopted strategies: the massless soil approach, the spring and dashpot boundary conditions, the Perfectly Matched Layer (PML) [23] and the Infinite Elements (IEs) [24]. The results reveal that the inertial part of the SSI, the so-called radiation damping [25], is a fundamental aspect which cannot be neglected in the simulation of dam seismic behaviour. Therefore, the massless approach turns out to be inadequate for this purpose. It is worth noting that no one has ever really proven that the massless approach is fully conservative in the seismic analysis of dams.

Once a particular deterministic model, i.e., a *model class* $\mathcal{C}$, has been defined, epistemic uncertainties related to numerical model parameters arise. They produce a variability of the model output leading to a mismatching between results and real observations [26]. The discussion above suggests a natural division of epistemic uncertainties into two categories based on their origin: model class and model parameter uncertainties, (Figure 2).

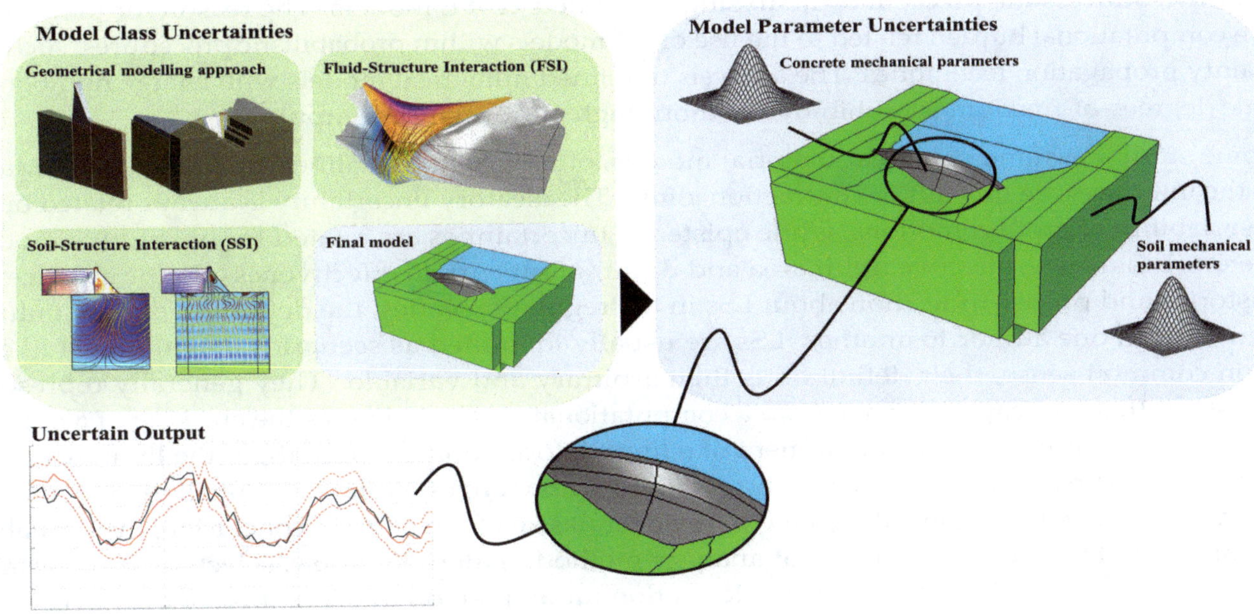

Figure 2. Classification of epistemic uncertainties in dam engineering.

Epistemic uncertainties should be reduced as much as possible in order to provide a reliable prediction of the system behaviour [27]. The process of reducing uncertainty is generally known as *model parameter calibration*, and it is performed by observing the actual behaviour of the system. Being an ill-posed inverse problem in the Hadamard' sense [28], the model parameter calibration requires a regularisation to be solved.

Both deterministic and probabilistic procedures can be adopted for this purpose. Deterministic procedures [29] aim to minimize objective functions expressing the relationship between observations and model output. Probabilistic procedures, such as the Bayesian inference [30], exploit prior information about the uncertain parameters (i.e., prior distributions) to regularise the inverse problem. Deterministic procedures require low computational performances and they enable deterministic error estimators related to the quality of the fitting to be calculated. On the other hand, probabilistic approaches are more computationally demanding than deterministic ones, but they allow deriving more comprehensive estimations of model parameters, i.e., Probability Density Functions (PDF) of model parameters and errors. The comparison between posterior distributions of errors related to different modelling approaches is the basic tools for the *model class selection*.

In this context, Sevieri et al. proposed a Bayesian approach for the reduction of the uncertainty related to elastic parameters of the materials based on static [31] and dynamic [26] measurements recorded by dam monitoring system. The authors proposed probabilistic hybrid predictive models for static and dynamic Quantity of Interest (QIs), such as displacements, frequencies, mode shapes, in which the dam behaviour is represented by the results of FE models. The computational burden is strongly reduced by using the general Polynomial Chaos Expansion technique (gPCE) [32] to surrogate the numerical results. These important examples show the possibility to apply probabilistic approaches to real dams.

3. Seismic Fragility Assessment of Existing Concrete Gravity Dams

Fragility functions are an important tool in earthquake engineering to compute the probabilities of different damage states as a function of seismic response of every type of structures and infrastructures. Originally created to characterize the seismic risk of nuclear power plants [33], fragility curves were developed for large systems such as buildings and bridges (e.g., [34]) and also for individual components (e.g., [35]).

Several recent studies focus on the seismic fragility assessment of existing concrete gravity dams. Most of them address the problem by analysing simplified 2D FE models. The reason for that is the massive computational burden related to the use of 3D models within probabilistic procedures, such as uncertainty propagation techniques. The analysis of refined numerical models with a large number of dynamic degrees-of-freedom is prohibitive without high performance computing.

Some studies assume nonlinear material models, otherwise the nonlinearities are concentrated within the interfaces, be they lift or contraction joints. The aleatory uncertainty is always related only to the variability of ground motions, while epistemic uncertainties are related to constitutive model parameters, basin level, effective uplift area and drain/grout curtain effectiveness. In the absence of case histories and precise indication about LSs in the regulation codes, the definition of dam failure is different from one author to another. LSs are usually identified as scenarios that are most likely based on common sense, their definition is then arbitrary and variable. They generally represent damage states that are easy to manage from a computational point of view, as the cracking of a certain section (e.g., dam base, neck), the attainment of a limit deformation or strength, or the limit extent of damaged areas. Thresholds are set in the form of limit stress ranges, limit crest-to-hell displacements or maximum damaged areas, mostly related to the overall size of the dam. In conclusion, the available studies are more focused on the choice of analysis method, material models, selection of uncertain parameters and methods for fragility curve derivation rather than on the definition of proper LSs.

One of the first probabilistic seismic assessment of existing concrete gravity dams is performed by de Araújo, J. M. and Awruch, A.M. [36], who combine aleatory uncertainties related to the record-to-record variability of ground motions and epistemic uncertainties related to the materials characteristics, by using a Monte Carlo simulation. The computational burden is reduced by modelling seismic excitation as a non-stationary stochastic process having two basic random variables, the acceleration amplitude and the phase angle. The dam concrete compressive strength is modelled as random variable, while other material properties, such as tensile strength, Young's modulus and adhesion of the dam-foundation interface are deduced through deterministic equations. Instead of calculating proper fragility curves, the authors derive the Cumulative Distribution Function (CDF) of safety factors against slide at the base, concrete crushing at the toe and tensile cracking at the heel for 50 simulations.

Tekie and Ellingwood [37] perform a probabilistic analysis of the Bluestone gravity dam in West Virginia (US). In this study, the aleatory uncertainties are only related to the ground motion. The deterministic numerical model is composed of rigid blocks with frictional interfaces, while dam-foundation interface is characterized by a perfectly plastic Mohr-Coulomb law. The epistemic uncertainties are related to the mechanical properties of the concrete and foundation, the drain and the grout curtain effectiveness, the water elevation, the effective uplift area. The effects of epistemic and aleatory uncertainties are assessed through a Monte Carlo simulation using the Latin Hypercube Sampling (LHS) method. Different LSs are considered in this work, (1) cracking of the neck; (2) compressive failure of the foundation material at the toe; (3) sliding at the dam-foundation interface and (4) deflection of the crest to the heel, that was limited to a percentage of the dam height. Fragility curves are log-normally distributed for different LSs.

Lin and Adams [38] present a set of seismic fragility curves based on expert judgement for concrete gravity dams located in eastern and western Canada. Two-dimensional dam models are used to derive fragility curves in terms of damage state according to [39]. Log-normal CDFs are then used to fit discrete points representing damage states.

Mirzahosseinkashani and Ghaemian [40] present a study on the Pine Flat dam, where the only record-to-record variability is considered, while epistemic uncertainties are neglected. The soil in the FE model is elastic and massless and the concrete of the dam body is simulated via a smeared crack model. Two LSs are considered, the one is attained for a given crack length at the base of the dam and the other for a given amount of the cracked area on the dam faces.

Lupoi and Callari [41] present a new probabilistic procedure for fragility assessment of existing concrete dams by considering epistemic and aleatory uncertainties separately. The LS function is linearized with respect to the model parameters (i.e., random variables). The reliability problem is thus characterized by a general cut-set formulation, and the failure probability is calculated by using a Monte Carlo simulation. The authors apply the proposed procedure to the case of the Kasho gravity dam, using an elastic FE model simulating the full fluid structure interaction. Only operational LSs are considered: (1) crest-to-heel displacements; (2) cracking at the neck; (3) base sliding governed by tangential stress and (4) cracking at the upstream face. Epistemic uncertainties are related to only two parameters: the concrete strength and the soil strength. Two aleatory uncertainties are considered: the seismic action and the water level.

Hebbouche et al. [42] adopt the procedure proposed by Tekie and Ellingwood to consider both epistemic and aleatory uncertainties in the fragility calculation. The authors assume a linear behaviour for the dam body, while the soil is modelled as perfectly plastic material with Mohr-Coulomb yield criterion. The dam–soil interface is governed by Coulomb's friction law. The LHS method is used to sample from six uncorrelated random variables: friction angle, cohesion, foundation dilation angle, Young's moduli of concrete and soil and concrete compressive strength. Four LSs are considered, (1) cracking at the neck; (2) sliding at the dam-foundation interface; (3) crest to heel displacement and (4) crushing at the toe. Six near-fault ground motions are used and scaled with regard to the pseudo-spectral acceleration (PSA) from 0.2 to 2 g.

Ghanaat et al. [43] provide fragility curves for different concrete gravity dams. Both material parameters and seismic input are assumed as random variables and the sampling approach is based on the LHS method. The authors analyse the seismic performance of the Mühleberg's gravity dam using a detailed 3D FE model having only a nonlinear contact surface between dam and foundation. The following uncertainties are considered: concrete elastic modulus, concrete damping, rock elastic modulus, cohesion and friction angle. They perform Incremental Dynamic Analysis (IDA) [44] with 30 three-components ground motion records. Given the high computational cost, each analysis is stopped when the first convergence failure occurs. In this way, any possible "resurrection" [44] at higher intensity is avoided. Finally, the fragility curve is obtained by fitting a log-normal CDF through the data points, via the least-square approach.

In Ghanaat et al. [45] a simplified version of the previous method is presented. In this case the number of trial analyses for the combination of epistemic and aleatory uncertainties is smaller with respect to the previous study. A different scaling approach and different directional factors are also used. Fragility curves are then calculated by considering a Weibull distribution rather than a log-normal one. The proposed procedure is applied to the highest non-overflow monolith of a gravity dam with shape similar to that of the Folsom dam. The 3D model has a nonlinear dam-foundation interface and an additional upper lift joint at the neck. Local and global failure modes are then considered. Elastic modulus and damping coefficients of concrete and rock are treated as random variables together with tensile strength, friction angle and cohesion of the nonlinear joints.

Finally, in Ghanaat et al. [46], a 3D FE model with linear elastic materials, massless soil, contraction and peripheral joints with nonlinear behaviour, is adopted for the seismic fragility analysis of a concrete dam. Thirty ground motions are selected and scaled in order to achieve five LSs. Each analysis is performed twice, treating the parameters both as random variables and as deterministic. The model parameters are: elastic modulus, compressive and tensile strength of concrete, maximum aggregate size, elastic modulus and tensile strength of rock, cohesion and friction angle of the interfaces.

Kadkhodayan et al. [47] analyse an arch gravity dam through nonlinear IDA, considering the nonlinearity concentrated within the contraction joints. The selected DM is the percentage of overstressed area on the dam faces, initially proposed by Ghanaat [48]. A linear relationship between DM and IM is assumed. Nine three-components ground motion records are selected and three potential IMs are considered: PGA, Peak Ground Velocity (PGV) and Spectral Acceleration at the structure's first-mode period ($S_a(T_1)$). Three LSs are then defined based on different damage levels.

Bernier et al. [49,50] derive seismic fragility curves for concrete gravity dams through the analysis of a 3D FE model with nonlinearity concentrated within the dam-foundation interface and the lift-joint at the neck of the dam. The parameters of the nonlinear interfaces, cohesion, tensile strength, friction angle and damping ratio are treated as epistemic uncertainties, whereas, the seismic variability is the only aleatory uncertainty. The vertical component of the ground motion is also considered, by scaling the horizontal component with a random factor ranging between 0.5 and 0.8. Two LSs are assumed: sliding at the dam-foundation interface and sliding at the lift–joint interface. After the seismic analyses, the authors compare three distributional models, e.g., normal, log-normal and Weibull, applying two fitting methods: the Sum of Squares due to Error (SSE) and the Maximum Likelihood Estimator (MLE). Finally, the effect of the spatial variation of the friction angle in the dam-foundation interface is investigated.

Hariri-Ardebili et al. [51] assess the seismic fragility of concrete gravity dams via the Multiple Stripe Analysis (MSA). Nine ground motions are selected and three different intensity levels are considered for each of them, thus resulting in 27 transient analyses. The authors perform a set of linear elastic analyses and a set of nonlinear analyses, where the nonlinearity is modelled by rotating smeared crack and Mohr-Coulomb base joints. Only the record-to-record variability is considered. The fragility curves are derived considering the following DMs, in the case of linear analyses: Demand Capacity Ratio (DCR) [52], Cumulative Inelastic Duration (CID) [53], Cumulative Inelastic Area (CIA) and Damage Spatial Distribution Ratio (DSDR). In the case of nonlinear analyses the following DMs are considered: Joint opening damage index, Joint sliding damage index and Crack-based damage index.

In Hariri-Ardebili and Saouma [54] the authors use the Cloud Analysis (CLA) [55] for the seismic assessment of concrete gravity dams. The authors analyse the tallest non-overflow monolith of a 122 m high dam via a 2D model, in which the interface between dam and foundation is the only source of nonlinearity. Uplift pressures are then automatically adjusted with regard to the crack lengths. Epistemic uncertainties related to the model parameters are neglected, but a large set of ground motions of 100 records is considered. Moreover, fragility curves and surfaces are developed for 70 different IMs in order to compare them and to identify the one with minimum dispersion.

In Hariri-Ardebili and Saouma [12] a numerical model similar to the one developed in Hariri-Ardebili and Saouma [54] is considered, this time adopting the concrete smeared crack model. Two different ground motions combinations are used, horizontal component only or horizontal and vertical components, in two different loading scenarios, empty and full reservoir. Twenty-one earthquake records are applied for each combination of loading scenarios and ground motion assumptions. For each case a log-normal CDF is fitted through the model outputs.

In Hariri-Ardebili and Saouma [56] the Endurance Time Analysis (ETA) method is adopted for the derivation of the fragility curve. Since only one ETA is performed, the aleatory uncertainty related to the record-to-record variation is neglected. A 2D model with nonlinearity concentrated within the interface between dam and foundation soil is used in this work. The elastic parameters of concrete and rock, and those of the lift joints, including the specific fracture energy, are treated as random variables. Two probabilistic Monte Carlo simulations, using LHS sampling, are applied to propagate the uncertainties through the numerical model. In one case the random variables are considered correlated, while in the second case they are considered uncorrelated. Log-normal CDFs are finally fitted through the deterministic results by using the MLE approach to derive the fragility curves.

In Hariri-Ardebili et al. [57] the Koyna dam is analysed by means of a 2D elastic FE model with massless soil. In a first analysis the dam concrete is assumed homogeneous, so the elastic parameters are modelled as random variables. Whereas, in a second analysis the dam concrete is assumed heterogeneous, and the elastic parameters are modelled as random fields. The structural safety is evaluated during a post-process step, in which several performance indices are considered. The authors vary the correlation length of the random field showing the effect on the model output in terms of crest and neck displacement, principal stress at the base and performance indices.

4. Illustrative Application: The Effect of Epistemic Uncertainties on the Seismic Fragility Analysis of Existing Concrete Gravity Dams

4.1. The Dam

An existing Italian large concrete gravity dam is selected as case study to show how epistemic uncertainties related to material mechanical parameters affect the seismic fragility analysis. The dam is 55.5 m tall and 199.20 m in length. As no vertical contraction joints are present and the dam is curved in plan, a 3D FE model is built to properly simulate the structural dam behaviour.

4.2. The Numerical Model

The original technical drawings and the orographic map of the region are elaborated to build a 3D Computer-Aided Drawing (CAD) of the dam-reservoir-soil system (Figure 3). ABAQUS 6.14 [58] is then used to perform FE analyses.

The mesh of the FE model is made up of 13,280 quadratic tetrahedral C3D10 elements for the dam body, while 7671 linear tetrahedral AC3D4 acoustic elements simulate the reservoir at its maximum level (52.9 m). As discussed in Section 2, both SSI and FSI must be modelled to achieve a proper refinement and subsequent accuracy of the solution. Therefore, 237 linear hexahedral CIN3D8 infinite elements reproduce the unboundedness of the soil domain, thus considering the radiation damping, while low reflecting boundary conditions are applied at the boundary of the acoustic domain, thus avoiding wave reflections. It is worth noting that the minimum mesh size is smaller than one tenth of the smallest wavelength of the seismic input, considering a significant frequency range of the seismic motion.

The model reproduces the foundation soil which is composed of two geological formations, an arenaceous mass and a marl mass, which have completely different mechanical characteristics.

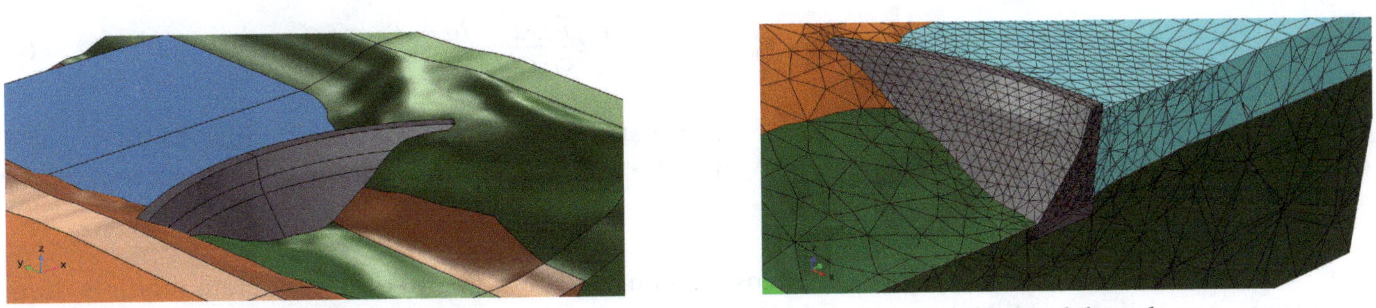

(a) Detailed view of the dam. **(b)** Model mesh.

Figure 3. Numerical model of the analysed dam.

Only the collapse LS of the dam body is considered. Special attention is then paid to the modelling of the dam concrete and its parametrization for the solution of the forward problem.

The concrete damage plasticity constitutive model [59] is suitable to simulate the behaviour of the dam concrete under cyclic loadings. In addition to the elastic parameters, this model requires the definition of plastic flow parameters and the concrete post-elastic behaviour both in compression and in tension. Plastic flow parameters commonly available in the scientific literature [60] were used in this study (Table 1). In particular, ψ is the dilation angle, ε is the eccentricity that defines the rate at which the flow potential approaches the asymptote, σ_{b0}/σ_{c0} is the ratio of initial equibiaxial compressive yield stress to initial uniaxial compressive yield stress and Λ is the ratio of the second stress invariant on the tensile meridian to that on the compressive meridian.

Table 1. Plastic flow parameters.

ψ [°]	ε	σ_{b0}/σ_{c0}	Λ
36.31	0.1	1.16	0.66

The post-elastic compressive behaviour (Figure 4a) is represented by three different branches: the *hardening branch*, the *first softening branch* and the *second softening branch*. Let us consider the concrete compressive strength $f_{C,c}$, the undamaged concrete elastic moduli E_C, the compressive stress value beyond which the material shows a non-linear behaviour $f_{C,c0} = 0.85 f_{C,c}$ and the compressive stress value beyond which the material strength exponentially decreases $f_{C,cu} = 0.3 f_{C,c}$ (these two latter correspond to $\epsilon_{C,c0} = f_{C,c0}/E_C$ and $\epsilon_{C,cu} = 0.005$ respectively). The post-elastic compressive behaviour can be then described in terms of stress σ_c and strain ϵ by the following equations:

- *Hardening branch*: if $\epsilon_{C,c0} < \epsilon_c \leq \epsilon_{C,c}$,

$$\sigma_c(\epsilon_c) = f_{C,c0} + (f_{C,c} - f_{C,c0}) \sqrt{\frac{2(\epsilon_c - \epsilon_{C,c0})}{\epsilon_{C,c} - \epsilon_{C,c0}} - \frac{(\epsilon_c - \epsilon_{C,c0})^2}{(\epsilon_{C,c} - \epsilon_{C,c0})^2}}. \quad (2)$$

- *First softening branch*: if $\epsilon_{C,c} < \epsilon_c \leq \epsilon_{C,cu}$,

$$\sigma_c(\epsilon_c) = f_{C,c} - (f_{C,c} - f_{C,cu}) \left(\frac{\epsilon_c - \epsilon_{C,c}}{\epsilon_{C,cu} - \epsilon_{C,c}}\right)^2. \quad (3)$$

- *Second softening branch*: if $\epsilon_c > \epsilon_{C,cu}$,

$$\sigma_c(\epsilon_c) = f_{C,cr} + (f_{C,cu} - f_{C,cr}) \exp\left(2 \frac{f_{C,cu} - f_{C,c}}{\epsilon_{C,cu} - \epsilon_{C,c}} \frac{\epsilon_c - \epsilon_{C,cu}}{f_{C,cu} - f_{C,cr}}\right). \quad (4)$$

where $f_{C,cr}$ is the residual value of the compressive strength, herein assumed equal to $0.1 f_{C,c}$ and $\epsilon_{C,c}$ is the strain value related to $f_{C,c}$. Once $f_{C,c}$ and E_C are sampled from their PDF, the compressive post-elastic behaviour is completely defined.

The parametrization of the post-elastic tensile behaviour (Figure 4b) of the concrete requires more attention. According to Hillerborg et al. [61] the fracture energy G_f is the energy required to open a unit area of crack and it can be thought related to the area under the post-elastic stress-fracture curve. Therefore, defining the tensile constitutive law by the stress σ_t and displacement u_t (or cracking displacements u_t^{ck} if one considers only post-elastic displacements) is a particularly convenient choice to reduce the mesh dependence of the problem solution. In this study, the tensile post-elastic behaviour of the concrete is described by the law

$$\sigma_t\left(u_t^{ck}\right) = \alpha_t \exp\left(-\beta_t u_t^{ck}\right). \quad (5)$$

where α_t and β_t are the shape parameters of the exponential function. Once a value of the tensile strength $f_{C,t}$ is sampled from their distributions, α_t and β_t are calculated in order to have G_f equal to 150 N/m. This value for G_f is commonly adopted in dam engineering [62]. The calibrated post-elastic tensile behaviour is finally used to determine the tensile damage law (Figure 4c) which describes the variation of the tensile damage variable d_t versus u_t^{ck}.

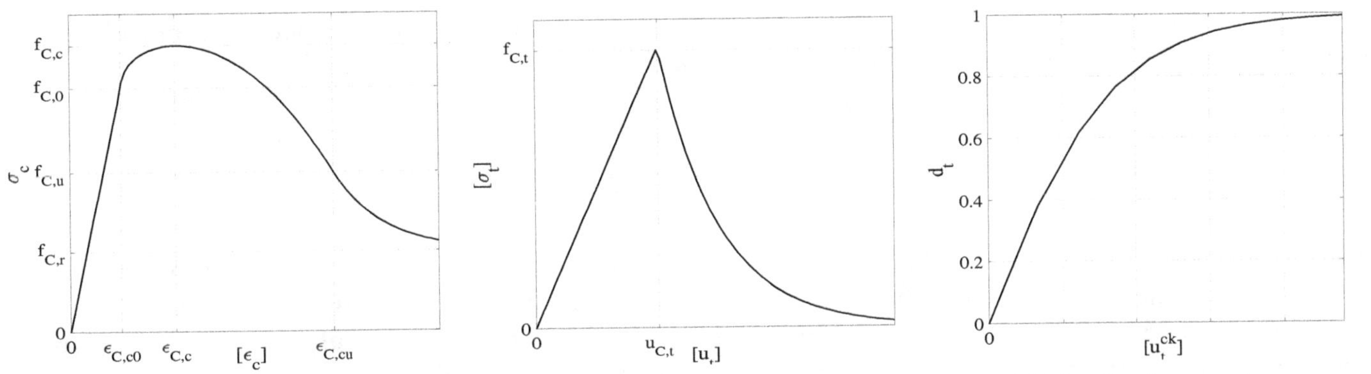

(a) Compressive behaviour. (b) Tensile behaviour. (c) Tensile damage law.

Figure 4. Constitutive model of the dam concrete.

The reduction of the concrete Young moduli due to damage related to the development of micro-cracks is expressed by a scalar value d_t. The tensile damage law is then simulated by a second order polynomial curve (Equation (6)) and its parameters are calibrated by assuming that the related plastic displacements u_t^{pl} (Equation (7)) must be positive and descending,

$$d_t\left(u_t^{ck}\right) = a_{d_t} + b_{d_t}u_t^{ck} + c_{d_t}u_t^{ck^2}, \tag{6}$$

$$u_t^{pl} = u_t^{ck} - \frac{d_t}{(1-d_t)}\frac{\sigma_t l_0}{E_0}. \tag{7}$$

A proper Matlab [63] code was developed to automatize the calculation of the tensile behaviour parameters for every sampled tensile strength value.

The values of mechanical characteristics of the dam concrete and the soil are deduced from the experimental campaigns conducted over the years. The results of in situ tests on dam concrete and foundation soil performed in 2012 are reported in Table 2 in the form of mean values and the standard deviations (s.d.). The Young modulus E, the Poisson's ratio υ the density ρ, the compressive and the tensile strengths, f_c and f_t respectively of concrete, arenaceous mass (orange in Figure 3) and marl mass (green in Figure 3), indicated with subscripts C, A and M, respectively, are reported below.

Table 2. Material test results.

	E_C [Mpa]	υ_C	ρ_C [kg/m^3]	E_A [Mpa]	υ_A	ρ_A [kg/m^3]	E_M [Mpa]	υ_M	ρ_M [kg/m^3]	$f_{C,c}$ [Mpa]	$f_{C,t}$ [Mpa]
Mean	34,620	0.16	2270	21,000	0.16	1800	7000	0.22	1800	16.7	2
s.d.	12,498	0.06	189.17	9545	0.07	-	3181.8	0.1	-	5.16	1.1

4.3. Uncertainty Quantification (UQ) in the Modal Analysis of Concrete Dams

In this section, the epistemic uncertainties related to the elastic parameters of the materials are propagated through the numerical model to evaluate how they affect the dam dynamic behaviour in terms of modal analysis (sensitivity analysis).

Mean values and standard deviations of the material Young moduli E and Poisson's ratios υ, derived during the 2012 experimental campaign (Table 2), are used to define the PDF of the bulk K and shear G moduli of concrete (i.e., K_C and G_C, respectively) as well as those of arenaceous mass (i.e., K_A and G_A) and marl mass (i.e., K_M and G_M). These random variables are assumed log-normally distributed in the present study.

The gPCE belongs to the family of the *spectral methods* for the propagation of uncertainties through deterministic models [32]. Let us assume that the uncertain model parameters are collected in $\boldsymbol{\theta}$, in this approach a polynomial expansion $\hat{u}_P^{\mathrm{gPCE}}$, also called *response surface*, of the uncertain model output $u(\boldsymbol{\theta})$ is provided by using orthogonal basis functions $\boldsymbol{\Phi}_{\mathbf{i}}(\boldsymbol{\theta})$, that is,

$$u(\boldsymbol{\theta}) \approx \hat{u}_P^{\mathrm{gPCE}}(\boldsymbol{\theta}) = \sum_{|\mathbf{i}| \leq P} \boldsymbol{u}_{\mathbf{i}} \boldsymbol{\Phi}_{\mathbf{i}}(\boldsymbol{\theta}) \tag{8}$$

where P is the degree of the polynomial expansion, $\mathbf{i}$ is a finite multi-index set and $\boldsymbol{u}_{\mathbf{i}}$ is the matrix of the combination coefficients. The selection of the orthogonal basis functions is based on the PDF of the random parameters $\boldsymbol{\theta}$. Whereas, the combination coefficients are calculated with regard to some reference solutions of the deterministic model by applying interpolation, regression or the Bayes rule. The Bayesian approach for the gPCE combination coefficient derivation proposed by Rosić and Matthies [64] is adopted in this study. Once a reliable gPCE is provided, the forward problem (i.e., the calculation of the model output statistics) is easily solved by processing the combination coefficients collected in $\boldsymbol{u}_{\mathbf{i}}$. In addition, the variance-based sensitivity analysis (Sobol' method, [65]) can be performed without any addition computational cost [66]. As already mentioned, the gPCE has been already applied in dam engineering both to propagate uncertainties and to build physic-based predictive models for the static and dynamic structural control [26,31].

The presence of the SSI in the FE model leads to a large number of numerical modes related to the soil motion only, while the propagation of the uncertainties changes the relative positions of modes once a set of new parameters is considered. These two issues are part of the so-called *mode matching problem* [26]. Being interested in the quantification of the uncertainty for specific dam modes, let us say those which mobilise the largest amount of dam mass, a selection criterion is needed to discard modes related to the soil only. The procedure proposed by Sevieri [60] is used to this purpose. Figure 5 shows the distributions of the frequencies related to the first three modes of the system which mobilise the highest amount of dam mass (Figure 6). The results show that the variation of the elastic parameters leads to a variability in the first three frequencies, which slightly increases toward higher modes. The mean values are 6.32 Hz, 7.8 Hz and 9.2 Hz, while the Coefficient of Variations (CoVs) are 0.1, 0.12 and 0.15, respectively.

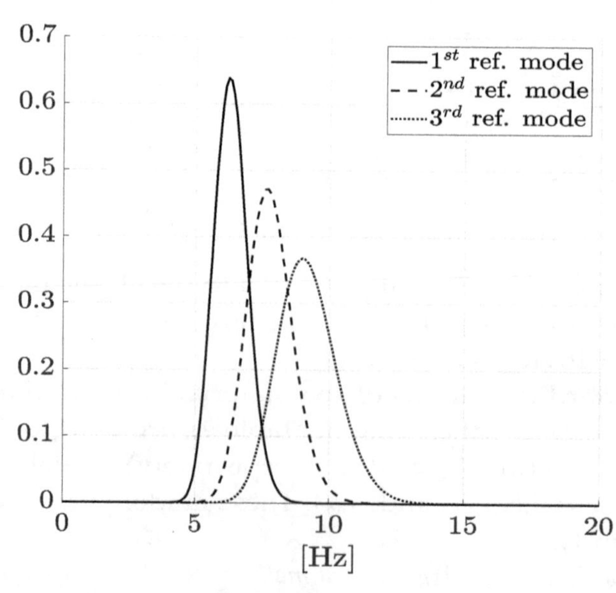

Figure 5. Distributions of the first three frequencies of the dam.

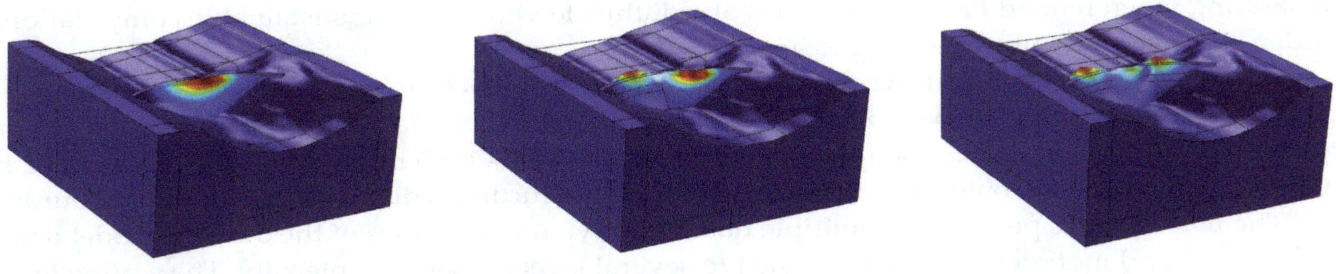

(**a**) First reference mode. (**b**) Second reference mode. (**c**) Third reference mode.

Figure 6. Reference modes for sensitivity analysis.

The influence of each random parameter on the first seventeen frequencies of the system is evaluated through the Sobol' Indices [65]. Figure 7 shows that each random parameter influences the result of the modal analysis. Therefore, none of them can be neglected in the seismic fragility analysis.

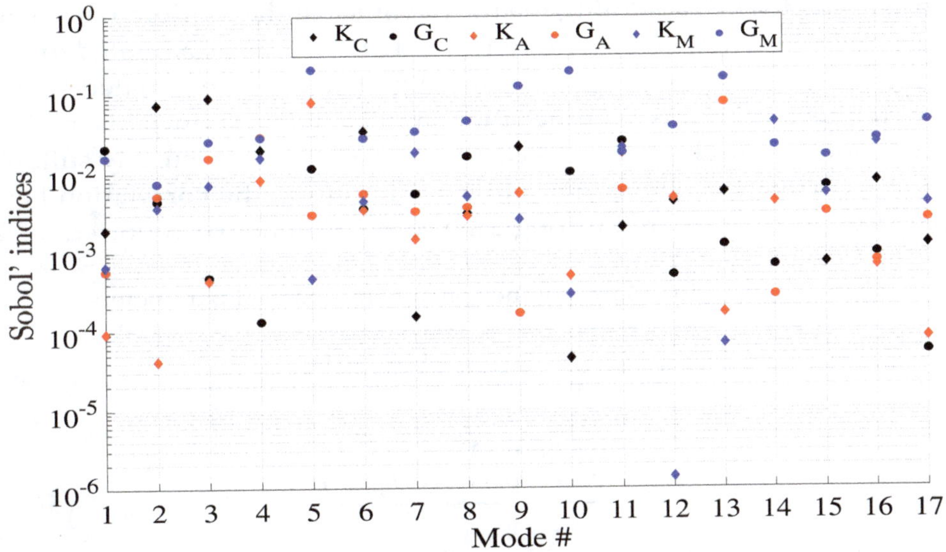

Figure 7. Sobol' indices for the modal analysis.

4.4. Fragility Analysis

A preliminary study allows the collapse LS to be calibrated. The LS is expressed as the attainment of a limit value of the difference between crest displacements recorded at the middle of the dam length and the corresponding point at the heel., $\Delta\delta^{\text{top-heel}}$. As it was demonstrated that in this case displacements greater than 0.003% of the dam height lead to a diffused damage of the dam body, this value can be related to the uncontrolled water release, and can also be associated with the achievement of the collapse LS (Equation (9)).

$$\Delta\delta^{\text{top-heel}} \geq 16.65\text{mm}. \tag{9}$$

In addition to the elastic parameters of materials, whose importance is highlighted in Section 4.3, concrete strength parameters are also modelled as random variables in the fragility analysis. They are considered normally distributed with statistics derived from Table 2.

Considering both epistemic and aleatory uncertainties in the fragility assessment of civil structures and infrastructures is a challenging task from the computational perspective. This is even more demanding when refined FE models are involved, thus leading to an unsustainable computational burden. Therefore, resolution strategies are needed.

In this study, the Multi-Record-IDA (MR-IDA) approach is adopted, as it enables the record-to-record variability to be considered by performing series of IDA with different selected ground motions. The Approximated Second Order Second Moment (ASOSM) resolution strategy proposed by Liel et al. [67] is used in conjunction to the MR-IDA [44] to further reduce the computational burden.

The IDA requires performing multiple nonlinear dynamic analyses of the dam FE model under a suite of ground motion records, each scaled to several levels of seismic intensity. The scaling levels are appropriately selected to study the structural behaviour from elastic to inelastic and finally to global dynamic instability, where the structure essentially experiences collapse. Tracing algorithms can be used to reduce the computational burden and to automatize the procedure. In this study the *Hunt & Fill* algorithm [44] is adopted.

Modelling SSI requires the seismic motion to be deconvolved before applying it to the boundary of the model. In this regard, Sooch and Bagchi [68] propose a suitably defined numerical procedure for ground motion deconvolution.

The first step of the ASOSM method requires the calculation of the fragility curve related only to the record-to-record variability, i.e., assuming the mean values of the random parameters. The effects of the epistemic uncertainties are then considered by calculating approximated gradients of the LS function for each random variable. The approximated gradients are evaluated by perturbing each random variable with plus and minus 1.7 times their standard deviation. This method also enables correlations between random parameters to be considered, so it is particularly suitable for studying the effect of a possible correlation between variables. Therefore, the correlation between concrete elastic/strength parameters is considered through the correlation matrix reported in Table 3.

Table 3. Correlation Matrix of the concrete mechanical parameters.

	K_C	G_C	$f_{c,C}$	$f_{t,C}$
K_C	1			
G_C	0	1		
$f_{c,C}$	0.83	0.83	1	
$f_{t,C}$	0.83	0.83	0.9	1

In the present study, 15 ground motions (Table 4) are selected. For each of them the magnitude Mw [69] is higher than 6 and the site class is A.

Table 4. Ground Motion Records.

#	Event	Year	Station ID	Mw	Epicentral Distance [km]	EC8 Site Class
1	Campano Lucano	1980	ST96	6.9	32	A
2	Friuli	1976	ST20	6.5	23	A
3	Campano Lucano	1980	ST98	6.9	25	A
4	Bingol	2003	ST539	6.3	14	A
5	South Iceland	2000	ST2558	6.4	5	A
6	Duzce 1	1999	ST1252	7.2	34	A
7	Tabas	1978	ST54	7.3	12	A
8	Umbria Marche	1997	ST238	6	21	A
9	Montenegro	1979	ST64	6.9	21	A
10	Basso Tirreno	1978	ST49	6	34	A
11	Golbasi	1986	ST161	6	29	A
12	Duzce 1	1999	ST3136	7.2	23	A
13	South Iceland	2000	ST2556	6.5	35	A
14	Izmit	1999	ST575	7.6	9	A
15	Friuli	1976	ST36	6	28	A

Figure 8 shows the results of the fragility analysis first considering only the record-to-record variability (dashed line) and then also the uncorrelated (dotted line) and correlated (continuous line) epistemic uncertainties.

The mean values of the three fragility curves are very close: 0.661 g if only the record-to-record variability is considered, and 0.652 g if also epistemic uncertainties are modelled. The fragility curve CoV is equal to 0.015 when only the record-to-record variability is modelled. Considering epistemic and aleatory uncertainties together leads to a higher variability of the fragility curve, the CoV is 0.0327 for the uncorrelated case and 0.0394 for the correlated case. Therefore, the fragility curve variability is particularly sensitive to the epistemic uncertainty effect, while mean values are fairly constant. In addition, in this case study, modelling epistemic uncertainties as correlated does not really affect the fragility curve derivation.

These results clearly show that epistemic uncertainties related to the mechanical parameters of the materials cannot be neglected in the seismic fragility analysis of concrete dams.

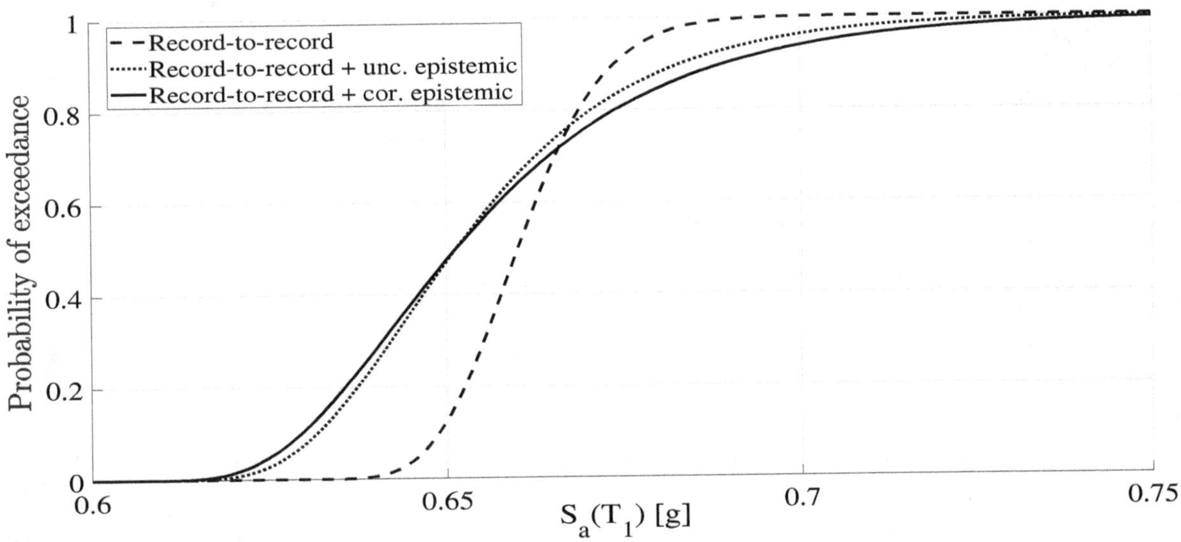

Figure 8. Collapse fragility curves.

5. Conclusions

This paper addresses fundamental issues related to the seismic risk assessment of existing concrete gravity dams.

The application of the Probabilistic Earthquake Engineering (PBEE) to existing dams is an interesting methodology which would allow overcoming the problems related to the use of the Load-and-Resistance-Factor Design approach in this field. The fragility analysis is a fundamental step for the quantification of the seismic risk associated to existing dams.

The focus of the paper is the quantification of the effect of the epistemic uncertainties related to the variability of the material mechanical parameters on the seismic fragility analysis of dams.

This paper discusses the main sources of epistemic uncertainty in dam engineering and presents a state-of-the-art review on the seismic fragility analyses of existing concrete dams. The effects of epistemic uncertainties are quantified through the analysis of a case study.

The results show that epistemic uncertainties strongly affect the simulation of dam dynamic behaviour and ultimately the quantification of its seismic performance. The comparison between fragility curves calculated under different assumptions show that epistemic uncertainties have a massive impact on the variation of the results.

Nevertheless, the application of the PBEE to real concrete dams still requires further studies aimed to solve practical issues, such as the development of damage-to-loss curves and the selection of significant damage scenarios.

Author Contributions: Conceptualization, G.S., A.D.F. and G.M.; Methodology, G.S., A.D.F. and G.M.; Software, G.S.; Validation, G.S., A.D.F. and G.M.; Formal analysis, G.S. and A.D.F.; Investigation, G.S.; Resources, G.S., A.D.F. and G.M.; Data curation, G.S.; Writing—original draft preparation, G.S. and A.D.F.; Writing—review and editing, G.S., A.D.F., and G.M.; Visualization, G.S. and A.D.F.; Supervision, A.D.F.; Project administration, A.D.F. and G.M.; Funding acquisition, A.D.F. and G.M.. All authors have read and agree to the published version of the manuscript.

Acknowledgments: This research work has been supported by the Directorate of Dams of the Italian Ministry of Infrastructures and Transport which is is kindly acknowledged.

References

1. International Commission on Large Dams (ICOLD). *World Register of Dams*; Technical Report; ICOLD: Paris, France, 2003.
2. Zhang, L.; Peng, M.; Chang, D.; Xu, Y. *Dam Failure Mechanisms and Risk Assessment*; John Wiley & Sons Singapore Pte. Ltd.: Hoboken, NJ, USA, 2016; p. 476.
3. Hall, J.F. The dynamic and earthquake behaviour of concrete dams: Review of experimental behaviour and observational evidence. *Soil Dyn. Earthq. Eng.* **1988**, *7*, 58–121. [CrossRef]
4. Hariri-Ardebili, M.A. Risk, Reliability, Resilience (R3) and beyond in dam engineering: A state-of-the-art review. *Int. J. Disaster Risk Reduct.* **2018**, *31*, 806–831. [CrossRef]
5. Galambos, T.V. Load and resistance factor design. *Eng. J. Am. Inst. Steel Constr.* **1981**, *18*, 74–82.
6. Moehle, J.; Deierlein, G.G. A framework methodology for performance-based earthquake engineering. In Proceedings of the 13th World Conference on Earthquake Engineering, Vancouver, BC, Canada, 1–6 August 2004.
7. Bozorgnia, Y.; Bertero, V.V. *Earthquake Engineering: From Engineering Seismology to Performance-Based Engineering*; CRC Press: Boca Raton, FL, USA, 2004.
8. European Committee for Standardization (CEN). *EN 1990:2002—Basis of Structural Design*; Technical Report; CEN: Brussels, Belgium, 2002.
9. Madsen, H.O.; Krenk, S.; Lind, N.C. *Methods of Structural Safety*; Dover Publications, Inc.: Mineola, NY, USA, 2006.
10. Porter, K.A. An Overview of PEER's Performance-Based Earthquake Engineering Methodology. In Proceedings of the 9th International Conference on Applications of Statistics and Probability in Civil Engineering, San Francisco, CA, USA, 6–9 July 2003.
11. Ciano, M.; Gioffrè, M.; Grigoriu, M. The role of intensity measures on the accuracy of seismic fragilities. *Probabilistic Eng. Mech.* **2020**, *60*, 103041. [CrossRef]
12. Hariri-Ardebili, M.A.; Saouma, V.E. Collapse Fragility Curves for Concrete Dams: Comprehensive Study. *J. Struct. Eng.* **2016**, *142*, 04016075. [CrossRef]
13. Federal Emergency Management Agency (FEMA). *FEMA 93—Federal Guidelines for Dam Safety*; Technical Report; FEMA: Washington, USA, 2004.
14. Hariri-Ardebili, M.A. Performance Based Earthquake Engineering of Concrete Dams. Ph.D. Thesis, University of Colorado, Boulder, CO, USA, 2015.
15. Der Kiureghian, A.; Ditlevsen, O. Aleatory or epistemic? Does it matter? *Struct. Saf.* **2009**, *31*, 105–112. [CrossRef]
16. Faber, M.H. On the Treatment of Uncertainties and Probabilities in Engineering Decision Analysis. *J. Offshore Mech. Arct. Eng.* **2005**, *127*, 243–248. [CrossRef]
17. Huang, Y.; Shao, C.; Wu, B.; Beck, J.L.; Li, H. State-of-the-art review on Bayesian inference in structural system identification and damage assessment. *Adv. Struct. Eng.* **2018**, *22*, 1329–1351. [CrossRef]
18. Andreini, M.; Falco, A.D.; Marmo, G.; Mori, M.; Sevieri, G. Modelling issues in the structural analysis of existing concrete gravity dams. In Proceedings of the 85th ICOLD Annual Meeting, Prague, Czech Republic, 3–7 July 2017; pp. 363–383.
19. De Falco, A.; Mori, M.; Sevieri, G. Simplified Soil-Structure Interaction models for concrete gravity dams. In Proceedings of the 6th European Conference on Computational Mechanics (ECCM 6) and the 7th European Conference on Computational Fluid Dynamics (ECFD 7), Glasgow, UK, 11–15 June 2018; pp. 2269–2280.
20. De Falco, A.; Mori, M.; Sevieri, G. Soil-Structure Interaction modeling for the dynamic analysis of concrete gravity dams. In Proceedings of the COMPDYN 2019, 7th International Conference on Computational Methods in Structural Dynamics and Earthquake Engineering, Crete, Greece, 24–26 June 2019.

21. De Falco, A.; Mori, M.; Sevieri, G. Bayesian updating of concrete gravity dams model parameters using static measurements. In Proceedings of the 6th European Conference on Computational Mechanics (ECCM 6) and the 7th European Conference on Computational Fluid Dynamics (ECFD 7), Glasgow, UK, 11–15 June 2018.
22. Westergaard, H.M. Water pressures on dams during earthquakes. *Trans. Am. Soc. Civ. Eng.* **1933**, *98*, 418–433.
23. Liu, Q.H.; Tao, J. The perfectly matched layer for acoustic waves in absorptive media. *J. Acoust. Soc. Am.* **1997**, *102*, 2072–2082. [CrossRef]
24. Astley, R.J. Infinite elements. In *Computational Acoustics of Noise Propagation in Fluids—Finite and Boundary Element Methods*; Marburg, S., Nolte, B., Eds.; Springer: Berlin/Heidelberg, Germany, 2008._8. [CrossRef]
25. Wolf, J.P. *Dynamic Soil-Structure Interaction*; Prentice-Hall: Upper Saddle River, NJ, USA, 1985; p. 466.
26. Sevieri, G.; De Falco, A. Dynamic Structural Health Monitoring for concrete gravity dams based on the Bayesian inference. *J. Civ. Struct. Health Monit.* **2020**, *380*, 1–16. [CrossRef]
27. Marsili, F.; Croce, P.; Friedman, N.; Formichi, P.; Landi, F. Seismic Reliability Assessment of a Concrete Water Tank Based on the Bayesian Updating of the Finite Element Model. *ASCE-ASME J. Risk Uncert. Eng. Syst. Part B Mech. Eng.* **2017**, *3*, 021004. [CrossRef]
28. Hadamard, J. *Lectures on Cauchy's Problem in Linear Partial Differential Equations*; Dover publications, INC: Mineola, NY, USA, 1923.
29. De Falco, A.; Girardi, M.; Pellegrini, D.; Robol, L.; Sevieri, G. Model parameter estimation using Bayesian and deterministic approaches: The case study of the Maddalena Bridge. *Procedia Struct. Integr.* **2018**, *11*, 210–217. [CrossRef]
30. Box, G.E.P.; Tiao, G.C. *Bayesian Inference in Statistical Analysis*; Wiley-Interscience: Hoboken, NJ, USA, 1992; p. 608.
31. Sevieri, G.; Andreini, M.; De Falco, A.; Matthies, H.G. Concrete gravity dams model parameters updating using static measurements. *Eng. Struct.* **2019**, *196*, 109231. [CrossRef]
32. Xiu, D. *Numerical Methods for Stochastic Computations*; Princeton University Press: Princeton, NJ, USA, 2010; p. 142.
33. Kennedy, R.P.; Ravindra, M.K. Seismic fragilities for nuclear power plant risk studies. *Nucl. Eng. Des.* **1984**, *79*, 47–68. [CrossRef]
34. Huo, Y.; Zhang, J. Effects of pounding and skewness on seismic responses of typical multispan highway bridges using the fragility function method. *J. Bridge Eng.* **2013**, *18*, 499–515. [CrossRef]
35. Choe, D.E.; Gardoni, P.; Rosowsky, D.; Haukaas, T. Probabilistic capacity models and seismic fragility estimates for RC columns subject to corrosion. *Reliab. Eng. Syst. Saf.* **2008**, *93*, 383–393. [CrossRef]
36. de Araújo, J.M.; Awruch, A.M. Probabilistic finite element analysis of concrete gravity dams. *Adv. Eng. Softw.* **1998**, *29*, 97–104. [CrossRef]
37. Tekie, P.B.; Ellingwood, B.R. Seismic fragility assessment of concrete gravity dams. *Earthq. Eng. Struct. Dyn.* **2003**, *32*, 2221–2240. [CrossRef]
38. Lin, L.; Adams, J. Lessons for the fragility of Canadian hydropower components under seismic loading. In Proceedings of the Ninth Canadian Conference on Earthquake Engineering, Ottawa, ON, Canada, 26–29 June 2007; pp. 1762–1771.
39. Applied Technology Council. *ATC-13: Earthquake Damage Evaluation Data for California*; Technical Report; Applied Technology Council: Redwood City, CA, USA, 1985.
40. Mirzahossein Kashani, S.; Ghaemian, M. Seismic fragility assessment of concrete gravity dams. In Proceedings of the 29th Annual USSD Conference, Nashville, TN, USA, 20–24 April 2009.
41. Lupoi, A.; Callari, C. A probabilistic method for the seismic assessment of existing concrete gravity dams. *Struct. Infrastruct. Eng. Maint. Manag. Life-Cycle Des. Perform.* **2012**, *8*, 37–41. [CrossRef]
42. Hebbouche, A.; Bensaibi, M.; Mroueh, H. Seismic Fragility and uncertainty Analysis of Concrete Gravity Dams under Near-Fault Ground Motions. *Civ. Environ. Res.* **2013**, *5*, 123–129.
43. Ghanaat, Y.; Hashimoto Philip, S.; Zuchuat, O.; Kennedy, R.P. Seismic fragility of Muhlberg dam using nonlinear analysis with Latin Hypercube Simulation. In Proceedings of the 31st Annual USSD Conference, San Diego, CA, USA, 11–15 April 2011.
44. Vamvatsikos, D.; Allin Cornell, C. Incremental dynamic analysis. *Earthq. Eng. Struct. Dyn.* **2002**, *31*, 491–514. [CrossRef]
45. Ghanaat, Y.; Patev, R.C.; Chudgar, A.K. Seismic fragility analysis of concrete gravity dams. In Proceedings of the 15th World Conference on Earthquake Engineering, Lisbon, Portugal, 24–28 September 2012.

46. Ghanaat, Y.; Patev, R.C.; Chudgar, A.K. Seismic fragility for risk assessment of concrete gravity dams. In Proceeding of the 2015 USSD Annual Conference, Louisville, KY, USA, 13–17 April 2015; pp. 645–660.
47. Kadkhodayan, V.; Meisam Aghajanzadeh, S.; Mirzabozorg, H. Seismic Assessment of Arch Dams Using Fragility Curves. *Civ. Eng. J.* **2015**, *1*, 14–20. [CrossRef]
48. Ghanaat, Y. Failure modes approach to safety evaluation of dams. In Proceedings of the 13th World Conference on Earthquake Engineering, Vancouver, BC, Canada, 1–6 August 2004.
49. Bernier, C.; Padgett, J.E.; Proulx, J.; Paultre, P. Seismic fragility of concrete gravity dams with modeling parameter uncertainty and spacial variation. *J. Struct. Eng.* **2014**, *142*.
50. Bernier, C.; Padgett, J.E.; Proulx, J.; Paultre, P. Seismic Fragility of Concrete Gravity Dams with Spatial Variation of Angle of Friction: Case Study. *J. Struct. Eng.* **2016**, *142*, 05015002. [CrossRef]
51. Hariri-Ardebili, M.A.; Saouma, V.E.; Porter, K.A. Quantification of seismic potential failure modes in concrete dams. *Earthq. Eng. Struct. Dyn.* **2016**, *45*, 979–997. [CrossRef]
52. U.S. Army Corps of Engineers (USACE). *Earthquake Design and Evaluation of Concrete Hydraulic Structures*; United States Army Crops of Engineers (USACE): Washington, DC, USA, 2007.
53. U.S. Army Corps of Engineers (USACE). *Time-History Dynamic Analysis of Concrete Hydraulic Structures*; United States Army Crops of Engineers (USACE): Washington, DC, USA, 2003.
54. Hariri-Ardebili, M.A.; Saouma, V.E. Probabilistic seismic demand model and optimal intensity measure for concrete dams. *Struct. Saf.* **2016**, *59*, 67–85. [CrossRef]
55. Jalayer, F.; Ebrahimian, H.; Miano, A.; Manfredi, G.; Sezen, H. Analytical fragility assessment using unscaled ground motion records. *Earthq. Eng. Struct. Dyn.* **2017**, *46*, 2639–2663. [CrossRef]
56. Hariri-Ardebili, M.A.; Saouma, V.E. Sensitivity and uncertainty quantification of the cohesive crack model. *Eng. Fract. Mech.* **2016**, *155*, 18–35. [CrossRef]
57. Hariri-Ardebili, M.A.; Seyed-Kolbadi, S.M.; Saouma, V.E.; Salamon, J.; Rajagopalan, B. Random finite element method for the seismic analysis of gravity dams. *Eng. Struct.* **2018**, *171*, 405–420. [CrossRef]
58. ABAQUS. *ABAQUS Version 6.14—Documentation Manual*; Technical Report; 2014. Avaliable online: http://ivt-abaqusdoc.ivt.ntnu.no:2080/texis/search/?query=wetting&submit.x=0&submit.y=0&group=bk&CDB=v6.14 (accessed on 29 January 2020).
59. Lee, J.H.; Fenves, G.L. Plastic-damage model for cyclic loading of concrete structures. *J. Eng. Mech.* **1998**, *124*, 892–900. [CrossRef]
60. Sevieri, G. The Seismic Assessment of Existing Concrete Gravity Dams: FE Model Uncertainty Quantification and Reduction. Ph.D. Thesis, University of Pisa & Technical University of Braunschweig, Braunschweig, Germany, 2019.
61. Hillerborg, A.; Modéer, M.; Petersson, P.E. Analysis of crack formation and crack growth in concrete by means of fracture mechanics and finite elements. *Cem. Concr. Res.* **1976**, *6*, 773–781. [CrossRef]
62. Omidi, O.; Valliappan, S.; Lotfi, V. Seismic cracking of concrete gravity dams by plastic-damage model using different damping mechanisms. *Finite Elem. Anal. Des.* **2013**, *63*, 80–97. [CrossRef]
63. The Mathworks Inc. *MATLAB*; MathWorks: Natick, MA, USA, 2016.
64. Rosić, B.; Matthies, H.G. Sparse bayesian polynomial chaos approximations of elasto-plastic material models. In Proceedings of the XIV International Conference on Computational Plasticity, Fundamentals and Applications, Barcelona, Spain, 5–7 September 2017; pp. 256–267.
65. Sobol', I. Sensitivity Estimates for Nonlinear Mathematical Models. *Math. Model. Comput. Exp.* **1993**, *1*, 407–414. [CrossRef]
66. Sudret, B. Global sensitivity analysis using polynomial chaos expansions. *Reliab. Eng. Syst. Saf.* **2008**, *93*, 964–979. [CrossRef]
67. Liel, A.B.; Haselton, C.B.; Deierlein, G.G.; Baker, J.W. Incorporating modeling uncertainties in the assessment of seismic collapse risk of buildings. *Struct. Saf.* **2009**, *31*, 197–211. [CrossRef]
68. Sooch, G.S.; Bagchi, A. A New Iterative Procedure for Deconvolution of Seismic Ground Motion in Dam-Reservoir-Foundation Systems. *J. Appl. Math.* **2014**, *2014*, 287605. [CrossRef]

11

Modelling and Characterizing a Concrete Gravity Dam for Fragility Analysis

Rocio L. Segura [1], Carl Bernier [2], Capucine Durand [3] and Patrick Paultre [1,*]

1 Department of Civil Engineering and Building Engineering, University of Sherbrooke, Sherbrooke, QC J1K 2R1, Canada; rocio.lilen.segura@usherbrooke.ca
2 Department of Civil and Environmental Engineering, Rice University, Houston, TX 77005, USA; carl.bernier@rice.edu
3 Institut des Sciences de la Terre, Université Grenoble Alpes, 38058 Grenoble, France; capucine.durand2539@gmail.com
* Correspondence: patrick.paultre@usherbrooke.ca.

Abstract: Most gravity dams have been designed and built during the past century with methods of analysis that are now considered inadequate. In recent decades, knowledge of seismology, structural dynamics and earthquake engineering has greatly evolved, leading to the evaluation of existing dams to ensure public safety. This study proposes a methodology for the proper modelling and characterisation of the uncertainties to assess the seismic vulnerability of a dam-type structure. This study also includes all the required analyses and verifications of the numerical model prior to performing a seismic fragility analysis and generating the corresponding fragility curves. The procedure presented herein also makes it possible to account for the uncertainties associated with the modelling parameters as well as the randomness in the seismic solicitation. The methodology was applied to a case study dam in Eastern Canada, whose vulnerability was assessed against seismic events with characteristics established by the current safety guidelines.

Keywords: seismic effects; dam safety; concrete dams; structural safety and reliability; finite elements

1. Introduction

Most gravity dams have been in operation for 30 to 50 years, and were built with the knowledge and technologies of the time. Certainly, concrete gravity dams are designed to remain operational under normal conditions, to sustain minimal damage under infrequent operative conditions and to prevent total loss of the reservoir under extreme events such as seismic actions. The importance of the behaviour of the dam under seismic loading then arises, given the human fatalities and the economic losses that can occur as a result. In recent decades, the knowledge of seismic engineering has evolved considerably, making it necessary to reassess the stability of dams under seismic loading.

Methods for analysing the structural response of a dam–reservoir–foundation system rely on deterministic or probabilistic approaches. Deterministic methods are often considered too safe, or even unsafe in some cases, because they neglect the different sources of uncertainty and because of the use of extreme load cases with very low probabilities of occurrence [1–3]. Procedures that provide a more rational way of assessing the safety of concrete gravity dams are required to establish priorities. Moreover, a probabilistic approach is required to manage the various sources of uncertainty that may impact the dam performance and decisions related thereto [1]. Within this probabilistic framework, a fragility analysis is a promising alternative, particularly suited to study the seismic vulnerability of structures and to estimate the level of damage likely to be caused by seismic events.

The implementation of fragility curves in the domain of dams is a relatively new practice. To the best of the authors' knowledge, the first study in the field was carried out by Ellingwood and Tekie [4].

Four limit states were considered in the seismic fragility analyses, and non-linear temporal dynamic analysis were performed including the dam–reservoir–foundation (DRF) interaction effects. Samples of the model were obtained using the method of Latin hypercube sampling (LHS) and were coupled with twelve historical accelerograms. These samples were analysed for six levels of seismic intensity, by calibrating the acceleration time compared to the spectral acceleration measured at the pitch period, to obtain the fragility curves. This study reveals that in all cases the log-normal model of fragility fit the results of the simulation very well. In the same manner, Mirzahosseinkashani and Ghaemian [5] and Ghanaat et al. [6,7] applied the same methodology proposed by Ellingwood and Tekie [4] and performed non-linear analysis to assess the vulnerability of a case study dam considering finite element models with different degrees of complexity. Similarly, the work performed by Lupoi and Callari [8] presents a probabilistic seismic assessment method able to manage the physical complexity of the DRF system and the uncertainties regarding structural data and external actions. The fragility curves were obtained via a standard Monte Carlo simulation procedure, and no probability was assumed to represent the fragility curves. Further studies on the subject were performed by Zhong and Lin [9], where considering the ground motion and material uncertainties, 180 non-linear seismic analyses were performed, based on which typical seismic damage modes were obtained. This research used samples obtained by thirty Monte Carlo simulations and six levels of seismic intensity.

The most recent studies were from Bernier et al. [10,11], Hariri-Ardebili and Saouma [12], Lallemant et al. [13] and Segura et al. [14]. In the first two studies [10,11], the finite element method is used to model a single block of a case study dam considering DRF interaction. Uncertainties in the ground motions and modelling parameters are included, and an LHS technique is used to propagate these sources of uncertainty. The fragility curves are developed using non-linear time-history analysis to evaluate two limit states: sliding at the dam base and at lift-joints. In the fist paper of Bernier et al. [10], the uncertainty related to the spatial variation of angle of friction is included in the fragility analysis through the incorporation of random fields modelling. In their second paper [11] the same methodology is used together with the conditional spectrum method to select ground motion times series. Hariri-Ardebili and Saouma [12] explored the development of seismic fragility curves for gravity dams with and without vertical ground motion. The derived fragility curves using scaled records were compared with those from probabilistic seismic demand analysis, showing acceptable consistency between the two methods. Similarly, the study performed by Lallemant et al. [13] provides a synthesis of the most commonly used methods for fitting fragility curves and highlights some of their significant limitations. Finally, in the study by Segura et al. [14], a similar model of the case study dam in Bernier et al. [10,11] is used to develop up-to-date fragility curves for the sliding limit states using a record selection method based on the generalised conditional intensity measure approach. These fragility functions are then combined with the recently developed regional hazard data to evaluate the annual risk, which is measured in terms of the unconditional probability of limit state exceedance. Further and more exhaustive information can be found in the state-of-the-art review of seismic fragility applied to concrete dams by Hariri-Ardebili and Saouma [15].

These studies show that the most reliable methods should be implemented where dam safety is concerned. Undoubtedly, one of the best methods for analysing gravity dams is non-linear dynamic time history analysis [16]. However, sampling techniques must be used in order to minimise the cost of the non-linear finite element analyses required to develop the fragility functions. While these previous studies have mainly focused on the methodology to develop fragility curves, they provide limited guidelines regarding the modelling and characterisation of concrete gravity dams for such analysis. Moreover, the simplifications regarding the modelling assumptions can lead to the improper evaluation of the seismic response of the dam. Among these simplifications, of note are the use of a very general or wide range of parameters due to the lack of rigorous material properties characterisation, the consideration of linear or unverified numerical models, the lack of a soil–fluid–structure interaction, the assumption of a massless foundation and the inadequate use of border conditions to describe the physical behaviour of the structure. In light of the abovementioned limitations, the main objective of this paper is to present a case study dam to illustrate a systematic methodology for modelling the DRF system and characterisation of the uncertainties associated with this type of structure to perform a seismic fragility analysis. Accordingly, the major contributions of this paper, in order of presentation, can be listed as follows: (i) the development, calibration and validation of a nonlinear finite element model (FEM) of a dam-type structure; (ii) the presentation of a procedure for applying static and dynamic loading in the DRF system, including the detailed deconvolution of the ground motion records; (iii) the design of experiments to generate samples of the system to perform a probabilistic analysis; (iv) the provision of insight into the parameters relevant for the uncertainty modelling and determination of their distributions; (v) illustration of the procedure for selection and scaling of the seismic input and (vi) the presentation of a fragility framework to generate fragility curves and assess the expected seismic performance under extreme events. The proposed methodology was applied to a case study structure in Eastern Canada.

2. Finite Element Model of the Case Study Dam

The present study is focused on a case-study concrete gravity dam spillway structure. The case study dam was built in the 1970s, has a length of 1638 m, a maximum height of 44.2 m and consists of three structures: west dike structure, bottom outlet spillway structure and main dam. This study is particularly interested in the bottom outlet spillway structure, the characteristics of which are summariSed in Table 1, and one of its blocks is considered representative of the structure. This block is a concrete gravity-type monolith, and its main geometric features and general configuration are shown in Figure 1.

Table 1. Main characteristics of the case study dam's bottom outlet spillway.

Soil foundation	Diorite rock, gneiss, granite, hybrid
Max. height	29.3 m
Crest length	192.6 m
Max. width	28.8 m
Crest width	12.2 m
Drain	Drained with injection curtain

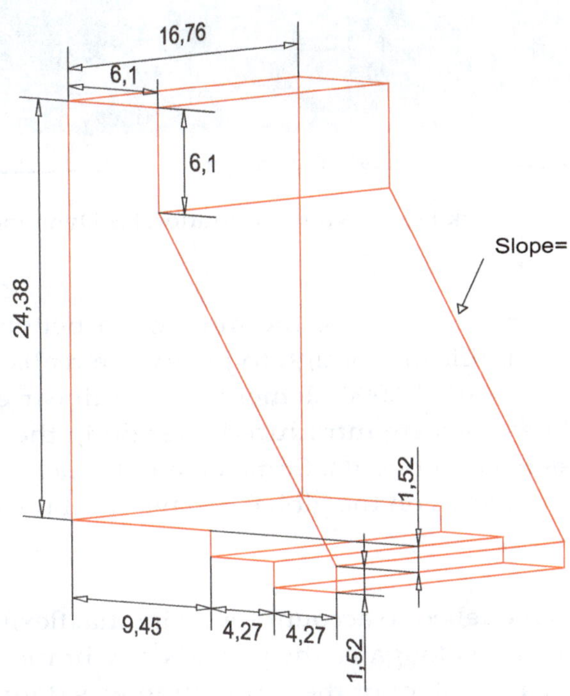

Figure 1. Block-6 main geometric characteristics.

2.1. *Model Configuration and Mesh Considerations*

The numerical model must adequately represent dynamic behaviour to minimise the epistemic uncertainties related to the modelling assumptions. The model was developed using the computer software LS-Dyna [17], following the recommendation of Noble [18] and the United States Bureau of Reclamation (USBR) [19]. The dimensions and meshing of the foundation and the reservoir in the finite element model were defined consistent with the recommendations of USBR [19], where the extent of the foundation is defined as a function of the block height H and the foundation mesh element sizes are, in general, controlled by seismic requirements. In the same manner, the reservoir is modelled with the same length as the foundation and a constant depth. The elements are sufficiently small to ensure the adequate transmission of the soil movement through the water. Further details are provided in Sections 2.1.2 and 2.1.3. Figure 2 presents the main features of the finite element model. To reduce the computational burden, an explicit time integration method [20] was used within LS-Dyna with single-point-of-integration hexahedral elements to estimate the seismic response of the dam.

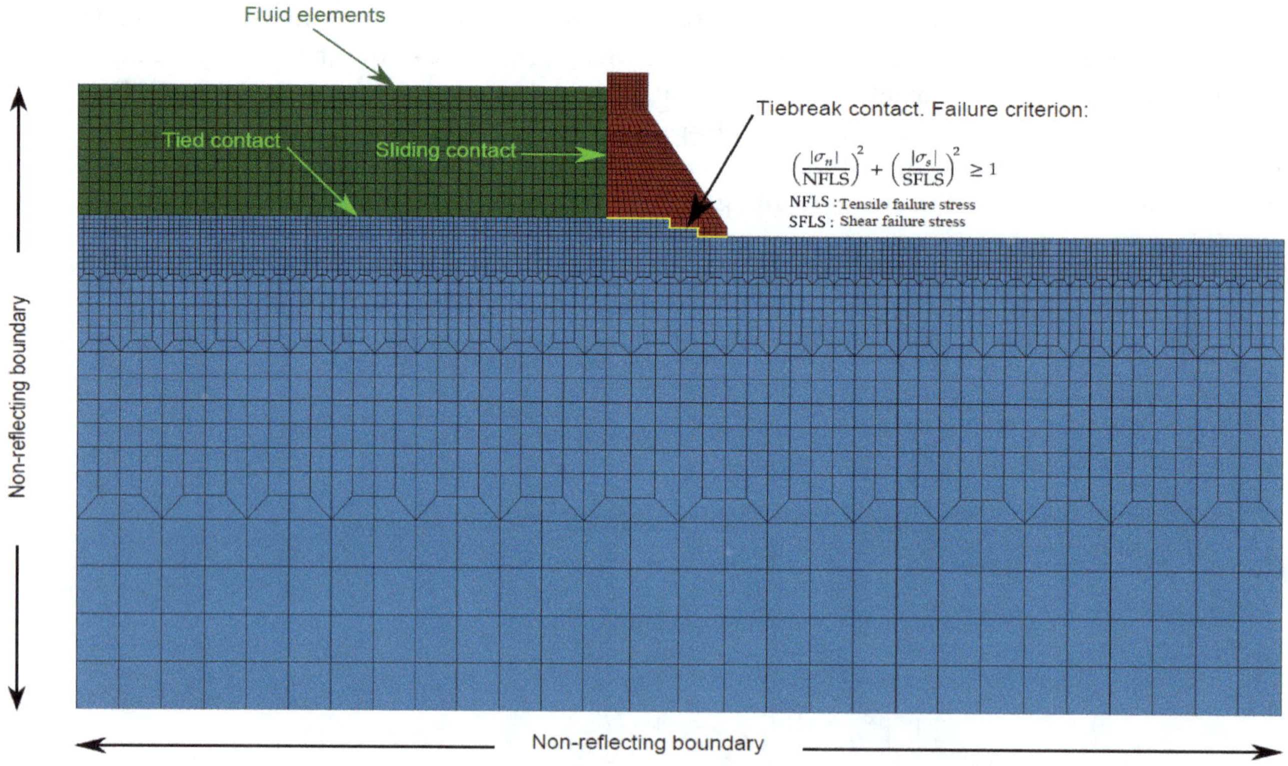

Figure 2. Block-6–reservoir–foundation LS-Dyna model.

2.1.1. Block-6

Due to the presence of contraction joints, the interaction between the blocks of the bottom outlet spillway structure is not significant enough to justify the realisation of a complex 3D model. Thus, only block-6 was modelled with 10,000 elements and a linear elastic material to model the concrete behaviour. The nonlinearities are introduced later on in the model in the form of contact surfaces between the dam system components. Some elements such as drainage galleries were not taken into account since their influences on the global behaviour of the structure are not significant.

2.1.2. Foundation

Part of the foundation was modelled to account for its inertia, flexibility and damping given that the structure–foundation interaction elongates the periods of vibration and provides an additional energy dissipation mechanism. The extent of the foundation was defined as a function of the block height H (25–28 m) [19]. Its length in the upstream–downstream direction is 160 m (6H), and its depth is 80 m (3H). The foundation mesh element sizes are generally controlled by seismic requirements. Consequently, the maximum element size depends on the foundation material properties and on the frequencies of the structure [19]. There should be no fewer than 10 elements per wavelength, and every attempt should be made to maintain uniform elements. The wavelength of interest is given by:

$$w = \frac{\sqrt{\frac{E_{rock}}{2(1+v_{rock})\rho_{rock}}}}{f_0}, \tag{1}$$

where w is the wavelength, $E_{rock}, v_{rock} and \rho_{rock}$ are the modulus of elasticity, the Poisson coefficient and the density of the foundation, respectively, and f_0 is the fundamental frequency of the structure–reservoir–foundation system. Considering from preliminary analysis that a system's fundamental frequency is approximately 10 Hz, the maximum foundation element dimension was fixed at 20 m. To limit the number of elements in the model, the foundation width was identical to that of the block-6, and the size of the elements was progressively increased, as shown in Figure 2. The final model of the foundation comprised 57,300 elements.

2.1.3. Reservoir

To account for the interaction between the reservoir and the structure, part of the reservoir was modelled with fluid elements having the physical properties of water. The reservoir was modelled on the same length as the foundation, had a constant depth of 21.94 m, and possessed 6500 elements. The compressibility of water was considered in order to adequately model the propagation of pressure waves in the reservoir, whereas the viscous effect was neglected since it had no influence in the fluid–structure interaction. The elements must be sufficiently small (<15 m) to ensure the adequate transmission of the soil movement through the water [19]. This condition is widely respected in the present model since the maximum dimension of a reservoir element remained less than 2 m. The elements of the reservoir were modelled using a Lagrangian formulation and NULL material in LS-Dyna that has no shear stiffness and no yield strength and behaves in a fluid-like manner. The NULL material is associated with an equation of state, and as recommended by Noble [18], a linear polynomial relation was used, where the pressure P applied to one element is given by:

$$P = K\left(\frac{\rho}{\rho_0} - 1\right), \tag{2}$$

where K is the water isostatic modulus of elasticity equal to 2.18 GPa, ρ is the water density at time t during the analysis, and ρ_0 is the initial water density.

2.2. *Boundary Conditions and Contact Interfaces*

In this study, as recommended by several studies [19,21] and detailed below, the loads were applied in two phases: a dynamic relaxation phase for static loads and a dynamic phase for the seismic loads. The boundary conditions were different depending on the loading phase. During the dynamic relaxation phase, a symmetric boundary condition was applied in the bottom and downstream–upstream faces, where the normal displacements were zero to simulate the constraints of the model in the space. In the dynamic phase, symmetric boundary conditions were removed, and the reactions obtained from the dynamic relaxation phase were also applied on the boundary faces for equilibrium. Additionally, non-reflective boundaries were added on the bottom, downstream and upstream faces of the model to account for the radiation damping and to dissipate the energy trapped in the foundation due to the finite length of the model (i.e., to simulate a semi-infinite behaviour). In addition, horizontal and vertical displacements were restricted at the bottom edges of the foundation to prevent rigid body movements of the model during dynamic analyses.

Regarding the contact surfaces between the different components of the system, sliding contact with zero friction was used to model the block–reservoir interface. For the reservoir–foundation interface, tied contact was applied, except near the upstream face of the block, where sliding contact with zero friction was used to maintain the reservoir load during the sliding of the block. Preliminary linear analyses—detailed further below—identified the block–foundation interface at the base of the block as high-tensile-stress areas, and therefore where cracking and sliding are likely to occur. Consequently, the model nonlinearity was constrained to these areas only, using tiebreak contact elements with a tension-shear failure criterion. The finite element code assumes the contact surface is broken when the following criterion is satisfied:

$$\left(\frac{|\sigma_n|}{NFLS}\right)^2 + \left(\frac{|\sigma_s|}{SFLS}\right)^2 \geq 1, \tag{3}$$

where σ_n and σ_s are the normal and shear stress, respectively, $NFLS$ is the tensile strength, and $SFLS$ is the cohesion. As long as the failure criterion is not reached, the tensile-compressive and shear stresses are transmitted at the contact level. If the normal stress is in compression, the first term of Equation (3) is not considered, and only cohesion is taken into account. In addition, no resistance is mobilised by friction before breaking the contact. Once the failure criterion is reached, only the friction contributes to the resistance of the contact.

2.3. Damping and Hourglassing Control

Energy dissipation within a complex dam–reservoir–foundation model may occur from a number of mechanisms. One major source is the use of nonreflecting boundary conditions; otherwise, too much energy could be trapped within the foundation domain. Another energy dissipation mechanism comes from the interaction of the dam with the reservoir through the generation of elastic waves in the reservoir. This phenomenon is known as radiation damping, which is considered in the model by also using a nonreflecting boundary condition in the upstream face of the reservoir. Regarding the block structure, to consider structural damping and potential nonlinear behaviour, a viscous damping of $\xi_{concrete} = 1.5\%$ was associated with the concrete material. For the foundation, and as recommended by the USBR [19], a viscous damping of $\xi = 5\%$ was also associated with the rock material.

The largest disadvantage of single-point integration elements is the need to control the zero energy modes that can arise, known as hourglassing modes [22]. The hourglass control method used in this study is the stiffness method, where a small elastic stiffness capable of stopping the formation of the anomalous modes but having a negligible effect on the stable global modes was added to the model. The hourglass modes control methods must be calibrated so that the energy associated with these modes remains less than 10% of the internal energy of the system and optimally less than 5% [23]. By setting the hourglassing parameters according to the recommendations of the USBR [23] in LS-Dyna, these parasitic modes were eliminated from the model, and the hourglassing energy criterion was always respected.

2.4. Validation of the Numerical Model

Dynamic Behavior

To evaluate the accuracy of the FEM, the dynamic properties of the block–reservoir–foundation system were compared to the results obtained from existing models of the structure developed by the case study dam owners. The validation of the dynamic characteristics was based on the fundamental period of the system and global damping. Because of the type of element used to model the reservoir in the LS-Dyna model, a modal analysis could not be performed. Therefore, a free vibration test was simulated to estimate the fundamental period and the damping of the system. Considering gravity loads only, a force was applied at the crest of the block in the upstream–downstream direction, and then suddenly withdrawn. The recorded horizontal displacement time series of a node at the crest of the block was then used to estimate the fundamental period of the system through the calculation of the Fourier spectrum. The global damping was approximated using the logarithmic decrement:

$$\xi = \frac{1}{2\pi m} \ln \left(\frac{u_n}{u_{n+m}} \right), \tag{4}$$

where u_n is the displacement measured at a given time and u_{n+m} is the displacement measured m cycles later. The damping was calculated as the average value using the first representative peak (peak n) and m numbers of cycles afterward.

To match the dynamic characteristics of the model, the material properties were selected as the calibration parameters. A first model, which will be referred to as model M0, was generated with initial material properties identical to those adopted by the dam owners and the typical viscous damping values used to calibrate numerical models using dynamic tests on an undamaged dam [24]. The material properties used for the M0 model are summarised in Table 2. This model was used to estimate the difference between the fundamental period of the system and the target value of $T_{1,target} = 0.096$ s, obtained by the dam owners.

Figure 3 shows the displacement time series of a crest node during the free vibration test simulation conducted with the M0 model. The fundamental period of the system with the M0 model was $T_{1,M0} = 0.110$ s and presented a difference of 14.7% compared to the target period. The overestimation of the fundamental period relative to the reference model could be explained by the excessive flexibility of the foundation. Indeed, unlike the model considered by the dam owners, the mass of the foundation was considered together with the absorbing boundary conditions. According to the USBR [19], these border conditions—while more realistic—may introduce additional flexibility to the foundation. Therefore, a second model M1 was tested to confirm this hypothesis, whose rock modulus of elasticity was increased to $E_{rock} = 50$ GPa according to the ASTM D4394-04 [25] for rock foundation sites. Table 3 provides a comparison of the dynamic properties obtained with the different models. By adapting the value of the foundation modulus of elasticity, it was possible to match the target value of the natural period and highlight the impact of this parameter on the dynamic behaviour of the studied system and therefore on the seismic response. Regarding the damping of the system, it matches well with the forced-vibration test performed by Proulx and Paultre [24] on undamaged dam-type structures. Nevertheless, given the unavailability of in situ experimental results, the exact natural period of the system and its damping remain unknown. Consequently, as part of the probabilistic framework conducted here, the dynamic properties of the structure will be varied by defining the concrete damping and the concrete and foundation modulus of elasticity as random variables.

Table 2. Material properties of the M0 model.

Elasticity modulus	$E_{rock} = 40$ GPa, $E_{concrete} = 21.4$ GPa
Damping	$\xi_{rock} = 5\%$, $\xi_{concrete} = 1.5\%$
Density	$\rho_{rock} = 2600$ kg/m^3 , $\rho_{concrete} = 2300$ kg/m^3
Poisson coefficient	$\nu_{rock} = 0.3$, $\nu_{concrete} = 0.164$
Reservoir	Westergaard added mass

Table 3. Dynamic properties of the finite element model (FEM).

Model	Fundamental Period (s)	Damping (%)
Reference	0.096	5.50
LS-Dyna M1	0.106	2.88
LS-Dyna M0	0.110	3.19

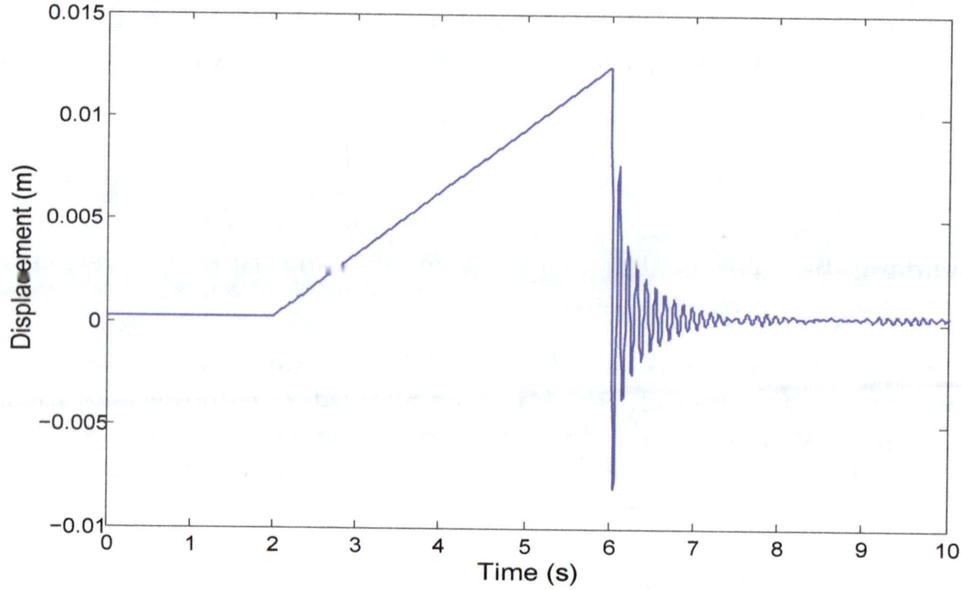

Figure 3. Free vibration test–displacement time history for model M0.

3. Loads

The loads considered in this study are summarised in Table 4. Loads resulting from ice thrust, sediment surge and thermal gradient were neglected, considering they would not affect the overall behaviour of the system under seismic actions. Only one loading case was considered to analyse the seismic response of the case study structure, which included the self-weight of the block, the hydrostatic and hydrodynamic loads exerted by the reservoir on the block, the uplift pressures at the concrete–rock contact and the horizontal and vertical seismic loads.

Table 4. Load conditions.

Load	Details
Self-weight - Plot-6	$\rho_{concrete} = 2300\,\text{kg/m}^3$
Reservoir weight	Reservoir level = MOL
Upstream hydrostatic thrust	Reservoir level = MOL
Uplift	Concrete–rock contact Reservoir level = MOL Drain efficiency: between 0% and 70%
Hydrodynamic load	Automatic with fluid elements Reservoir level = MOL
Horizontal seismic load	Representative accelerograms
Vertical seismic load	Representative accelerograms

MOL: maximum operating level.

3.1. Static and Dynamic Loads

The weight of the block was integrated in all the analyses, and it was determined directly by LS-Dyna as well as the reservoir and the hydrostatic load. For all the analyses, the maximum operating level (MOL) of the reservoir was considered (i.e., a constant reservoir elevation of 21.94 m with respect to the downstream base of block-6). Even if the foundation is associated with a mass, its self-weight was not taken into account in the calculations to avoid settlement during the analysis. Concerning the uplift loads, they were only applied at the concrete–rock contact, and they were calculated at each node of the structure base according to the prescribed expressions by the United States army corps of engineers (USACE) [26].

The dynamic loads considered in the analysis comprised the seismic load in the horizontal (upstream–downstream) and vertical directions. In addition, the model of the reservoir with fluid elements made it possible to automatically account for the hydrodynamic thrust due to the seismic shocks. The dynamic effects on the uplift distribution were neglected, assuming that the uplift profile would remain constant during seismic loading [19,26].

3.2. Application of Static and Dynamic Loads

The application of gravity and static loads in an explicit analysis can produce unexpected results if the load is applied too quickly, because the solution may diverge. To avoid this problem, as previously mentioned, the static loads were applied during a dynamic relaxation phase, and the seismic loads were applied in a dynamic analysis phase. The dynamic relaxation then reduced the nodal velocity at each time step, which is equivalent to a highly damped dynamic analysis [22]. Within the context of this study, the application of the gravity and statics loads took a total duration of 8 s. The self-weight of the dam was applied progressively from 0 to 4 s, whereas the weight of the reservoir and uplift pressures were applied gradually, after the self-weight, between 4 and 8 s. At the end of the dynamic relaxation phase, the reaction forces at each node belonging to the foundation faces were recorded to be used in the next phase to maintain a quasistatic state of the model.

Regarding the dynamic phase, the most straightforward method for applying ground motions to finite element models is to apply accelerations at the base [19]. In contrast, earthquake motion cannot be specified directly at the model truncations, as this would render any absorbing boundary ineffective [27]. Instead, effective earthquake forces were computed from the earthquake excitation and applied either directly at the absorbing boundaries. Given the presence of an absorbing boundary condition during the dynamic phase, it is necessary to use nodal forces instead of base accelerations at the same location as a non-reflecting boundary condition. The seismic time series were applied as a stress time series using the following expressions:

$$\sigma_h = 2\rho_{rock} v_h \sqrt{\frac{E_{rock}}{2\rho_{rock}(1+\nu_{rock})}}, \tag{5}$$

$$\sigma_v = 2\rho_{rock} v_v \sqrt{\frac{E_{rock}(1-\nu_{rock})}{\rho_{rock}(1+\nu_{rock})(1-2\nu_{rock})}}, \tag{6}$$

where σ_h is the horizontal stress time series, ρ_{rock} is the foundation density, v_h is the horizontal ground velocity (obtained by integration of the horizontal acceleration), σ_v is the vertical stress time series, v_v is the vertical ground velocity (obtained by integration of the vertical acceleration) and E_{rock} and ν_{rock} are the modulus of elasticity and the Poisson coefficient of the foundation, respectively. Since the non-reflecting boundary condition and earthquake ground motions were to be applied at the same location, care was taken to ensure that the nodal forces applied were of the proper magnitude to ensure that the incident waves propagating through the foundation medium produced the proper excitation at the base of the structure [27].

3.3. Deconvolution of the Ground Motions

Given that the foundation mass and inertia effect were considered in the analysis, ground motions needed to be deconvolved. Deconvolution methods allow for the adjustment of the amplitude and frequency contents of a seismic ground motion applied at the base of the foundation to achieve the desired target acceleration time series at the structure–foundation interface. Following the method proposed by Sooch and Bagchi [28] and presented in Figure 4, the ground motion was initially applied at the base of the foundation, and it was assumed to be the same as the free-field ground acceleration. The acceleration time series at the top surface (i.e., block–foundation interface) was then estimated by solving the wave propagation problem using the finite element model. This estimated or reproduced ground motion at a reference point on the block–foundation interface was then compared to the original free-field ground motion after transforming both signals into the frequency domain using Fourier analysis. The synthesised and free-field signals at the top of the foundation were then compared in the frequency domain, and a correction factor for each frequency was computed. The modified ground motion was then transformed back into the time domain, and the wave propagation analysis for the foundation system was repeated iteratively with the modified ground motion applied at the base of the foundation. This procedure was repeated until the difference between the spectra of the output and the target ground motion was less than 5% over the range of periods of interest ($0.2T_1 - 2T_1$) and 10% elsewhere. As shown in Figure 5, the comparison of the acceleration response spectra further confirms the effectiveness of the deconvolution.

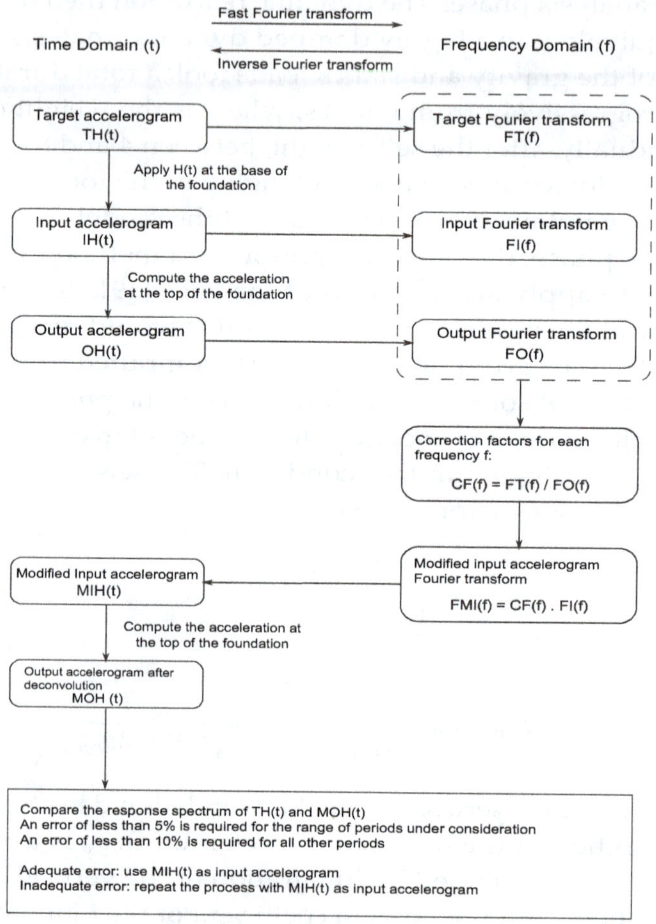

Figure 4. Deconvolution methodology.

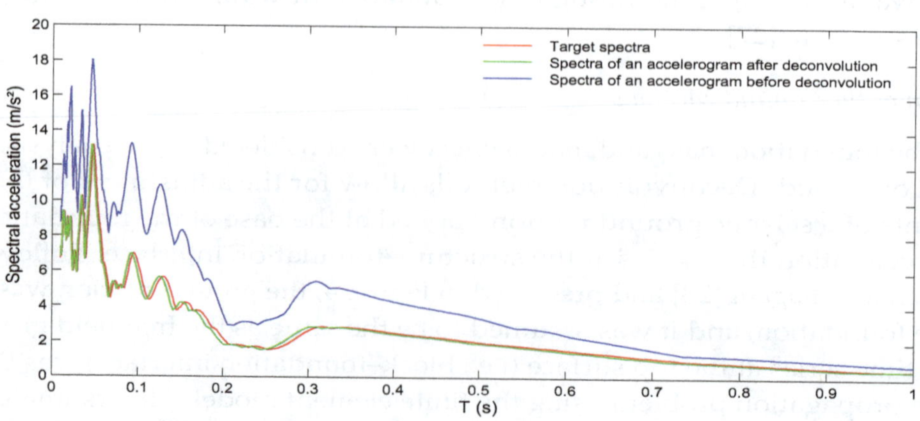

Figure 5. Comparison between response spectrum (before and after deconvolution) and target spectrum.

4. Uncertainty Modelling

Each parameter of the analysis was defined either as a fixed value or as a random variable associated with a probability density function (PDF). The preliminary analyses carried out in the finite element model showed that some parameters (e.g., moduli of elasticity) had to be defined as random variables in order to account for the uncertainty about the estimated value. The lack of knowledge relative to the parameters quantifying the resistance of the structure, loading conditions and seismic input were considered in this study. All the parameters considered as random variables and their associated PDFs are listed in Table 5, while all the material properties considered constant throughout the analysis are summarised in Table 6. The 11 uncertain parameters listed in Table 5 were sampled using Latin hypercube sampling (LHS) to generate $N = 22$ model samples to propagate sources of uncertainty in the fragility analysis. When using an LHS experimental design, the minimum number of design runs required for optimally spaced design points is typically at least twice the number of parameters considered [29,30]. This sampling method was chosen because of its ability to efficiently generate representative samples by dividing the range of possible values for each variable into N equiprobable intervals.

Table 5. Uncertain parameters.

Modelling Parameter	Probability Distribution	Distribution Parameter		Units
Directionality factor	Uniform	$L = 0.5$	$U = 0.8$	-
Concrete damping	Log-Normal	$\lambda = -2.99$	$\zeta = 0.35$	%
Concrete–concrete cohesion	Uniform	$L = 0.5$	$U = 2.5$	MPa
Concrete–rock cohesion	Uniform	$L = 0.1$	$U = 1.5$	MPa
Concrete–concrete angle of friction	Uniform	$L = 45$	$U = 55$	°
Concrete–rock angle of friction	Uniform	$L = 30$	$U = 45$	°
Concrete–concrete tensile strength	Uniform	$L = 0.3$	$U = 2.3$	MPa
Concrete–rock tensile strength	Uniform	$L = 0.05$	$U = 0.8$	MPa
Concrete elasticity modulus	Uniform	$L = 18.5$	$U = 26.9$	MPa
Rock elasticity modulus	Uniform	$L = 40$	$U = 50$	MPa
Drain efficiency	Uniform	$L = 0$	$U = 70$	%

Table 6. Sample's constant parameters.

Material Properties	Constant Value
Concrete density	2300 kg/m^3
Rock density	2600 kg/m^3
Concrete Poisson's coefficient	0.164
Rock Poisson's coefficient	0.300
Foundation damping	5%

4.1. Material Properties

To test the accuracy of the finite element model, it was necessary to specify the densities, moduli of elasticity, damping and Poisson coefficients of the concrete and foundation material (Table 2). However, these values were re-evaluated here for estimation of the impact of their uncertainty on the dynamic behaviour of the structure.

4.1.1. Modulus of Elasticity

As already pointed out, the seismic response of the system changes according to the value of the concrete and foundation modulus of elasticity. It is therefore appropriate to define these parameters as a random variables, especially given the lack of empirical or experimental data to calibrate them accurately. From a series of tests carried out at the dam site, it was determined that the concrete static modulus of elasticity could be associated with extreme values between 14.8 and 21.5 GPa. As recommended by [26,31], the concrete modulus of elasticity is amplified by 25% in the presence of dynamic loadings. Therefore, the extreme values of the modulus of elasticity were 18.5–26.9 GPa. Regarding the foundation, because no test was carried out to characterise the modulus of elasticity, the values adopted in models M0 and M1 (40–50 GPa) were considered herein as the extreme values of the uniform distribution associated with this variable.

4.1.2. Damping

The concrete damping is difficult to estimate, especially since the possible cracking of the structure under the effect of the seismic load will cause additional energy dissipation. Given that the concrete is associated with an elastic linear behaviour in the numerical model, the increase in damping with cracking is not automatically taken into account. The average damping should therefore be increased with respect to the 1.5% value obtained from an experimental test for relatively small vibrations [24] and used during model verification to anticipate cracking of the structure and the nonlinear mechanisms that can develop during an earthquake. As proposed by Ghanaat et al. [6], for the concrete damping, a log-normal distribution with a median of 5% and a standard deviation of 0.35 was adopted. Taking these values into account, the system damping was estimated with the free-decay method, and it was verified that the total system damping remained under 7% [32]. With respect to the foundation damping, given that the knowledge on this parameter is limited, the viscous damping of the foundation was assumed to be 5% for all samples [19].

4.1.3. Shear and Tensile Strength

In the absence of sufficient data, a uniform PDF was also used to describe the parameters characterising the shear and tensile strength. For determining the shear and tensile strength parameters at the rock–concrete contact and based on the construction plan of the dam, the foundation rock at the dam site was determined to be composed of diorite, gneiss and granite. The maximum and minimum values for the cohesion and the angle of friction were estimated from literature test results for the same type of rock foundation [33]. As a result, a range of values of 0.1–1.5 MPa and 30–45° were assigned to the uniform PDF for the cohesion and the angle of friction, respectively. Similarly, for the tensile strength, the adopted extreme values varied between 0.05 and 0.8 MPa.

4.2. Load Parameters

4.2.1. Drain Efficiency

The efficiency of the drainage system is defined by the proportion of the flow that can be captured by this system to reduce the hydraulic load. To account for the uncertainties in the uplift pressures, the drain efficiency was defined as a random variable with a uniform PDF. From previous studies of the dam owners at the base of block-6 and using the extreme values of a water column height to calculate the uplift load, it was concluded that the drains allow for a decrease in the uplift pressures between 15% and 75%, while the USBR [19] recommends limiting the effectiveness of the drains to 67%. Consequently, the efficiency of the drain system was sampled by considering values between 0% and 70% to also include the undrained case.

4.2.2. Directionality Factor

The directionality factor $f_{h/v}$ is defined as the ratio of the vertical to the horizontal peak ground acceleration (PGA), as expressed by Equation (7):

$$\mathrm{PGA}_v = f_{h/v}\,\mathrm{PGA}_h, \tag{7}$$

where PGA_v and PGA_h represent the peak accelerations in the vertical and horizontal directions, respectively. Traditionally, this factor is assumed to be equal to 2/3, but its value can vary greatly, having a significant influence on the response of the structure. In fact, from the available recorded ground motions, it can be seen that this ratio generally varies between 0.3 and 1.0 depending on the magnitude, the epicentral distance, the tectonic environment and the spectral period [34–36]. Due to the lack of seismic data at the dam site, the range of values was slightly reduced to 0.5–0.8.

4.3. Ground Motions

To perform the fragility analysis for assessing the seismic vulnerability of the case-study structure, a representative set of ground motion time series (GMTS) that efficiently depicts the aleatory uncertainty was needed. Each of the 22 samples generated with the LHS procedure was randomly coupled with a seismic event. Thus, 22 ground motion time series in the vertical and horizontal direction were selected.

4.3.1. Levels of Seismic Intensity

The response of each sample was analysed for several levels of seismic intensity through an incremental dynamic analysis (IDA). Most of the studies found in the literature concerning seismic fragility assessment for gravity-dam-type structures use the peak ground acceleration (PGA) or the spectral acceleration at fundamental period of the structure as the seismic intensity measure [10,14,15]. Nevertheless, given that the fundamental period depends on the concrete and foundation modulus of elasticity, defined herein as random variables, the spectral acceleration at the fundamental period varies according to the sample considered for the analysis. Therefore, in this study, the PGA in the horizontal direction was chosen to characterise the seismic solicitation at seven target intensity levels: $\text{PGA}_{target} = \{0.1g, 0.2g, 0.25g, 0.3g, 0.4g, 0.5g, 0.7g\}$ to efficiently cover return periods corresponding to 200–20,000 years. Thus, the 22 selected horizontal accelerograms were scaled according to these seven intensity levels, while the vertical accelerograms were scaled taking into account the directionality factor associated with the sample in question.

Due to the limited availability of strong ground motion records for the considered region, synthetic accelerograms developed by Atkinson [37] for Eastern Canada were used. The accelerograms were simulated for the following cases: magnitude 6 and epicentral distances between 10 and 15 km and between 20 and 30 km, and magnitude 7 and epicentral distances of 15–25 km and 50–70 km. For each of these four sets, three random components of the earthquake were simulated at 15 random locations around the epicentre. Seismic solicitations were then selected considering accelerograms from the four sets.

4.3.2. Selection and Scaling of Accelerograms

Current seismic guidelines for dimensioning recommend the use of seismic events compatible with the uniform hazard spectrum (UHS) for the studied site. The UHS provided by the Geological Survey of Canada (GSC) at the dam site for soil type A ($VS30 \geq 1500\,\text{m/s}$), referred to as the target spectrum, was used for this study, and it is shown in Figure 6a. Taking into consideration the different simulated sets of accelerograms in terms of magnitude and epicentral distance, GMTS were selected for soil type A, keeping those whose response spectrum closely matched the target spectrum. For that purpose, the method proposed by Atkinson [37] was used to compare the shape of the response spectra of the target and the selected ground motion over the range of periods of interest, $T \in [0, 1]$ s. Finally, the 22 accelerograms presenting the smallest values of the standard deviation within the considered range of periods were selected. Each simulated record, with initial PGA_h, was then multiplied by the factor $\text{PGA}_{target}/\,\text{PGA}_h$ to match the peak ground acceleration at each intensity level. Figure 6b shows the selected horizontal accelerograms for the seismic intensity level $\text{PGA} = 0.23g$, corresponding to the UHS PGA. It is possible to note in this figure that the spectra are relatively well dispersed, which will allows for the representation of the random nature of the seismic solicitation.

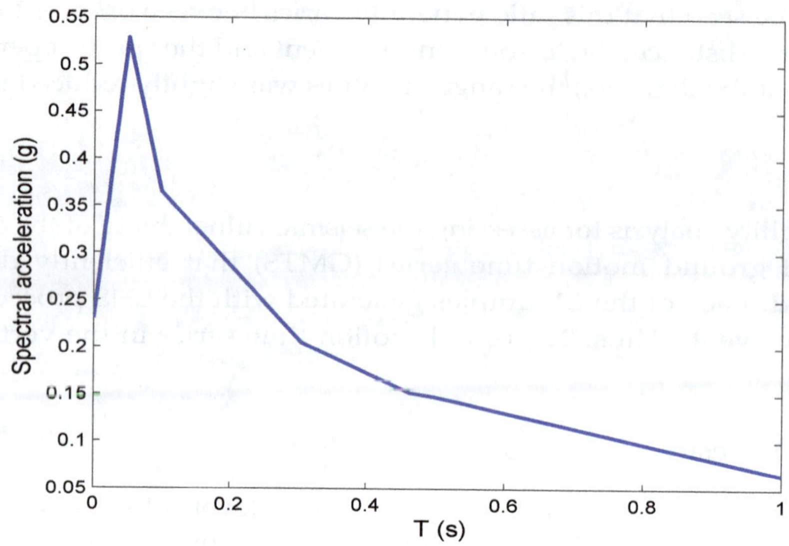

(a)

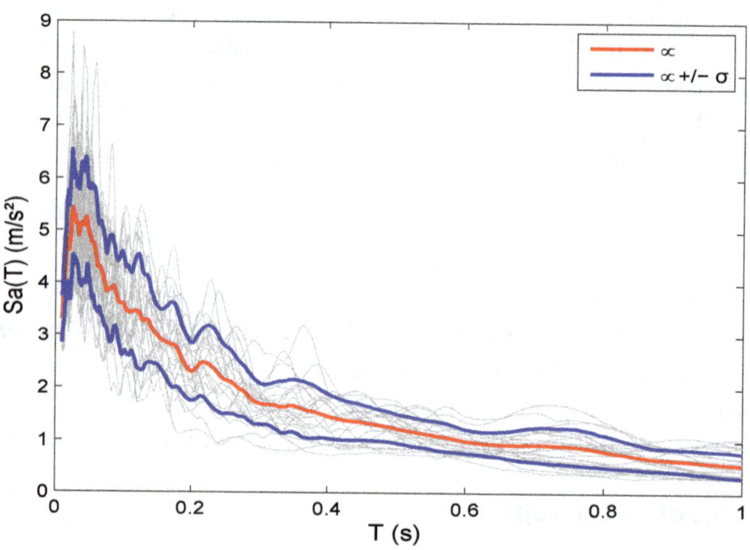

(b)

Figure 6. Spectra of selected horizontal accelerograms. **(a)** Target spectrum; **(b)** Selected horizontal accelerograms.

Each horizontal accelerogram is associated with a vertical accelerogram from the two remaining simulated components, and the one with the closest response spectrum to the UHS was kept. Following the same procedure as the horizontal component, in addition to the directionality factor $f_{h/v}$ associated with the sample, the vertical acceleration values were multiplied by the factor $f_{h/v} \times \mathrm{PGA}_{target} / \mathrm{PGA}_h$.

5. Fragility Analysis

With the FE model and the uncertainty characterisation presented above, it was possible to perform a fragility analysis to evaluate the vulnerability of the case study dam. Fragility analysis is a technique for assessing and displaying, in probabilistic terms, the capability of an engineered system to withstand a specified event [3]. In other words, as presented in Equation (8), the fragility $P_f(y)$

defines the conditional probability that a system reaches a structural limit state (LS) for a given ground motion intensity measure (IM):

$$P_f(y) = Pr[LS|IM = y]. \tag{8}$$

In this equation, y is the demand variable that defines the ground motion intensity. Two fundamental steps are needed before performing the fragility analysis: the (i) identification of all limit states relevant to the system performance and (ii) rational assessment of all sources of uncertainty likely to affect the response of the structure, as described in the previous section.

5.1. Limit States: Preliminary Analyses and Nonlinearities

Before initiating fragility analyses, a series of preliminary linear analyses were conducted. The objective was to determine where tensile stresses are likely to appear in order to limit the nonlinearities to those places only and to identify the typical damage modes that could lead to the potential collapse of dams after a seismic event. Linear analyses were conducted on the 22 samples assuming a peak ground acceleration of PGA = 0.25g—the highest seismic intensity level studied near the PGA of the target spectrum. All samples developed tensile stresses at the upstream and downstream foot during seismic loading, varying between 0.25 and 4.1 MPa. Figure 7a illustrates the development of tensile stresses at the upstream and downstream foot. Similarly, tensile stresses were also identified on several samples at the upstream or downstream faces of the block, as shown in Figure 7b, but given that no specific area could be identified, only the block–foundation contact at the base of the block was modified to introduce contact elements with a shear-tension failure criterion, as detailed above.

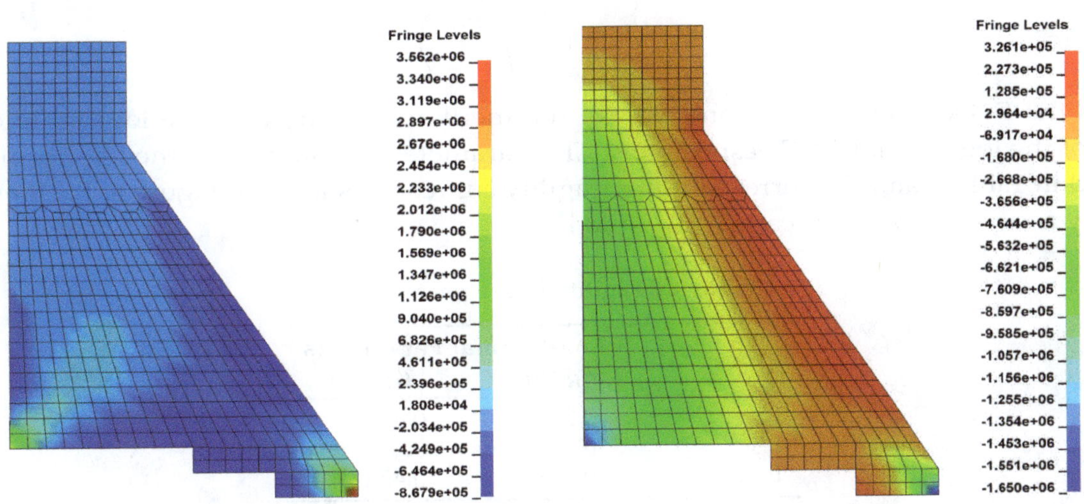

(b) (a)

Figure 7. (**a**) Main tensile stress areas; (**b**) Example of tensile stress zone at the downstream face.

In the same manner, the results from these analyses identified sliding as the critical failure mode for the case-study dam, and other failure modes would only occur after sliding was already observed. As a consequence, the only damage state contemplated in this study was sliding at the concrete–rock interface. Each limit state, presented in Table 7, was defined based on the damage states proposed in Tekie and Ellingwood [38].

Table 7. Limit states.

Limit State	Displacement (mm)
LS0—Minor damage	Incipient sliding
LS1—Moderate damage	25
LS2—Severe damage	50

5.2. Fragility Curves

Within this 7-intensity-level IDA framework, 22 samples of the finite element model were generated with LHS and were paired with the 22 sets of ground motion records selected as explained in Section 4.3.2. Nonlinear dynamic analyses were performed for the 22 samples at each considered intensity level, resulting in a total of 154 analyses. From each simulation, the maximum relative base displacement was computed. Table 8 summarises the results of the IDA analyses in terms of the fraction of the number of samples where sliding exceeded the limit state divided by the total number of samples. The results from Table 8 (i.e., the seven fragility point estimates) allow the development of fragility curves for the three considered limit states.

Table 8. Incremental dynamic analysis (IDA) results—Fraction of samples exceeding the limit state. PGA: peak ground acceleration.

Limit State	Seismic Intensity Levels—PGA (g)						
	0.1	0.2	0.25	0.3	0.4	0.5	0.7
LS0	0/22	2/22	5/22	10/22	15/22	17/22	20/22
LS1	0/22	0/22	1/22	3/22	6/22	10/22	19/22
LS2	0/22	0/22	0/22	0/22	3/22	5/22	11/22

A well-known mathematical equation was then fitted to these fragility estimates in order to develop fragility curves for the considered limit states. The log-normal cumulative distribution function (CDF), together with the least square estimate fitting technique, was found to be adequate for depicting the fragility of the three considered limit states, as given by:

$$P_f = \Phi\left(\frac{\ln y/m_R}{\beta_R}\right), \quad (9)$$

where y is the PGA level, and m_R and β_R are the median capacity and the logarithmic standard deviation of the log-normal CDF, respectively. The parameters m_R and β_R of the log-normal CDF are synthesised in Table 9, and the corresponding fragility curves are shown in Figure 8. Further details on the fragility analysis procedure can be found in [10].

Table 9. Fragility curve parameters.

Limit State	Log-Normal Parameters	
	m_R (g)	β_R
LS0	0.336	0.456
LS1	0.499	0.374
LS2	0.698	0.454

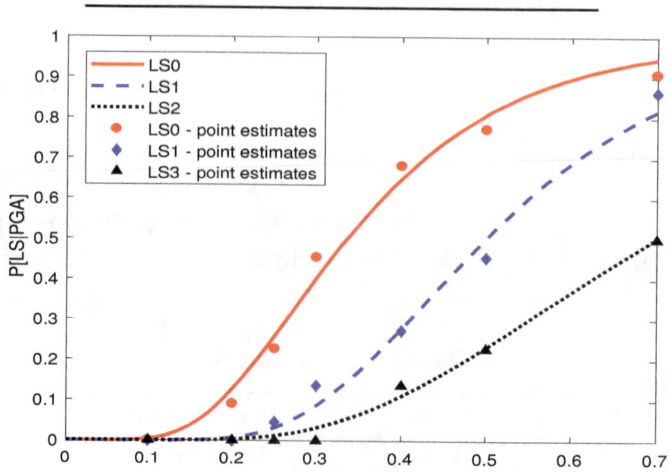

Figure 8. Fragility curves.

6. Expected Seismic Performance

For structures of special importance, such as dams and nuclear power plants, it is typical to consider the high confidence low probability of failure (HCLPF) criteria. This notion represents the level for which the probability of exceedance of a limit state is sufficiently low to consider that it will not be reached throughout the life of the structure [39] and usually corresponds to a probability of less than 5%. Accordingly, to evaluate the expected seismic performance of the case-study structure for the HCLPF level, the spectral acceleration for each limit state was extracted from the fragility curves for a probability of exceedance of 5%. Moreover, given that the PGA at the dam site for seismic events with a return period of 2475 could be computed, the expected seismic performance of the structure for events with these characteristics, usually used in the safety guidelines for dimensioning and verification, could also be estimated. Considering a $PGA_{target} = 0.23g$ from the UHS, the probability of exceedance was extracted from the fragility curves for each limit state.

The results presented in Table 10 show that the sliding at the concrete–rock interface of block-6 could be initiated for a PGA greater than $0.16g$. Additionally, the sliding remained low (<25 mm) for PGA values below $0.27g$, while it is unlikely to reach such a level of damage during an earthquake with a return period of 2475 years. Finally, the probability of reaching a severe damage limit state that could lead to the loss of control of the reservoir becomes significant when the PGA reaches values close to $0.33g$.

Table 10. Expected seismic performance results. HCLPF: high confidence low probability of failure.

Limit State	PGA for HCLPF	PGA_{target}
LS0	$0.16g$	20%
LS1	$0.27g$	2%
LS2	$0.33g$	0.7%

7. Conclusions

The main objective of this study was to assess the seismic vulnerability of a dam-type structure through a fragility framework and to present all the required analyses and verifications of the numerical model prior to obtaining the fragility curves. The methodology was applied to the bottom outlet spillway structure of the case study dam and more particularly to block-6 of the latter. For that purpose, a finite element model of the block–reservoir–foundation system was developed to describe the real behaviour of the structure as closely as possible. Following the results of the preliminary analyses, the study of the fragility of the system was restricted to a single damage state related to the sliding at the base of the block. Fragility curves were generated, taking into account the results of nonlinear dynamic analyses that were calculated with the structure samples obtained with LHS and coupled to GMTS, describing the seismic scenario at the dam site.

The methodology proposed herein makes it possible to account for the uncertainties associated with the modelling parameters as well as the randomness in the seismic solicitation. The results obtained from the fragility curves indicate that the probability of observing an incipient sliding at the base of the block, when solicited with PGA values used by the current guidelines for dimensioning and verification, is 20%. In contrast, it is unlikely that more severe damage to the structure for this intensity level will be observed.

Author Contributions: Data curation, C.D.; Methodology, C.B.; Supervision, P.P.; Validation and Writing—original draft, R.L.S.

Acknowledgments: The authors also thank Douglas Sparks, Martin Roberge and Benjamin Miquel, all of Hydro-Quebec, for their cooperation during this research project. Computational resources for this project were provided by Compute Canada and Calcul Quebec.

References

1. Lave, L.B.; Resendiz-Carrillo, D.; McMichael, F.C. Safety Goals for High-Hazard Dams: Are Dams Too Safe? *Water Resour. Res.* **1990**, *26*, 1383–1391.
2. Schultz, M.T.; Gouldby, B.P.; Simm, J.D.; Wibowo, J.L. *Beyond the Factor of Safety: Developing Fragility Curves to Characterize System Reliability*; Technical Report ERDC SR-10-01; U.S. Army Corps of Engineers: Washington, DC, USA, 2010.
3. Tekie, P.B. Fragility Analysis of Concrete Gravity Dams. Ph.D. Thesis, Johns Hopkins University, Baltimore, MD, USA, 2001.
4. Ellingwood, B.; Tekie, P.B. Fragility Analysis of Concrete Gravity Dams. *J. Infrastruct. Syst.* **2001**, *7*, 41–48. [CrossRef]
5. Mirzahosseinkashani, S.; Ghaemian, M. Seismic fragility assessment of concrete gravity dams. In Proceedings of the 29th Annual USSD Conference, Nashville, TN, USA, 20–24 April 2009; pp. 173–181.
6. Ghanaat, Y.; Hashimoto, P.S.; Zuchuat, O.; Kennedy, R.P. Seismic Fragility of Muhlberg Dam using Nonlinear Analysis with Latin Hypercube Simulation. In Proceedings of the 31st Annual USSD Conference, San Diego, CA, USA, 11–15 April 2011; pp. 1197–1212.
7. Ghanaat, Y.; Patev, R.C.; Chudgar, A.K. Seismic Fragility Analysis of Concrete Gravity Dams. In Proceedings of the 15th World Conference on Earthquake Engineering, Lisbon, Portugal, 24–28 September 2012.
8. Lupoi, A.; Callari, C. A probalistic method for the seismic assessment of existing concrete gravity dams. *Struct. Infrastruct. Eng.* **2011**, *8*, 985–998.
9. Zhong, H.; Lin, G.; Li, H.J.; Li, X.Y. Seismic Vulnerability Analysis of a Gravity Dam Based on Typical Failure Modes. In Proceedings of the 15th World Conference on Earthquake Engineering, Lisbon, Portugal, 24–28 September 2012.
10. Bernier, C.; Padgett, J.E.; Proulx, J.; Paultre, P. Seismic fragility of concrete gravity dams with modeling parameter uncertainty and spatial variation. *Am. Soc. Civ. Eng.* **2014**, *142*. [CrossRef]
11. Bernier, C.; Monteiro, R.; Paultre, P. Using the Conditonal Spectrum Method for Improved Fragility Assessment of Concrete Gravity Dams in Eastern Canada. *Earthq. Spectra* **2016**, *32*, 1449–1468. [CrossRef]
12. Hariri-Ardebili, M.A.; Saouma, V.E. Collapse Fragility Curves for Concrete Dams: A Comprehensive Study. *J. Struct. Eng.* **2016**, *142*, 04016075. [CrossRef]
13. Lallemant, D.; Kiremidjian, A.; Burton, H. Statistical procedures for developing earthquake damage fragility curves. *Earthq. Eng. Struct. Dyn.* **2015**, *44*, 1373–1389. [CrossRef]
14. Segura, R.; Bernier, C.; Monteiro, R.; Paultre, P. On the seismic fragility assessment of concrete gravity dams in eastern Canada. *Earthq. Spectra* **2018**, *35*, 211–231. [CrossRef]
15. Hariri-Ardebili, M.A.; Saouma, V.E. Seismic fragility analysis of concrete dams: A state-of-the-art review. *Eng. Struct.* **2016**, *128*, 374–399. [CrossRef]
16. Shinozuka, M.; Feng, M.; Lee, J.; Naganuma, T. Statistical Analysis of Fragility Curves. *J. Eng. Mech.* **2000**, *126*, 1224–1231. [CrossRef]
17. Livermore Software Technology Corporation. LS-Dyna User's Manual. In *Reference Manual*; Livermore Software Technology Corporation: Livermore, CA, USA, 2013.
18. Noble, C.R. Finite Element Techniques for Realistically Simulating the Seismic Response of Concrete Dams. Ph.D. Thesis, University of California, Davis, CA, USA, 2007.
19. USBR. *State-of-Practice for the Nonlinear Analysis of Concrete Dams at the Bureau of Reclamation*; Technical Report; United States Bureau of Reclamation: Washington, DC, USA, 2013.
20. Paultre, P. *Dynamics of Structures*; John Wiley & Sons, Inc.: Hoboken, NJ, USA, 2011.
21. Depolo, D.; Kennedy, E.; Walker, T.; Tom, R. Analysis and design of large-scale civil works structures using LS-Dyna. In Proceedings of the 11th International LS-Dyna Users Conference, Detroit, MI, USA, 6–8 June 2010.
22. Hallquist, J. *LS-Dyna Theory Manual*; Livermore Software Technology Corporation: Troy, MI, USA, 2006.
23. USBR. *Nonlinear Analysis Research—Dam Safety Technology Development Program*; Technical Report DSO-08-02; United States Bureau of Reclamation: Washington, DC, USA, 2008.

24. Proulx, J.; Paultre, P. Experimental and numerical investigation of dam-reservoir-foundation interaction for a large gravity dam. *Can. J. Civ. Eng.* **1997**, *24*, 90–105.
25. ASTM. *Standard Test Method for Determinig the In Situ Modulus of Deformation of Rock Mass Using the Rigid Plate Loading Method*; Norme D4394-04; American Society for Testing and Materials: West Conshohocken, PA, USA, 2004.
26. USACE. *Gravity Dam Design*; Technical report EM 1110-2-2200; United States Army Corps of Engineers: Washington, DC, USA, 1995.
27. Lokke, A.; Chopra, A.K. Direct finite element method for nonlinear analysis of semi-unbounded dam-water-foundation rock systems. *Earthq. Eng. Struct. Dyn.* **2017**, *46*, 1267–1285. [CrossRef]
28. Sooch, G.S.; Bagchi, A. Effect of Seismic wave scattering on the Response of Dam-Reservoir-Foundation Systems. In Proceedings of the 15th World Conference on Earthquake Engineering, Lisbon, Portugal, 24–28 September 2012.
29. JMP. JMP, Data analysis software. In *User Manual 1989–2007*; SAS Institute Inc.: Cary, NC, USA, 2011.
30. Ghosh, J.; Padgett, J.E.; Dueñas Osorio, L. Surrogate modeling and failure surface visualization for efficient seismic vulnerability assessment of highway bridges. *Probabilistic Eng. Mech.* **2013**, *34*, 189–199. [CrossRef]
31. NRC. *Earthquake Engineering for Concrete Dams: Design, Performance and Research Needs*; National Academy Press: Washington, DC, USA, 1990; p. 165.
32. Gupta, H.K.; Rastogi, B.K. *Dams and Eartquakes*; Elsevier: Amsterdam, The Netherlands, 1976.
33. EPRI. *Uplift Pressures, Shear Strengths, and Tensile Strengths for Stability Analysis of Concrete Gravity Dams*; Technical Report TR-100345; Electric Power Research Institute: Palo Alto, CA, USA, 1992.
34. Campbell, K.W. A Study of the Near-Source Behavior of Peak Vertical Acceleration. *Eos Trans. Am. Geophys. Union* **1982**, 63, 1037.
35. Mohammadioun, G.; Mohammadioun, B. Vertical/Horizontal Ratio for Strong Ground Motion in the Near Field and Soil Non-linearity. In Proceedings of the 11th World Conference on Earthquake Engineering, Acapulco, Mexico, 23–28 June 1996.
36. Kunnath, S.K.; Abrahamson, N.; Chai, Y.; Erduran, E.; Yilmaz, Z. *Development of Guidelines for Incorporation of Vertical Ground Motion Effects in Seismic Design of Highway Bridges*; Rapport Technique CA/UCD-SESM-08-01; University of California: Davis, CA, USA, 2008.
37. Atkinson, G.M. Earthquake time histories compatible with the 2005 National Building Code of Canada uniform hazard spectrum. *Can. J. Civ. Eng.* **2009**, *36*, 991–1000. [CrossRef]
38. Tekie, P.B.; Ellingwood, B. Seismic fragility assessment of concrete gravity dams. *Earthq. Eng. Struct. Dyn.* **2003**, *32*, 2221–2240. [CrossRef]
39. Hadjian, A. Earthquake reliability of structures. In Proceedings of the 13th World Conference on Earthquake Engineering, Vancouver, BC, Canada, 1–6 August 2004.

12

Progressive Failure Analysis of a Concrete Dam Anchored with Passive Rock Bolts

Rikard Hellgren *, Richard Malm and Anders Ansell

Division of Concrete Structures, Department of Civil and Architectural Engineering, KTH the Royal Institute of Technology, 100 44 Stockholm, Sweden; richard.malm@byv.kth.se (R.M.); anders.ansell@byv.kth.se (A.A.)

* Correspondence: rikard.hellgren@byv.kth.se

Abstract: Passive rock bolts are commonly used to anchor concrete dams, and they may have a significant impact on stability-evaluations. However, these bolts are often omitted from dam safety analysis due to uncertainties regarding their condition and the size of displacements required in the dam-rock interface to mobilize significant bearing forces in the passive rock bolts. This paper address the latter question by studying the failure process of a small concrete dam anchored with rock bolts. Failure simulations were performed with the increased density method in a finite element model consisting of a dam, the corresponding part of the rock and rock bolts. Two types of approaches are used to simulate the anchorage of the rock bolts; a method where the anchorage to the rock is simulated using a fixed boundary condition; and a method where the anchoring of the bolts are modelled using springs. Depending on the method of analysis, the rock bolts contribute with 40–75% of the load-carrying capacity of the dam. The rock bolts increase the load-bearing capacity of the dam, partly through anchorage forces, but also by keeping the contact surface between rock and concrete together and thereby increase the shear capacity of the interface.

Keywords: passive rock bolt; concrete dam; progressive failure

---

1. Introduction

Passive rock bolts are commonly used to anchor concrete dams, and they may have a significant impact on stability-evaluations. However, from a dam safety perspective, there are two major concerns with such bolts. The first is the condition of the bolts with respect to degradation. In ageing hydraulic structures, it is not possible to determine the structural status of rock bolts due to their inaccessible placement [1]. This leads to large uncertainties regarding the strength of the bolts. The second concern is the failure process of a dam anchored with passive rock bolts that remain unloaded until movement occurs in the joint. For rock bolts under concrete dams, this means that a displacement of the dam must take place before the bolts are activated.

Dam safety authorities handle these uncertainties in different ways. The Swedish dam regulations do not permit the consideration of rock bolts in the stability assessments of new dams [2]. In some cases for old existing dams, it may be possible to consider bolts in a highly conservative manner. Engineers should neglect the contribution from rock bolts for dams in the highest consequence class. For dams in lower classes, rock bolts may be included, but the reinforcement stress must be limited to 140 MPa for steel with a characteristic yield strength of at least 370 MPa. The Norwegian dam regulation [3] contains a similar stress limitation, stating that the stress in the bolts should be limited to the lowest value of the yield stress divided by 2.1, or 180 MPa. In all the above countries, a safety factor of at least 1.0 shall be achieved without the contribution of bolts, even if bolts are included.

During the condition assessment and stability evaluation of older dams, contributions from rock bolts may be significant [4–6], and to exclude or limit their contribution may change the status of a dam from stable to unstable. To strengthen a stable dam is not effective, neither from an economic or an environmental perspective. However, to include the bolts in stability calculations requires a better understanding of the failure modes of dams anchored with rock bolts.

This paper aims to highlight the influence of rock bolts on the global dam safety and to present a methodology for including rock-bolts in FE-simulations, that easily can be adapted by dam engineers to different types of concrete dams. The failure process of a small concrete dam anchored with passive rock bolts is studied using a finite element method model consisting of a dam, the corresponding part of the rock and rock bolts. The rock bolts are included as beam elements using an explicit rock bolt model. Two types of approaches are used to simulate the anchorage of the rock bolts; a method where the anchorage to the rock is simulated using a fixed boundary condition; and a method where the anchoring of the bolts are modelled using non-linear springs. The results from the two approaches are compared with a focus on the displacement required to activate forces in the rock bolts. Further, the distribution in load-carrying between the bolts and the contact force in the rock-dam interface during the failure progression is described.

2. Method

To study how rock bolts affect the failure process of a small concrete dam, the research in this paper was performed as a case study, where one monolith from an 8 m high buttress dam in northern Sweden was used as the studied object. The dam is a small buttress monolith anchored with a total of 20 rock bolts. The front plate has a width of 8 m and a constant thickness of 1.2 m, with a 1 m wide dam toe and the buttress wall has a constant thickness of 3.0 m. Each rock bolt is drilled 3 m into the rock and embedded 2 m in the concrete. The bolts are ribbed steel rebars with a diameter of 25 mm. Eight of the bolts are located under the front plate in the upstream dam toe, and the remaining twelve are placed in two rows beneath the buttress wall. The dimensions of the monolith and the position of the bolts are shown in Figure 1. The bridge on the dam crest was not included in the analysis to simplify the geometry.

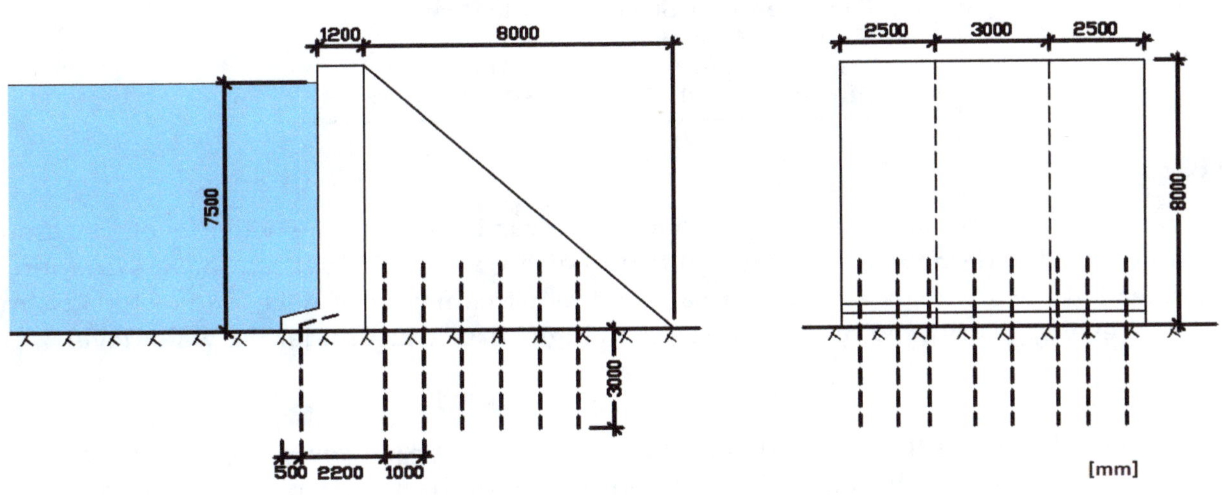

Figure 1. Sketch of the buttress monolith used as a case study, and the location of the 20 rock bolts.

2.1. *Analyzes*

Analytical stability calculations and simulations were performed for three cases to compare different assumptions regarding the bolts:

Case 1: Without bolts
Case 2: Bolts with a fixed attachment.
Case 3: Bolts with deformable attachment.

Case 1 serves as a reference and makes it possible to estimate the influence of the rock bolts compared to a dam without bolts. Cases 2 and 3 include rock bolts, but the analyses were performed with different assumptions for the bolt attachment. In Case 2, the bolts were anchored with a fixed stiff horizontal support without consideration to deformation in the grout and rock. In the hand calculation, that means the capacity of the bolt in the radial displacement was calculated using Equation (12). In the simulation, a restrained horizontal boundary condition was applied on the embedded part of the rock bolt element. In Case 3, the deformation in the grout and rock during shear loading was considered. The capacity of the bolts in the analytical stability evaluation was calculated using Equation (11). In the simulation, the rock bolt model presented in Section 2.4.2 was used as the horizontal support for the bolts.

2.2. *The Dam*

2.2.1. Material Properties

For the concrete and the rock, a linear material model was used, with all the properties for the simulations presented in Table 1. The values in Table 1 are based on [2,7,8].

Table 1. Material properties used for the case study.

Variable	Value	Unit
Density, concrete	23	kN/m^3
Elastic modulus, concrete	30	GPa
Density, rock mass	26.5	kN/m^3
Elastic modulus, rock	30	GPa
Adhesion grout-rock	2.0	MPa
Adhesion grout-steel	1.2	MPa
Adhesion concrete-steel	3.0	MPa
Elastic modulus, Steel	200	GPa
Ultimate strain, Steel,	0.15	-
Possion's ratio, Steel	0.30	-
Yield strength, Steel	370	MPa
Ultimate strength, Steel	600	MPa

2.2.2. Loads

Hydrostatic pressure, uplift pressure and ice load was included. The water level was assumed to be 0.5 m below the dam crest, and the uplift pressure was applied with a linear variation from full water head at the upstream edge to zero at the downstream edge. An ice load of 100 kN/m was applied as an evenly distributed pressure on a 0.6 m high surface just under the water level.

2.2.3. FE-Model

The FE-model includes the dam and part of the surrounding rock, see Figure 2. All the numerical analyses were performed with Abaqus [9], version 2018 with the standard implicit solver. The dam and rock were modelled with 8-node linear brick elements with reduced integration and hourglass control (element C3D8R in Abaqus).

The average element size was 0.2 m for the dam and 0.25 m for the rock, meaning that a total of 32,640 elements were used for the dam and 40,960 elements for the rock. In the model, the boundary condition was applied to the rock by prohibiting displacements perpendicular to each side at all outer boundaries of the rock, except the top surface, see Figure 2. The rock was included in the model with the length 20.0 m in the upstream-downstream direction, the width 8 m in the direction perpendicular to the cross-section of the dam, and the height 4.0 m. The purpose of including the rock in the model is to account for the friction at the interface and the dam is held in place by frictional forces in the interface between the dam and the foundation. The properties of the interface between dam and rock was considered according to the Swedish regulations for dam safety [2]. In the tangential direction a friction contact model was introduced where the friction coefficient was set to 1.0 and was kept constant throughout the analysis. The contact in the normal direction was modelled as a hard contact in compression but allowed separation due to tensile forces; no cohesive bond in the concrete-rock was considered.

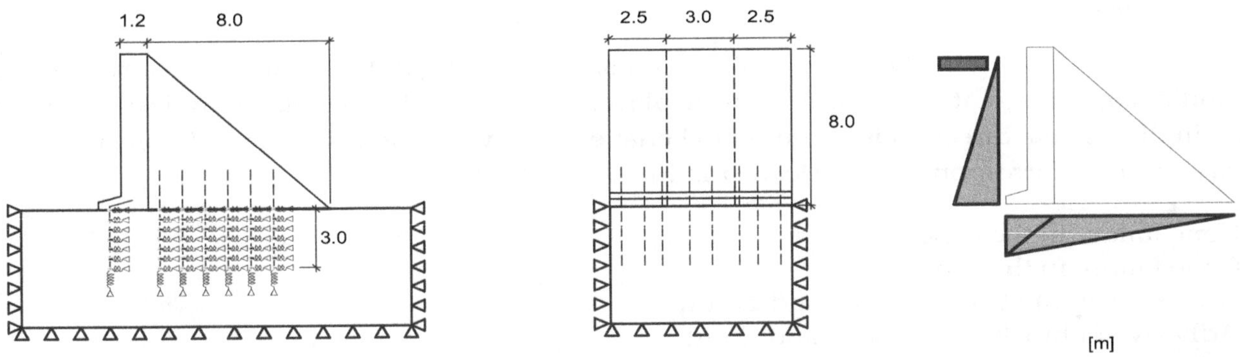

Figure 2. Static system with boundary conditions used in the numerical calculations and applied loads.

2.3. *Stability Analyses of Dams*

2.3.1. Analytical Stability Analyses

The analytical stability calculations were performed according to the Swedish guidelines, where a strong emphasis is put on the sliding and overturning failure modes [2]. In the dam safety assessments, it must be shown that concrete dams are safe against sliding and overturning failure modes. The stability analyses are performed with hand calculation methods, based on the force equilibrium of a rigid body.

The analytical calculations require that force resultant and lever arm are calculated for all loads, including self-weight. For sliding, the safety factor is defined as the ratio between the maximum friction force in the interface between concrete and rock and the acting forces parallel to the sliding surface (i.e., the horizontal forces).

$$S_{fs} = \frac{\mu_{int} \sum V}{\sum H} \tag{1}$$

where, μ_{int} is the coefficient of friction in the interface, $\sum H$ is the sum of all horizontal forces (i.e., parallel to the sliding plane) and $\sum V$ are the sum of all vertical forces.

The safety factor for overturning is defined as the ratio between the stabilizing moments M_s and the de-stabilizing moments, M_d.

$$S_{fo} = \frac{M_s}{M_d} \tag{2}$$

The moments were calculated with the dam toe as the axis of rotation.

2.3.2. Numerical Failure Analyses

There are two basic methods for simulating the failure of a structure; strength reduction and overload [10]. In the strength reduction method, normal loads are applied to the structure and thereafter are the strength reduced until failure occurs. In the overload method, normal loads are applied to the structure, and these loads are thereafter increased until a failure occurs. To achieve safety factors that are comparable with the analytical calculations, the failure analyses were performed with the increased density approach presented by [10,11]. In the increased density approach, the first step of analysis includes all design loads. In subsequent steps, all the destabilizing loads are gradually increased with a load factor of λ until failure occurs. The load factor that occurs at the point of failure is defined as the factor of safety of the structure, S.

$$S = \lambda S_{0(\lambda=1)} \tag{3}$$

where $S_{0(\lambda=1)}$ is the design load. The non-linearities considered in the analyses are the interface properties between rock and dam; the non-linear material behaviour of the steel; and the non-linear anchoring of the bolts caused by failure in the grout (incl. its interface to rock or bolt).

2.4. *Design of Rock Bolts for Dam Stability*

2.4.1. Analytical

Two kinds of movement occur in bolted joints; the opening of the joint in the normal direction (perpendicular to the joint plane) and shear displacement in the interface plane. In design, a rock bolt loaded in the normal direction is assumed to behave as the weakest link system. In tension, the load capacity of a bolt corresponds to the lowest failure load of [1]:

- Cone failure in the rock
- Cone failure in the concrete
- Adhesive failure between rock and grout.
- Adhesive failure between steel and grout.
- Adhesive failure between concrete and steel.
- Steel failure.

The cone failure in rock and concrete occurs when a volume theoretically shaped as a cone is pulled out. There are several ways to calculate the capacity of the cone. If the shear capacity of the material is neglected, the failure load can be calculated as the weight of the assumed cone according to Equation (4).

$$R_{\mathrm{Rock}} = V_{\mathrm{Rock}}\gamma_{\mathrm{Rock}} = \frac{\pi L_{\mathrm{E}}^{3}\sin(\beta)}{3}\gamma_{\mathrm{Rock}} \tag{4}$$

where R_{Rock} is the load that mobilises a rock volume V_{cone} with the angle β and height equal to the embedment depth of the rock bolt in the rock, L_{ER}; and γ_{rock} is the unit weight of the rock.

The adhesive bond failures occur when the stress reaches the maximum allowed bond stress for the contact surface, as mentioned above, the failure can occur in three interfaces where the load capacity for each interface is calculated as.

$$R_{\mathrm{Rock.Grout}} = \pi\phi_{H}L_{ER}\tau_{m.RG} \tag{5}$$

$$R_{\mathrm{Grout.Bar}} = \pi\phi_{B}L_{ER}\tau_{m.GB} \tag{6}$$

$$R_{\mathrm{Concrete.Bar}} = \pi\phi_{Bar}L_{EC}\tau_{m.CB} \tag{7}$$

Here, ϕ_H is the diameter of the rock bolt, ϕ_H the diameter of the hole and L_{EC} is the embedded length in the concrete. For the maximum allowed stress in the interface, $\tau_{m.XY}$, the subscript RG, GB and CB denotes the interface Rock-Grout, Grout-Bolt and Concrete-Bolt, respectively.

The capacity of the bolt is equal to the ultimate bond strength times the cross-sectional area.

$$R_{\mathrm{Steel}} = A_{\mathrm{Steel}}f_{yd} = \frac{\pi\phi_{\mathrm{Bar}}^{2}}{4}f_{yd} \tag{8}$$

The anchoring capacity of the rock bolt for an orthogonal load is the lowest value of the failure modes listed above.

$$R_{\mathrm{Rock\ Bolt}} = min(R_{\mathrm{Rock}}, R_{\mathrm{Rock.Grout}}, R_{\mathrm{Steel.Grout}}, R_{\mathrm{Concrete.Steel}}, R_{\mathrm{Steel}}) \tag{9}$$

For a rock bolt loaded in the shear plane, displacements activate forces in the bolt and the failure then depends on the interaction between steel, rock and grout and their non-linear material behaviour.

As forces build-up, the bolt holds the joint together, increasing the normal stress in the interface. This extra normal stress is added to the shear capacity of the joint due to friction, and its effect can exceed the shear capacity of the bolt itself. The resistance of the rock bolt to withstand shear deformations depends on the stiffness and compressive strength of the surrounding materials, i.e., grout, rock and concrete. The contribution of the bolt to the shear capacity of a joint may exceed the shear capacity of the steel material. This contribution consists of a dowel effect and a friction effect [8,12,13]. The dowel effect is due to the contribution from shear forces in the rock bolt while the friction effect is due to the increased friction in the joint caused by increased normal forces in the interface.

If the surrounding materials have high stiffness, then no bending of the bolt will occur. Instead, localized shear deformation occurs in the contact zone. This localized shear deformation is called the dowel effect and is illustrated in the left sketch in Figure 3. In hard rock, the mobilized shear stress becomes high enough to cut the bolt at low degrees of deformation before any significant tensile stresses have been developed. According to [14], this deformation could be as small as 1–3 mm. The dowel capacity of a bolt is equal to the shear strength times the cross-sectional area.

$$R_{\text{Steel.Shear}} = \frac{\pi\phi^2_{\text{Bar}}}{4}\frac{f_{yd}}{\sqrt{3}} \tag{10}$$

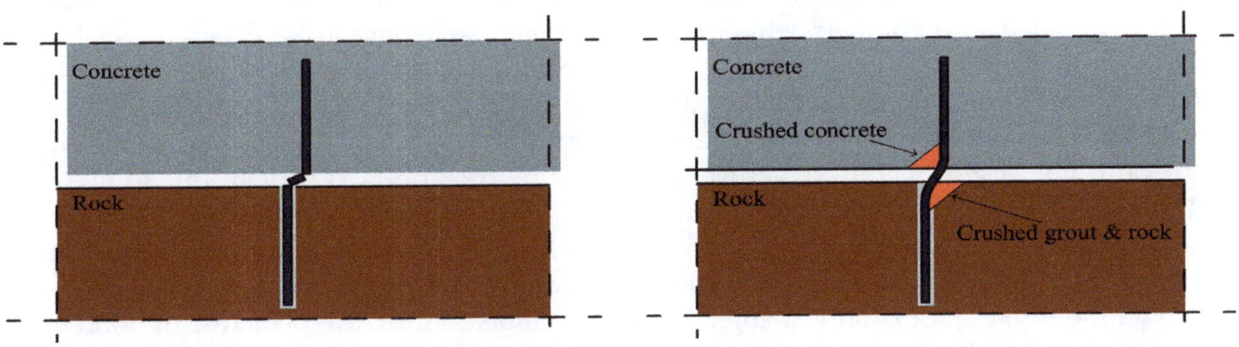

Figure 3. Illustration of the dowel effect (left) and crushing of surrounding material (right).

If the stiffness and compressive strength of the surrounding materials is low, then the bolt will gradually deform and crush the surrounding materials, and a tensile load component will develop, forming a crank shape [15]. In that case, the bolt will fail due to the fact that the tensile strength is exceeded giving a high deformation at failure. This deformation is usually in the order of the bolt diameter, which means that full strength is not mobilized until deformations of around 20 to 30 mm have occurred [15]. In addition, the contact zone between concrete and rock is not a smooth surface in reality. The roughness of the contact zone will give rise to a dilatation which will lead to tensile stresses in the bolt. The friction capacity of a rock bolt can be assumed to be equal to the tensile capacity. Hence, it can be determined from Equation (9).

If the interface between the rock and the dam is intersected by rock bolts, these will result in additional shear strength. When the sliding plane is subjected to shear deformation, the roughness will give rise to a dilatation which leads to tensile stresses in the bolt. The vertical component of the tensile stress in the bolt gives rise to a corresponding additional normal stress on the weakness plane, which leads to an increased shear strength for the weakness plane. Since the degree of mobilized tensile and shear stresses are often unknown in analytical calculations, it is generally assumed that the bolt either works as a dowel according to Equation (10) so that Equation (1) is rewritten as

$$S_S = \frac{\mu \sum V + \sum_i R_{\text{rock.bolt.shear}}}{\sum H} \tag{11}$$

or that only tensile stresses are developed according to Equation (9).

$$S_S = \frac{\mu(\sum V + \sum_i R_{\text{rock.bolt.tension}})}{\sum H} \tag{12}$$

where, $\sum R_{\text{rock.bolt.tension}}$ or $\sum R_{\text{rock.bolt.shear}}$ is the contribution from all rock bolts crossing the sliding plane.

In an overturning failure mode, analytical calculations are performed where it typically is assumed that the dam will rotate around the downstream toe. A moment balance equation is defined for the stabilizing and overturning moments. The rock bolts will introduce additional resistance to the stabilizing moment. As described earlier, the load capacity of the rock bolts can be defined as the minimum of all potential failure modes. In the conventional design, the rock bolts are however designed to result in a steel failure, due to its ductile behaviour.

2.4.2. Numerical Rock Bolt Model

Two approaches are used when analysing rock bolt by numerical methods; explicit [13,16,17], and implicit [18–21] methods. In the implicit method, the rock bolts contribution to the joint is smeared over the rock mass by adding it directly in the constitutional law of the rock or joint. This approach is used to simulate larger civil engineering problems where it is inconvenient to model each rock-bolt. Contrary, the rock bolt is modelled as an own part in the explicit method. Therefore, the explicit method allows a more detailed study of their behaviour. To simulate the behaviour of grouted rock bolts under shear using explicit models, the bolt, grout, and the nonlinear material behaviour and interactions between interfaces is often treated with a high level of detail, see for example [17,22,23]. These types of analyses are therefore computational demanding and are in most cases limited to the simulation of a single bolt. For large concrete structures anchored by rock bolts, it is desirable to use a simpler rock bolt model. Using a detailed simulation of a rock bolt to a dam monolith with 20–50 anchors involves high computational costs. As an alternative, a simplified approach is used in this paper, where the bolt is modelled with an explicit model where the bolt is represented with a beam element and the attachment to the rock with springs. For the axial behaviour, the principle of bond stress-slip relationship from [24,25] was used. The attachment to the rock is modelled using non-linear springs. The springs were modelled with connector elements with a predefined force-displacement curve in Abaqus. The anchorage of bolts are modelled with a bond stiffness-displacement $k(u)$ relationship consisting of four parts, according to Equation (13) where $k(u)$ is the foundations stiffness (N/m) and u the displacement.

The input for the model is the maximal force k_m, the residual force stress k_R; the three displacements $u_1 - u_3$; and the parameter α.

$$k(u) = \begin{cases} k_m \left(\frac{u}{u_1}\right)^\alpha, & \text{if } u \leq u_1 \\ k_m, & \text{if } u_1 < u \leq u_2 \\ k_m - (k_m - k_r)\frac{u-u_2}{u_3-u_2}, & \text{if } u_2 < u \leq u_3 \\ k_r, & \text{if } u_3 < u \end{cases} \tag{13}$$

The stiffness is recalculated to a force-displacement relationship at each node of the bolt element,

$$F(u) = k(u)h \tag{14}$$

where $F(u)$ is the force in the spring and h the length of the mesh element. This force-displacement relationship was used to model both the displacement in the axial and radial direction of the bolt. The rock bolt is considered in the modelled where a three-dimensional beam element with both transitional and rotational degrees of freedoms represents the bolts. A spring is attached to the bolt at every node, and on the other end of the spring, a fixed displacement boundary condition is applied.

The rock and the springs are therefore not connected, and the purpose of including the rock in the model is to account for the friction at the interface. The properties for the axial stiffness was chosen in accordance with [26], who performed calibrations of the spring stiffness in the axial direction on pull out test performed by [14]. The radial stiffness was calibrated based on shear tests on 16 mm rock bolts installed in granite blocks 250 × 250 × 250 mm^3 in size [14]. A test with an untensioned bolt installed perpendicular to the shear plane was chosen for the calibration. Figure 4a shows the layout of the FE-model and the input data is presented in Table 2. After a mesh size convergence study, an element size of one-tenth of the embedment length was used for all three parts. The boundary condition was applied to the rock by prohibiting displacements perpendicular to each side at all outer boundaries of the rock, except for the top surface, as seen in Figure 4a. The same contact properties were used for the interface as described above for the simulations of the dam where the friction coefficient was changed to correspond to the material data given in Table 2. A vertical pressure was applied on the top surface with magnitude in accordance with Table 2.

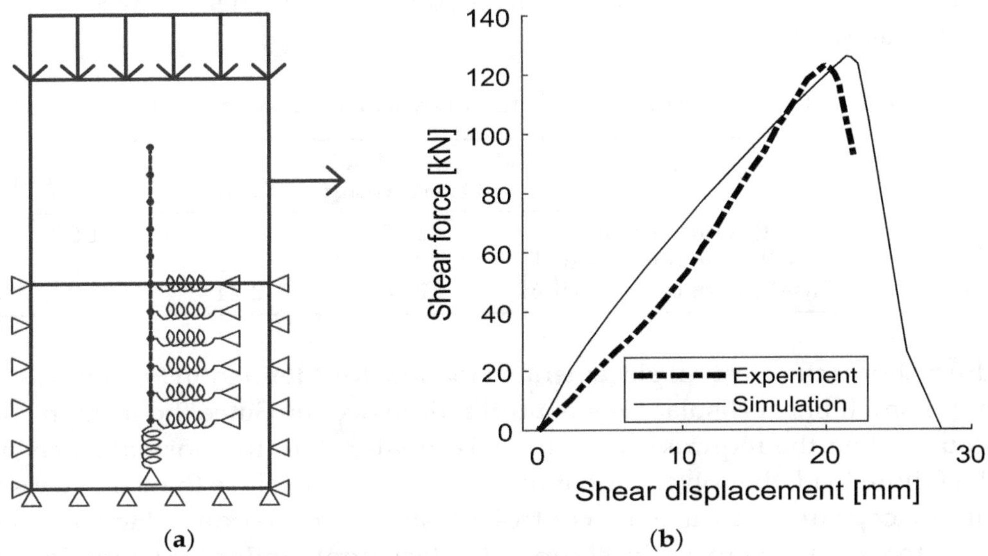

Figure 4. (**a**) Sketch of the principal design of the simulation models and shear tests, (**b**) Force displacement curves from tests performed by [14] and corresponding simulations.

Table 2. Material properties for the shear test from [14], that was used to calibrate the radial spring stiffness.

Variable	Value	Unit
Diameter rock bolt	16	mm
Diameter drilled hole	32	mm
Yield stress	370	MPa
Ultimate stress	600	MPa
Ultimate strain	0.05	-
Rock compressive strength	162	MPa
Rock elastic modulus	69	GPa
Friction angle along the shear plane	31	-
Applied normal stress	4.0	MPa

Figure 4b shows the force-displacement curve for the simulation and test with the best fit. Table 3 presents the forces used for the two directions.

Table 3. The properties used for the attachment spring.

Model	Unit	Axial	Radial
k_m	N/mm	$\pi\phi_B\tau_{m.GB}/u_1$	120
k_r	mm	0	0
u_1	mm	6	20
u_2	mm	35	25
u_3	mm	36	26
α		0.4	0.0

3. Results

Table 4 presents the safety factors from analytic stability calculations and simulations. The bolts significantly increase the factor of safety in all analyses. For overturning, the safety factor is almost the double for Case 2 and 3 with bolts compared to Case 1 without bolts. For sliding, the inclusion of rock bolts increases the safety factor by 70% and 120% for treating the bolts as dowels or tension, respectively. The difference between the safety factor for cases 1 and 2 or 3 is even more significant in the simulations. Overall, the passive bolts account for somewhere between 40% and 75% of the load capacity of the dam.

Table 4. Calculated factor of safety for analytical stability calculations and simulations.

	Analytic		Simulations
	Overturning	Sliding	
Case 1, without bolts	1.35	1.03	1.00
Case 2, with bolts, dowel	2.46	1.73	3.22
Case 3, with bolts, tension	2.46	2.24	3.23

Figure 5 shows the crest displacement vs load factor (defined in Section 2.3.2) for the simulation. Case 1 shows small linear displacements until failure occurs. Since the friction coefficient is set to 1.0, failure occurs when the required shear force is greater than the normal force. Whe the simulation reaches the failure load, the solver cannot find an equilibrium since the dam starts to slide. This sliding is difficult to capture with a load controlled static simulation. Hence, this non-convergence was interpreted as an increase in sliding displacement under constant loads, i.e., what would be characterized as a failure. The crest displacement up to this load level is small and identical for all the models.

At the failure load level for Case 1, the bolts are activated in Cases 2 and 3, and thereby, the load can be further increased. In these models, the behaviour differs between these two cases. In Case 2, the forces in the bolts can increase immediately, while Case 3 require larger displacements in the rock-dam interface before the load in the bolts increases. Already for small displacements, significant strains develop in the bolts. This leads to a significant increase in the bearing capacity of the bolts even for a small displacement of the dam. As the steel yields, the displacements increases and a large displacement occurs before the ultimate capacity of the dam is reached. Case 3 shows a similar failure process but with larger displacements. Since the springs allow displacement of the horizontal support of the bolts, larger deformations of the dam are required for the same load. In the initial state, displacements occur in the spring before sufficient resistance is developed so that the steel starts to deform. Thereafter the load-deformation curve follows a path similar to that of Case 2. Both models fail to converge at approximately the same load level when the ultimate stress for the bolts are reached in all bolts, and this non-convergence was interpreted as a failure. The displacement required to increase the load factor from 1.0, the failure of the unanchored dam, to 1.35 (the required safety factor for sliding according to the Swedish guidelines [2]) is 0.55 mm and 0.89 mm for Case 2 and Case 3, respectively. To corresponding displacements to achieve a bearing capacity that fulfils the load factor required for the overturning failure mode is 0.74 mm and 1.2 mm.

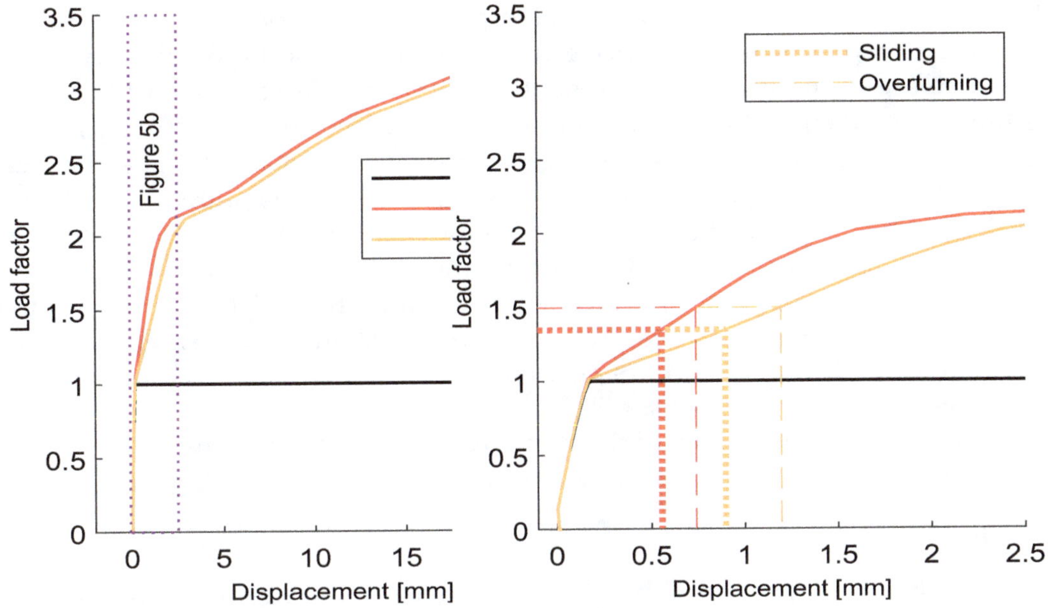

(a) **(b)**

Figure 5. Load factor as a function of crest displacement, a load factor up to 1.0 includes all design loads while only the destabilizing loads are amplified for greater load factors. Case 1 is without bolts; Case 2—bolts with a fixed attachment and Case 3—bolts with deformable attachment. (**a**) The complete load-displacement curve, and (**b**) load-displacement curve for the initial part of the curve with the the required safety factor for sliding (1.35) and overturning (1.5) according to the Swedish guidelines for dam safety marked [2].

Figure 6 shows the vertical and horizontal reaction forces, i.e., the contact forces in the surface between rock and dam and the total reaction force for all the bolts. At load factor zero (only the dead load applied) the shear force on the surface is approximately zero. As the water pressure increases, the shear force increases. For the normal forces, the relationship is the opposite. At load factor zero, the total normal force is equal to the dead load, and the normal force thereafter decreases linearly as the uplift pressure increases.

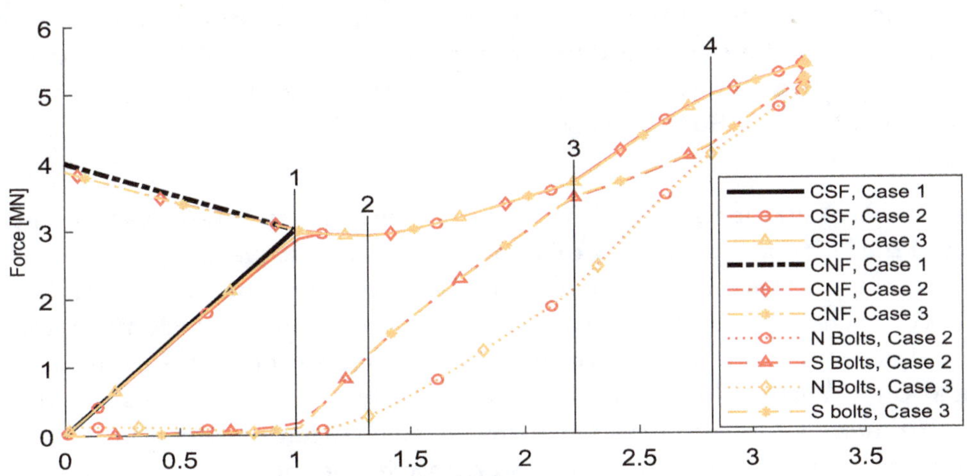

Figure 6. Shear and normal reaction forces as a function of load factor presented as forces in the bolts and on the surface of the dam-rock interface. The notation CNF is the contact normal force, and CSF is the contact shear force. N and S are normal and shear forces in the bolts, respectively. Case 1 is without bolts; Case 2 bolts with a fixed attachment and Case 3 bolts with deformable attachment.

During the progressive failure, the following notable events occur at the points marked in Figure 6.

1. The failure load for Case 1, the model without bolts. From the red dashed line with triangles in Figure 6, representing the horizontal forces in the bolts in the fixed model, it can be seen that there are horizontal forces present in the bolts before this load level. This is not the case for the orange dashed line with asterisks representing the shear forces in the bolt in the Case 3 model, where the bolt forces first occur after this load level. The difference in shear force in the bolts between the Case 2 and Case 3 models is at this point approximately 100 kN.
2. The initial linear slope in the force-displacement curve for Case 3 ends shortly after a minimum in the interface forces. After this point, the normal forces in the bolts are large enough to pull the dam towards the rock and create a frictional effect. The increasing load is carried by a combination of increasing forces in the concrete-rock interface and in the bolts.
3. A change in failure mode occurs as the increasing load is carried by an increasing proportion by normal force in the bolts.
4. Another change in behaviour of the dam, from here to the failure, is the increase in load capacity mainly due to increased forces in the bolts.

4. Discussion

Previous studies [4,6] and Table 4 shows that rock bolts may have a significant impact on the stability of small concrete dams. Inclusion of rock bolts in stability calculations can make the difference between not satisfying the design criteria and being able to withstand the design loads. Vice versa, excluding the rock bolts from the safety assessment of small dams, means that significant a part of the load-bearing system is neglected. This can also mean that an incorrect failure mode is assumed and that the dam is thereby strengthened in a suboptimal way. For example, post-tensioned anchorage may be installed to prevent an overturning failure where a backfill to increase the sliding resistance would be the measure that most increases the overall dam safety. Including the effect of rock bolts in the stability, analyses are therefore of great practical use with decreasing costs, from both an economical and an environmental aspect.

However, the bolts do not contribute to the load-bearing until a displacement occurs in the dam-rock interface. The results presented in Figure 5 indicates that a significant part of the additional bearing capacity is achieved already at relative small displacements. The displacement required to increase the load factor from 1.0 (the failure of the unanchored dam) to a load factor that fulfils all stability requirements is 1.2 mm. If dilatation and unevenness in the dam-rock interface were considered, this displacement would be even smaller. A common definition of a dam failure is: "Collapse or movement of a part of a dam or its foundation, so that the dam cannot retain water. In general, a failure results in the release of large quantities of water, imposing risks on the people or property downstream" [27]. Such release of water is highly unlikely to occur as a result of this magnitude of displacement.

The interaction between forces in the interface and the bolts are presented in Figure 6. For a dam, the friction effect of the bolts, i.e., the ability of a bolt to hold the interface together and thereby increase the friction force in the interface is significant. The bolts prevent the dam from lifting, leading to a maintained and even increased restraining friction force in the concrete-rock interface. After activation of the bolts, the normal force and corresponding shear force in the interface increase, although the resultant of the applied vertical load decreases as the uplift pressure increases. In the simulations, the up-lift pressure increases as the load factor increases. In reality, the ability of the bolts to hold the interface together may reduce or prevent water flow in the joint. The positive effect of rock bolts under the front plate is therefore likely to be underestimated in the simulations. The shear forces in the bolts occur at smaller displacements, and during the loading process, they are significantly higher than the normal forces in the bolts. However, the results indicate that the bolts are carrying a combination of tensile and shear forces. This implies that analytical calculations should preferably be based on the load-carrying capacity of combined normal and shear forces.

In the case study, a flat interface between rock and concrete and a constant friction coefficient was used, with cohesion neglected. In reality, the friction coefficient decreases from a static to a kinematic coefficient after the initial movement breaks the bonds created by asperities in the interface. The same reasoning can be used for cohesion. After the movement required to activate the bolts, the cohesive bond in the interface will be lost. The combination of cohesive bond failure along with the decrease in friction angle is not trivial and, for the purpose of studying the rock bolts impact, that issue was excluded. To simulate the behaviour of a grouted rock bolt, a model that includes deformation both in grout and in rock should be used. Neglecting these deformations leads to an inaccurate estimation of the failure behaviour for an anchored structure, where the bolt forces under small displacements are overestimated.

5. Conclusions

The effect of passive rock bolts contribution to the stability of small concrete dams has been studied. This has been studied through traditional analytical stability calculations and FE-simulations of the failure progress that allows a more detailed study of the distribution in load carrying capacity between the bolts and the contact force in the rock-dam interface, during the failure progression.

The analytical stability calculations confirm the background to this study and show that the inclusion of rock bolts can make the difference between a dam that is deemed not safe and one that with great margin fulfil the design criteria. Depending on the method of analysis, the rock bolts contribute with 40–75% of the total load-carrying capacity of the dam. Current design criteria hence neglect a significant part of the load-bearing system for small dam anchored with rock bolts.

Simulation of the failure process of an anchored concrete dam shows that the forces in the bolts are negligible up to the load level at which a failure occurs for the unanchored anchored dam. Thereafter, a displacement occurs in the contact surface, and the rock bolts are activated. Initially, the horizontal forces in the rock bolts increase but relatively early, the rock bolts help to hold the contact surface together so that the frictional force in the interface can be increased. The rock bolts increase the load-bearing capacity of the dam, partly through anchorage forces, but also by keeping the contact surface between rock and concrete together and thereby increase the shear capacity of the interface. The simulation shows that the contribution from passive rock bolts at 0.74 mm and 1.2 mm displacement in the dam-rock interface increases the load-carrying capacity sufficient to fulfil the required safety for sliding and overturning failure, respectively. Such displacement is unlikely to lead to an uncontrolled release of water. The current requirement for a safety factor of at least 1.0 without bolts is a sound criterion for the serviceability limit state, but it is overly conservative not to include the bolts in the ultimate failure capacity.

This study address one of the two major concerns regarding passive rock in dam safety presented in the introduction. The concern about the condition of inaccessible rock bolts under concrete dams with regards to degradation remains, and should be given further attention. In the case study, a flat interface between rock and concrete, and a constant friction coefficient was used, with cohesion neglected. In further studies, the failure of a dam with regard to the interaction between cohesive bond failure, dilatation, friction coefficient decrease and rock bolt should be studied.

Author Contributions: The authors confirm contribution to the paper as follows; conceptualization, R.M. and R.H.; methodology, R.H. and R.M.; software, R.H.; validation, R.H., R.M. and A.A.; formal analysis, R.H.; investigation, R.H, R.M.; resources, R.M. and A.A.; data curation, R.H.; writing original draft preparation, R.H.; writing review and editing, R.H., R.M. and A.A.; visualization, R.H.; supervision, R.M. and A.A.; project administration, R.M.; funding acquisition, R.M. All authors have read and agreed to the published version of the manuscript.

References

1. Ekström, T.; Hassanzadeh, M.; Janz, M.; Sederholm, B.; Stojanovic, B.; Ulriksen, P. *Condition Assesment of Rock Anchors in Dams*; Technical Report; Elforsk: Stockholm, Sweden, 2013.
2. RIDAS. *Swedish Hydropower Companies Guidelines for Dam Safety, Application Guideline 7.3 Concrete Dams*; Technical Report; Svensk Energi: Stockholm, Sweden, 2017.
3. NVE. *Guidelines for Concrete Dams*; Technical Report; Norges Vassdrags- og Energidirektorat: Oslo, Norway, 2005.
4. Larsson, C. *Investigation and Testing of Rock Bolts at Hotagen Power Station*; Technical Report; Elforsk: Stockholm, Sweden, 2009.
5. Hellgren, R.; Rios Bayona, F.; Malm, R.; Johansson, F. Pull-out tests of 50-year old rock bolts. In Proceedings of the International Symposium "Appropriate technology to ensure proper Development, Operation and Maintenance of Dams in Developing Countries", Johannesburg, South Africa, 18 May 2016; pp. 263–272.
6. Berzell, C. Load Capacity of Grouted Rock Bolts in Concrete Dams. Master's Thesis, KTH Royal Institute of Technology, Concrete Structures, Stockholm, Sweden, 2014.
7. Westberg Wilde, M.; Johansson, F. *Probabilistic Model Code for Concrete Dams*; Technical Report; Energiforsk: Stockholm, Sweden, 2016.
8. Malm, R.; Johansson, F.; Hellgren, R.; Ríos Bayona, F. *Load Capacity of Grouted Rock Bolts Due to Degradation*; Technical Report; Energiforsk: Stockholm, Sweden, 2017.
9. Dassault Systemes. Abaqus CAE (version 2017) 2017. Available online: https://www.3ds.com/products-services/simulia/products/abaqus/abaquscae/ (accessed on 27 May 2019).
10. Malm, R. *Guideline for FE Analyses of Concrete Dams*; Technical Report; Energiforsk: Stockholm, Sweden, 2016.
11. Nordström, E.; Malm, R.; Johansson, F.; Ligier, P.L.; Øyvind, L. *Betongdammars Brottförlopp Energiforsk Rapport 2015:122*; Failure of Concrete Dams—Literature Review and Potential for Development; Technical Report; Energiforsk: Stockholm, Sweden, 2015.
12. Bjurstrom, S. Shear strength of hard rock joints reinforced by grouted untensioned bolts. In Proceedings of the Third Congress of the International Society for Rock Mechanics, Denver, CO, USA, 1–7 September 1974; Volume 2, pp. 1194–1199.
13. Spang, K.; Egger, P. Action of fully-grouted bolts in jointed rock and factors of influence. *Rock Mech. Rock Eng.* **1990**, *23*, 201–229. [CrossRef]
14. Bjurström, S. *Bolted Hard Jointed Rock*; Technical Report; The Swedish Fortifications Agency: Stockholm, Sweden, 1973.
15. Stille, H. Keynote Lecture: Rock Support in Theory and Practice. In *Rock Support in Mining and Underground Construction*; CRC Press: Boca Raton, FL, USA, 1972; pp. 421–437.
16. Chen, S.H.; Qiang, S.; Chen, S.F.; Egger, P. Composite Element Model of the Fully Grouted Rock Bolt. *Rock Mech. Rock Eng.* **2004**, *37*, 193–212. [CrossRef]
17. Jalalifar, H.; Aziz, N. Experimental and 3D Numerical Simulation of Reinforced Shear Joints. *Rock Mech. Rock Eng.* **2010**, *43*, 95–103. [CrossRef]
18. Chen, S.H.H.; Egger, P. Three dimensional elasto-viscoplastic finite element analysis of reinforced rock masses and its application. *Int. J. Numer. Anal. Methods Geomech.* **1999**, *23*, 61–78. [CrossRef]
19. Stille, H.; Holmberg, M.; Nord, G. Support of weak rock with grouted bolts and shotcrete. *Int. J. Rock Mech. Min. Sci.* **1989**, *26*, 99–113. [CrossRef]
20. Chen, S.H.; Yang, Z.M.; Wang, W.M.; Shahrour, I. Study on rock bolt reinforcement for a gravity dam foundation. *Rock Mech. Rock Eng.* **2012**, *45*, 75–87. [CrossRef]
21. D'Amato, M.; Braga, F.; Gigliotti, R.; Kunnath, S.; Laterza, M. Validation of a modified steel bar model incorporating bond-slip for seismic assessment of concrete structures. *J. Struct. Eng.* **2012**, *138*, 1351–1360. [CrossRef]
22. Grasselli, G. 3D behaviour of bolted rock joints: experimental and numerical study. *Int. J. Rock Mech. Min. Sci.* **2005**, *42*, 13–24. [CrossRef]
23. Kilwic, A.; Yasar, E.; Celik, A.G. Effect of grout properties on the pull-out load capacity of fully grouted rock bolt. *Tunn. Undergr. Space Technol.* **2002**, *17*, 355–362. [CrossRef]
24. Bahrani, N.; Hadjigeorgiou, J. Explicit reinforcement models for fully-grouted rebar rock bolts. *J. Rock Mech. Geotech. Eng.* **2017**, *9*, 267–280. [CrossRef]
25. Fib. *Model Code for Concrete Structures 2010*; Technical Report; Wilhelm Ernst & Son: Lausanne, Switzerland, 2010.

26. Sjölander, A.; Hellgren, R.; Ansell, A. Modelling aspects to predict failure of a bolt-anchored fibre reinforced shotcrete lining. In Proceedings of the Eight International Symposium on Sprayed Concrete—Modern Use of Wet Mix Sprayed Concrete for Inderground Support, Trondheim, Norway, 11–14 June 2018; pp. 278–292.
27. ICOLD. *Dam Failures—Statistical Analysis Bulletin 99*; Technical Report; International Commission on Large Dams (ICOLD): Paris, France, 1995.

13

A Probabilistic Approach to the Spatial Variability of Ground Properties in the Design of Urban Deep Excavation

Jacob B. Herridge [1], Konstantinos Tsiminis [2], Jonas Winzen [3], Arya Assadi-Langroudi [1,*], Michael McHugh [4], Soheil Ghadr [5] and Sohrab Donyavi [1]

1 Department of Computing and Engineering, University of East London, London E16 2RD, UK
2 Fairhurst, Newcastle Upon Tyne NE4 6DB, UK
3 bei AECOM, 45127 Essen, Germany
4 MESLEC Ltd., Elstree WD6 3DP, UK
5 Faculty of Engineering, Civil Engineering Department, Urmia University, Urmia 5756151818, Iran
* Correspondence: A.AssadiLangroudi@uel.ac.uk

Abstract: Uncertainty in ground datasets often stems from spatial variability of soil parameters and changing groundwater regimes. In urban settings and where engineering ground interventions need to have minimum and well-anticipated ground movements, uncertainty in ground data leads to uncertain analysis, with substantial unwelcomed economical and safety implications. A probabilistic random set finite element modelling (RSFEM) approach is used to revisit the stability and serviceability of a 27 m deep submerged soil nailed excavation built into a cemented soil profile, using a variable water level and soil shear strength. Variation of a suite of index parameters, including mobilized working loads and moments in facing and soil inclusion elements, as well as stability and serviceability of facing and the integrated support system, are derived and contrasted with field monitoring data and deterministic FE modelling outputs. The validated model is then deployed to test the viability of using independent hydraulic actions as stochastic variables as an alternative to dependent hydraulic actions and soil shear strength. The achieved results suggest that utilizing cohesion as a stochastic variable alongside the water level predicts system uncertainty reasonably well for both actions and material response; substituting the hydraulic gradient produces a conservative probability range for the action response only.

Keywords: stochastic; probabilistic; excavation; movement; field; groundwater; soil nail; spatial; variability

1. Introduction

Rapid and sustained urban sprawl is expected to continue to accommodate an estimated 70% of the world population in cities by 2050 [1]. Exploitation of the urban sub-surface has attracted increasing attention in the past 50 years as a solution to a shortage of space and services per head, and is reflected, nationally and internationally, in many current planning policies, including [2] in the UK. Ground movements arising from deep excavation need to be strictly limited to permissible levels; that is, to be controlled through design and deployment of top-down support of excavation systems that allow prevention and mitigation of excessive ground movements from the outset. In an urban context and where the subsurface is already congested (by buried utilities and underground structures), any further disturbance can lead to variations in groundwater conditions, with immediate implications on some soil engineering properties. Spatial variability of shallow disturbed ground and limited space for intrusive ground investigation, as well as extreme climates that reportedly cause a broad range of problems, including fatigue [3] and mineral dissolution [4], add to the uncertainty and cast doubt on the reliability of the estimated stability and serviceability of support systems. At a practical level, these systems are commonly designed through deterministic analysis of engineered grounds using ground properties that are statistically moderated, and in the context of some standards, further factored down by an arguably arbitrary partial factor.

Ideally, support systems should be adaptive in allowing improved integration with the variable ground and groundwater conditions during the course of construction and also the service life. Use of arbitrary safety factors fall short in satisfying that integration. This paper studies the scopes for using an alternative statistical design approach, with an attempt to incorporate likely scenarios of actions and material properties into design within the context of an urban deep basement case study. The framework may be adopted as an alternative to arbitrary partial and safety factors. Stochastic discretization of random fields is used to determine and incorporate into finite element (FE) analysis sets of random variables to revisit a monitored 28 m deep submerged soil nail system. Effects of actions and serviceability are determined and reported in lower-bound to upper-bound ranges, and alongside, the most likely indexes (for structure performance) are contrasted, with outcomes of commonly practiced deterministic FE and FHWA-compliant [5] design. Two sets of stochastically dependent (water level and true cohesion) variables are initially used to verify the framework. Further analysis is presented using two sets of stochastically independent (water level and hydraulic gradient) variables to test the viability of different input variables.

2. Risk and Probabilistic Analysis in the Geotechnical Context

Naturally occurring materials, such as soil, are inherently variable. The uncertainty increases in space and upon ground disturbance, or in dynamic constantly changing total environments. National codes of practice allow statistical evaluation methods for determination of characteristic ground properties. The common universal practice is to factor down the ground resistance index parameters, statistically moderated ground properties, or a combination of the two using sets of partial factors that are predetermined and rather arbitrary. Whilst previous experience can be used as a basis for adopted factors of safety [6], this conservative approach falls short in addressing the impacts of spatial variability on computer reliability.

The probabilistic approach in geotechnical design has not received much interest, with stimulus from two directions: The mathematical structure of probabilistic approaches is often not appealing to practicing engineers; and the general perception is that probabilistic analysis incurs more time and cost. Early models include the first order second moment (FOSM) from Taylor Series and based on Bayes rule [7,8]; the method did not receive much attention mainly due to its inconsistency with non-linear functions [9]. Monte Carlo simulation in geotechnics was introduced and discussed in a series of seminal works [10,11], with limitations revisited more recently in [12]. Early attempts into the use of random set (RS) theory in geomechanics include the work of [13–15] in the design of linings for tunnels into rock. The RS framework was first used in conjunction with finite element (FE) analysis in [16].

3. Benchmark Problem and Methods

3.1. Benchmark Problem

The benchmark problem used in this contribution is a 28 m deep vertical cut in 2 to 2.5 m high lifts, from ground surface at 104.7 m above ordnance datum (mAOD) to 76.65 mAOD, and supported with a top-down soil nail system in a heavily built setting [17]. Figure 1 shows a cross-section diagram of the supported excavation, and adjoining features. Figure 1 shows the variation of the water level at the face of the excavation and also the hydraulic gradient, during the course of excavation. A 40 m long section was adopted for the purpose of this paper, where the land terrain led to a maximum depth of excavation. The cut was excavated into seven soil layers and supported with three flexible reinforced shotcrete facings, divided with two 2 m wide soil benches, and 13 rows of soil nails, 6 to 14 m long, installed at a 10° to 25° inclination. Soil nail elements included 28 to 32 mm steel bars installed centrally in 700 to 850 mm drilled holes that were eventually filled with a sand-cement grout. The soil nails were spaced 2 to 2.5 m horizontally and 1.5 to 3 m vertically. The cut was benched at 10.04 and 17.04 mbgl. A series of 4 to 6 m long, 4-inch drain perforated PVC pipes were installed at 10° elevations, at 3.10, 6.10, and 8.35 m from the bed of excavation. Drains were connected by series of 0.3 m wide vertical geocomposite blankets installed between shotcrete facing and retained ground from 87.0 to 76.65 mAOD. An evenly distributed 49 kPa permanent surcharge (20 m run) was applied 12 m from the excavation crest, representing the contact pressure from the neighbouring six-storey residential buildings. Piezometric readings derived from the ground investigation report [17] indicate a southward flow of groundwater, with a natural hydraulic gradient ranging from 0.1 to 0.2 (i.e., 1 to 2 m drop in every 10 m running perpendicular to project line—see Figure 1). The water level varied from 84.5 to 91.2 mAOD prior to construction. The system was designed for a mean 88 mAOD water level. Although commonly practiced, this is a substantial simplification, which reportedly led to excessive seepage and lateral displacement during the course of the staged excavation. Lateral ground movements were recorded using laser inclinometers at reflectors installed at 97 mAOD, five times each month to the end of the excavation. The northern wall section (covered by nine reflectors) was used for the purpose of this study, showing a cumulative lateral displacement of 115 mm, and a 61 mm horizontal displacement six months post-construction.

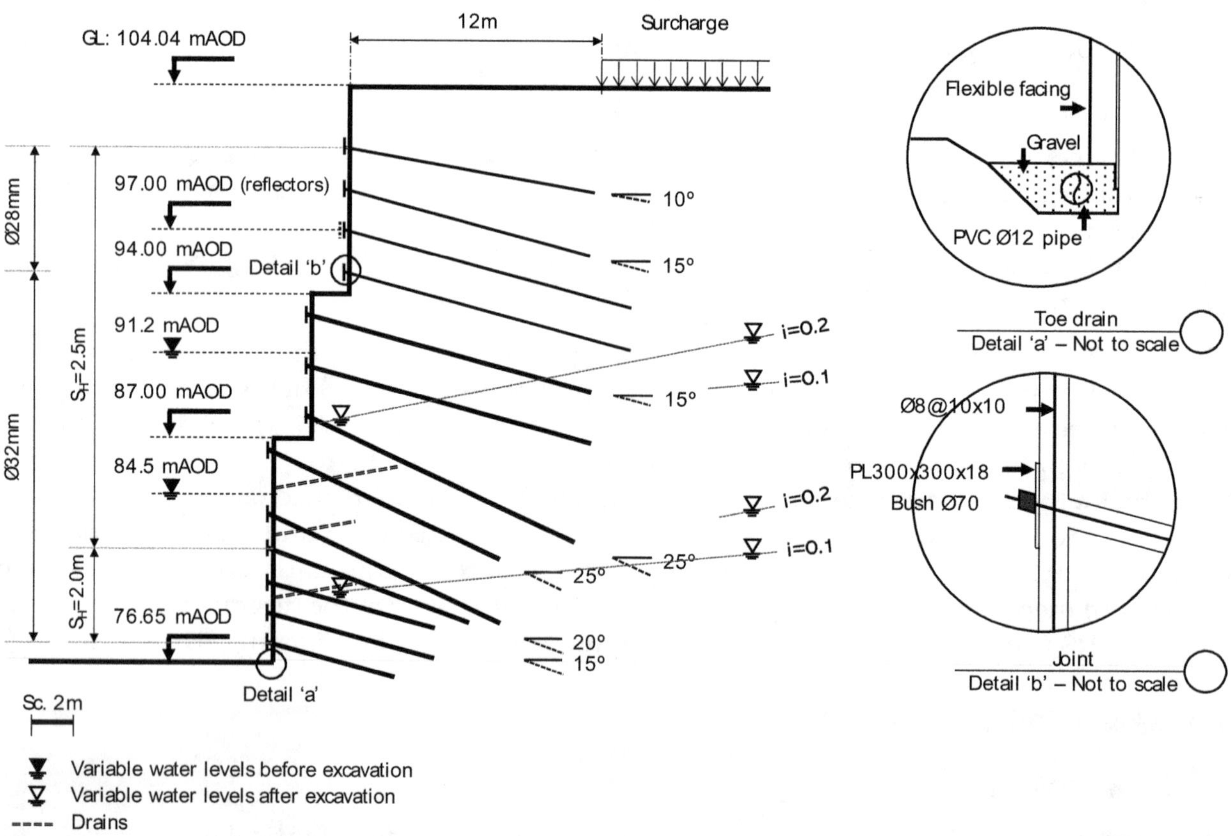

Figure 1. Cross-section diagram of the benchmark soil nailed system and adjoining features: S_H is horizontal spacing, GL is ground level, PL is short for plate, mAOD is metres above ordnance datum, and Ø is diameter.

3.1.1. Materials

The benchmark site is located in North Tehran and is composed of Quaternary (Pleistocene) permeable dark reddish to yellowish brown cemented poorly sorted alluvium (Bn fm). Table 1 summarizes the ground stratification, and mean (i.e., characteristic) long-term strength, stiffness, and hydraulic properties of soil layers in the benchmark site. The influence of construction and groundworks on soil properties is indisputable but was not considered either in the original design or in the present probabilistic analysis. Whilst this counts as a limitation, it allows study of the influence on actions and ground properties of seasonal groundwater regimes only.

Table 1. Stratification and factored (i.e., design) engineering properties used in deterministic analysis.

Layer	Soil Type	Depth: m	Ψ: °	ϕ': °	C′: kPa	γ: kN/m^3	E: kPa	ν	K_h: m/s	K_v: m/s
1	Made Ground	0.0–1.2	0^1	20	10	17	9800	0.4	1.0×10^{-4}	1.0×10^{-8}
2	Silty clayey sand	1.2–4.2	14	28	20	18	24,500	0.3	1.2×10^{-4}	1.2×10^{-8}
3	Alluvium	4.2–9.2	14	28	25	18	24,500	0.35	3.1×10^{-4}	3.1×10^{-8}
4	Silty clayey sand	9.2–18.2	14	34	20	18	34,300	0.3	4.2×10^{-7}	4.2×10^{-7}
5	Clay	18.2–21.2	0	25	35	19	34,300	0.4	1.9×10^{-7}	3.28×10^{-8}
6	Gravelly sandy clay	21.2–29.2	14	30	25	19	49,000	0.3	1.31×10^{-7}	4.36×10^{-6}
7	Silty clayey gravel	29.2–35.0	14	36	20	19	58,800	0.3	3.6×10^{-9}	1.0×10^{-9}

Ψ = dilation angle, ϕ' = drained friction angle, C′ = drained cohesion, γ = bulk unit weight, E = modulus of elasticity, ν = Poison's ratio, K_h = horizontal coefficient of permeability, K_v = vertical coefficient of permeability.

Granular soils form the top 18 m of the ground profile in the project site. The granular layers are generally cemented with hardly soluble calcite nodules; the clay fraction increases with depth and begins to control the overall soil behaviour from about 18 mbgl (meters below ground level) down to 29 mbgl. For the entire profile, the plasticity index ranges within the narrow range of 10% to 20% (i.e., low plasticity), with a Standard Penetration Number of >50, highlighting the role of calcite cementation. Piezometer readings over 7 months ahead and during the construction (May to November) showed a fluctuating water level from 11.5 to 18 mbgl (91.2 to 84.5 mAOD) on the project line; this was attributed in part to the natural terrain of the land and seasonal precipitation. The soil nail system was originally designed for a conservative 85.7 mAOD water level. Variations of the water level have direct implications on layer 4 (cemented sand) and level 5 (low-plasticity clay), and extend through the capillary rise effect to layers 2 and 3. The impacts of capillary rise were evident from the variable water content within soil layers 2 and 3.

3.1.2. Plaxis 2D Model

The nonlinear finite elements (FE) method was used to allow simulation of staged (sequenced) construction, soil–structure interaction, and steady-state flow, and also to allow measurement of mobilized stresses and deformations by the end and during the course of construction. Deterministic analysis was undertaken using the Plaxis 2D software. The soil layers were modelled using 15-node triangular elasto-plastic plain-strain finite elements. The local element factor was reduced to 0.1 for areas around drains to allow steady-state flow analysis. The Mohr-Coulomb failure criterion was adopted. Reinforced shotcrete flexible facing was modelled using elastic beam elements, with interface elements featuring a reduced 5% shear capacity (R_{int} = 0.95) compared to the surrounding soil. Elasto-plastic plate elements were preferred over geogrid elements to incorporate into the analysis the impact of the bending stiffness of steel bars and the nail-facing joint as a rigid connection. The axial and bending stiffness of plates was normalized against the horizontal nail spacing and are presented in Table 2 for drill holes of a 0.08 m diameter. Standard fixities were applied to horizontal and vertical boundaries. Drain elements contained five nodes and were assigned a head to represent every stage of excavation. The Phi/C (strength) reduction method was adopted to analyse the stability of the system at each stage and at the end of construction. Tolerance was limited to 1%; a maximum of 50 steps were deemed sufficient in determining the global factor of safety.

Table 2. Material properties for flexible facing and soil nail elements.

Material	E_{eq}: kPa	S_h: m	D_{dh}: m	EI: kN·m^{-2}·m^{-1}	EA: kN·m^{-1}	Model
Nail elements (top nine rows)	2.10×10^8	2.50	0.08	1.38×10^2	3.81×10^5	Elastoplastic
Nail elements (lower four rows)	2.10×10^8	2.00	0.08	1.72×10^2	4.79×10^5	Elastoplastic
Reinforced shotcrete facing	-	-	-	2.50×10^3	3.00×10^6	Elastic

E_{eq} = equivalent modulus of elasticity for grouted soil nails, S_h = horizontal spacing, D_{dh} = diameter of drill holes, EI = bending stiffness, EA = normal or axial stiffness.

3.2. Probabilistic Framework

Random fields were employed to formulate the variability of actions and material properties in a probabilistic framework. Three variables were paired into sets of stochastically dependent and stochastically independent parameters to first study the general impact on the serviceability and stability of uncertain input parameters, and then to seek the difference between two variable pairs and impacts on design.

To study the impact of water level variation and subsequent alterations to the apparent cohesion of cemented soils (i.e., design scenario 1: Stochastically dependent variables), random set realizations were defined using two datasets of equal probability (Table 3). Water level and cohesion were varied from the lower-bound to the upper-quartile in the first dataset; and from the lower-quartile to the upper-bound in the second dataset. A further two input data sets were defined using a stochastically independent hydraulic gradient and water level (i.e., design scenario 2) to revisit the stability and performance of the support system (Table 3).

Table 3. Input random sets for two design scenarios.

	Dataset	C′: kPa	WL: mAOD	I
Scenario 1	Set 1	20~24	79.75~86	
	Set 2	22~26	85~88	
Scenario 2	Set 1		79.75~86	0.1~0.15
	Set 2		85~88	0.15~0.2

C′ = drained cohesion, WL = water level (post excavation), I = hydraulic gradient, mAOD = metres above ordnance datum.

The water level was varied to match the captured fluctuations through the construction time, between 79.75 to 88 mAOD (i.e., 16 to 24.3 mbgl); the discharge rate at the face of excavation, representing a hydraulic gradient range of 0.1 to 0.2, was also varied. The hydraulic gradients were assigned to the model by introducing a head different between the phreatic line—cut intercept and the vertical model boundary. A head difference of 1V:10, 1.5V:10, and 2F.0V:10 H simulated hydraulic head gradients of 0.1, 0.15, and 0.2, respectively. For the water level to adapt itself during excavation with the location of drain points, a set of representative fixed hydraulic head values were assigned to drain points and the model's boundary for every phase of submerged excavation. The flow of the groundwater was stemmed from the natural hydraulic gradient, submerged excavation, and activation of drains. On the FE model, this condition was met by drawing inclined water levels and performing groundwater flow calculation. Interface elements were activated during the groundwater flow calculation to prevent flow through the wall.

The total number of realisations required to run a meaningful random set analysis is outlined through Equation (1) [18]:

$$N_c = 2^n \prod_{i=1}^{N} n_i \tag{1}$$

where N is the number of basic variables and n_i the number of sources available for each variable. Therefore, with a total of two basic variables, 16 realisations are required for random set analysis, outlining the data that feeds the cumulative probability distribution function in the form of probability boxes (i.e., *p*-box). It does mean that increasing the number of variables increases the output exponentially, therefore a thorough sensitivity analysis to identify critical parameters is generally advised. Focusing solely on univariates from the data sets, Figure 2 presents four *p*-boxes, highlighting the upper and lower bounds for each variable.

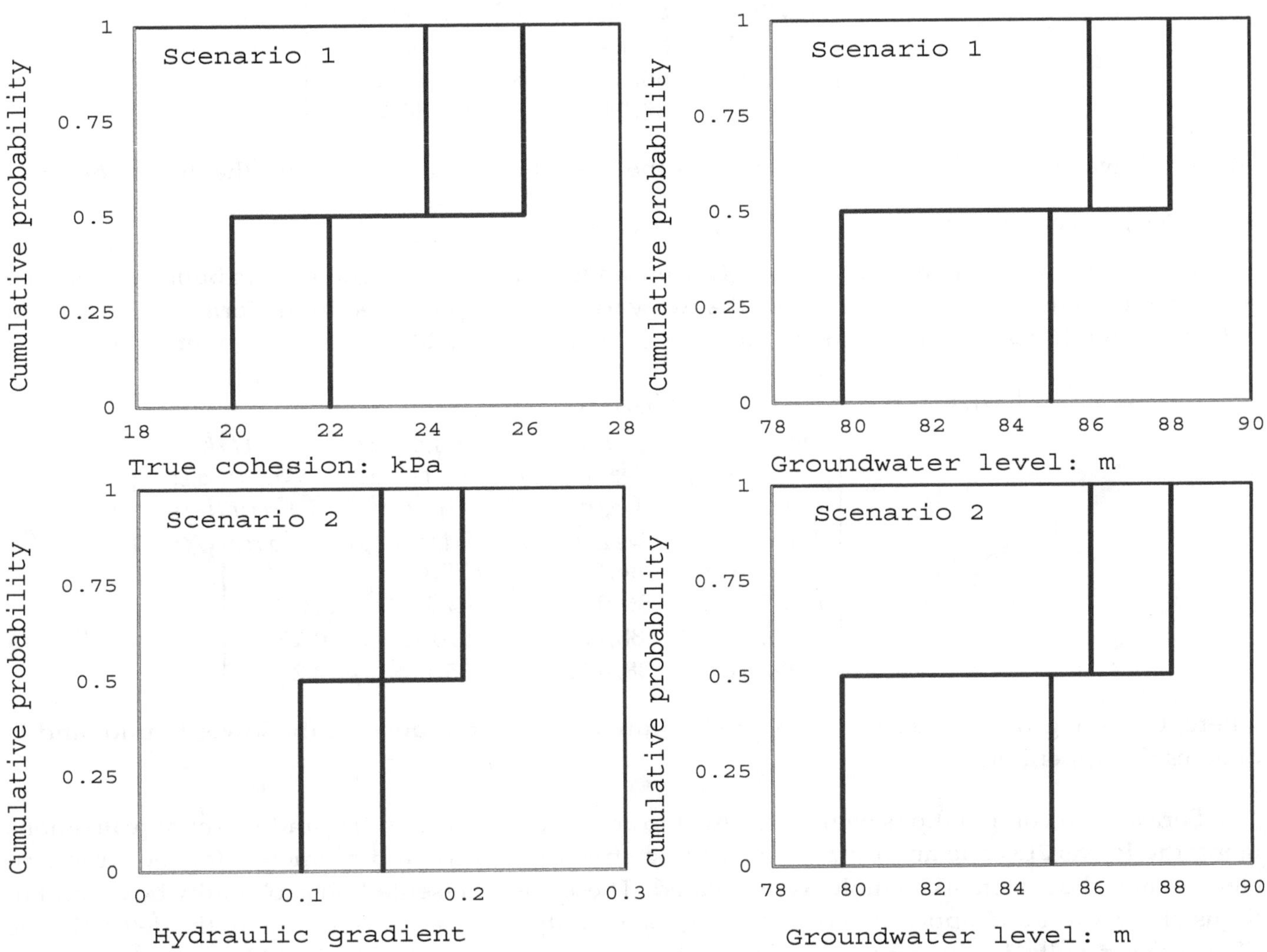

Figure 2. Range of input variables in RS *p*-Box; scenario 1: Stochastically dependent variables: Water level and cohesion, scenario 2: Stochastically independent variables: Water level and flow rate.

For $u = (x_1, \ldots, x_n)$, where x_i is the variables, n is the total number of variables, and u is the vector of set value parameters, a random relationship can be defined in the Cartesian form of $x_1 \times \ldots \times x_n$. This renders a vector of combinations of basic variables. In this research, two basic variables are deployed to form four combinations of basic variables; these are discussed in Sections 3.2.1 and 3.2.2 for two design scenarios.

3.2.1. Design Scenario 1

Drained cohesion (C′) and water level (G) were adopted as variables and used to build two random sets. The combined relationship between the two variables was presented in the form of a vector and defined by the 16 realisations arranged into four combinations, as illustrated in Equation (2) (below):

$$
\begin{aligned}
G\times C &= \{(G_1,C_1)_1,(G_2,C_1)_2,(G_1,C_2)_3,(G_2,C_2)_4\} \\
&= \begin{bmatrix} G_{1,LL}, i_{1,LL} & G_{2,LL}, i_{1,LL} & G_{1,LL}, i_{2,LL} & G_{2,LL}, i_{2,LL} \\ G_{1,LU}, i_{1,LU} & G_{2,LU}, i_{1,LU} & G_{1,LU}, i_{2,LU} & G_{2,LU}, i_{2,LU} \\ G_{1,UL}, i_{1,UL} & G_{2,UL}, i_{1,UL} & G_{1,UL}, i_{2,UL} & G_{2,UL}, i_{2,UL} \\ G_{1,UU}, i_{1,UU} & G_{2,UU}, i_{1,UU} & G_{1,UU}, i_{2,UU} & G_{2,UU}, i_{2,UU} \end{bmatrix} \\
&= \begin{bmatrix} 79.75,20 & 86,20 & 79.75,22 & 86,22 \\ 76.75,24 & 86,24 & 79.75,26 & 86,26 \\ 85,20 & 88,20 & 85,22 & 88,22 \\ 85,24 & 88,24 & 85,26 & 88,26 \end{bmatrix}
\end{aligned} \tag{2}
$$

where G is water level, C is cohesion, L denotes the lower bound, and U denotes the upper bound.

3.2.2. Design Scenario 2

Hydraulic gradient (i) and water level (G) were adopted as variables and used to build two random sets. The combined relationship between the two variables was presented in the form of a vector and defined by the 16 realisations arranged into four combinations, as illustrated in Equation (3):

$$
\begin{aligned}
G\times i &= \{(G_1,i_1)_1,(G_2,i_1)_2,(G_1,i_2)_3,(G_2,i_2)_4\} \\
&= \begin{bmatrix} G_{1,LL}, i_{1,LL} & G_{2,LL}, i_{1,LL} & G_{1,LL}, i_{2,LL} & G_{2,LL}, i_{2,LL} \\ G_{1,LU}, i_{1,LU} & G_{2,LU}, i_{1,LU} & G_{1,LU}, i_{2,LU} & G_{2,LU}, i_{2,LU} \\ G_{1,UL}, i_{1,UL} & G_{2,UL}, i_{1,UL} & G_{1,UL}, i_{2,UL} & G_{2,UL}, i_{2,UL} \\ G_{1,UU}, i_{1,UU} & G_{2,UU}, i_{1,UU} & G_{1,UU}, i_{2,UU} & G_{2,UU}, i_{2,UU} \end{bmatrix} \\
&= \begin{bmatrix} 79.75,0.1 & 86,0.1 & 79.75,0.15 & 86,0.15 \\ 76.75,0.15 & 86,0.15 & 79.75,0.2 & 86,0.2 \\ 85,0.1 & 88,0.1 & 85,0.15 & 88,0.15 \\ 85,0.15 & 88,0.15 & 85,0.2 & 88,0.2 \end{bmatrix},
\end{aligned} \tag{3}
$$

where, G is the groundwater level, i is the hydraulic gradient, L denotes the lower bound, and U denotes the upper bound.

Series of FE computations were performed to accommodate the entire random set of realizations using the Plaxis 2D code and outputs, including stresses, strains, and moments, for the structural element and the system as a whole were collated. These were presented on probability boxes and in terms of the lower and upper bounds, allowing the FE outputs to be contrasted with the deterministic FE analysis results, in-situ measurements, or equivalent parameters calculated in compliance with common codes of practice (i.e., ultimate and serviceability limit states).

3.3. *Ultimate and Serviceability Limit States*

The benchmark soil nail support system was designed using the mean ground properties summarized in Table 1 and for mean water level conditions. Ultimate limit states were determined in compliance with the British Standard BS 8006-2:2011+A1:2017 [19] and US Department of Transportation Federal Highway Administration FHWA-NHI-14-007 code [5] of practice and the methods used are revisited in this section. Limit states were used as index parameters. The most probable limit states derived from random set FE analysis were compared and contrasted on probability boxes with index parameters to examine the effectiveness of the probabilistic framework and implications of use of factored mean values representative of variable ground conditions.

This section summarises the limit state index parameters used in this study and methods deployed for their measurement.

3.3.1. Serviceability of Flexible Facing

The serviceability of the shotcrete lining was calculated here using [20]. To assess the serviceability of the facing, the admissible normal force needs to be greater than the working normal force, along the facing. The admissible (or allowed) normal force is formulated in Equation (4):

$$N_{max} = \frac{f_c d}{F_s}\left(1 - 2\,\frac{e(x) + e_a}{F_s d}\right) \tag{4}$$

where N_{max} is the maximum normal force, f_c is the uniaxial compressive strength of facing (here about 17 MPa), e_a is the eccentricity tolerance (here about 3 cm), $e(x)$ is the eccentricity (M/N = working bending moment (FHWA equations)/working normal force (FHWA equations or FE computational analysis), d is the thickness of the facing, and F_s is the partial factor (1.65). The bending moment is calculated for the wall profile segments, considering the facing as an indeterminate continuous beam, made up of segments. The values of N_{max} provide the data to determine the serviceability of facing through Equation (5):

$$g(x) = N_{max} - N \tag{5}$$

where $g(x)$ is the serviceability function and for $g(x) < 0$, the shotcrete wall would be likely to fail through cracking.

3.3.2. Bending Moments along Flexible Facing

To define the maximum bending moment along the shotcrete wall facing, the formulations offered for granular soils in the design code [21] for deep excavations were adopted and presented in Equations (6)–(11). For the soil fiction angle, $\emptyset$, the mean value for retained soils from the ground level to the bed of excavation was considered in the analysis. For standing groundwater, CIRIA [22] allows the use of the triangular water hydrostatic pressure envelope, with water pressure, U_w, gaining value with the depth of the submerged soil, h_w:

$$P[kN/m^2] = 0.65\gamma \times H \times K_a \tag{6}$$

$$K_a = \tan{(45 - 0.5\varnothing)}^2 \tag{7}$$

$$U_w = h_w{\cdot}\gamma_w \tag{8}$$

$$w = \frac{0.65\gamma{\cdot}H^2{\cdot}K_a}{H - \frac{1}{3}\,H_1 - \frac{1}{3}\,H_{n+1}} \tag{9}$$

$$w = \frac{0.65\gamma{\cdot}H^2{\cdot}K_a + \gamma_w\left(h_w - \frac{2}{3}H_{n+1}\right)}{H - \frac{1}{3}\,H_1 - \frac{1}{3}\,H_{n+1}} \tag{10}$$

$$M = \frac{wL^2}{10} \tag{11}$$

where M is the bending moment, w is the maximum uniformly distributed lateral load, H is the total height of the retained soil at the end of excavation, γ is the mean bulk density of the soil behind the wall, H_1 is the highest soil nail's length of embedment, and H_{n+1} is the distance between the lower-most nail and the bed of excavation. To find the value of the maximum bending moment, L can be substituted for the maximum vertical length of the nail influence area (S_v) along the inner wall, in compliance with the recommendations of [23].

3.3.3. Axial Mobilized Force in Soil Nails

As per the recommendations of [5], soil nails were designed to carry the maximum possible axial force mobilized ($T_{mob.}$) at each elevation. The maximum mobilized axial force in soil nails was formulated using Equations (12) and (13) for the end of construction conditions and outer and inner walls, respectively. $T_{mob.}$ excludes the effect of surcharge pressure at the ground level:

$$T_{mob.} = (0.65 - 0.75)\gamma \cdot H \cdot K_a \cdot S_h \cdot S_v \tag{12}$$

$$T_{mob.} = (0.32 - 0.37)\gamma \cdot H \cdot K_a \cdot S_h \cdot S_v \tag{13}$$

where γ is the soil unit weight, K_a is the lateral pressure coefficient, S_h is the horizontal spacing of nails, S_v is the vertical spacing of nails, and $S_h \times S_v$ is the influenced area surrounding each nail. The mobilized axial force should be lower than the failure load (ultimate resistance), which can be determined for each reinforcement bar by multiplying the steel yield strength by the bar's cross-sectional area. This is a conservative approach as steel bars sit within grouted drill holes.

3.3.4. Lateral and Vertical Displacements

The actual lateral ground movements of the support system, originally designed using the mean factored material properties (Table 1) and actions, were measured in situ using laser deflectometers during and after construction. Survey data taken from nine reflectors installed at 97 mAOD and along the section of interest in this study were used as a benchmark for an assessment of the efficiency of the adopted probabilistic framework. The vertical displacement of the support system evaluated using the random set framework was contrasted with values from the analytical methods recommended in [21] code and in Equation (14):

$$\delta_v = \left(\frac{\delta_v}{H}\right)_i \times H \tag{14}$$

where δ_v is the ground vertical movement, H is the total height of the retained soil, and $\left(\frac{\delta_v}{H}\right)_i$ is the coefficient of the soil properties.

4. Results

4.1. Design Scenario 1: Stochastically Dependent Variables

FHWA compliant calculated limit states are projected on belief and plausibility distribution functions [18] in Figure 3 for stochastically dependent cohesion and water level variables.

Limit states values pertinent to four realizations for each combination were sorted to derive the focal point extremes (i.e., minimum and maximum values). Middle values were discarded, and extremes were presented on the probability box in form of one step in the cumulative distribution function. As such, every step on the probability boxes represents extreme values for four random set combinations. Extreme values were sorted to form the upper and lower bound stepped line for each limit state. In Figure 3, the line marking the 0.5 cumulative probability represents the most likely values for each limit state, and the dashed line represents the index parameter either determined by the FHWA code of practice or measured in-situ. Lower and upper bounds are presented with stepped thick lines, with the space between representing possible values in the face of uncertain cohesion and water level. Table 4 summarises the design's maximum axial force (T_{max}) in soil nails 8 (submerged inner wall), 6 (middle wall), and 3 (outer wall) based on mean soil and water parameters. In Figure 4a–c, the design axial force for all walls falls within the probabilistically determined likely range and is relatively closer to the lower-bound line. This suggests that while the conventional design practice gives a reliable estimation of the soil nail axial force, variability of the water level can yield greater orders of axial loads, which is technically unexpected. While limit states are not reached and the system remains stable, the excessive axial forces in soil nails can impact the predicted ground lateral movements.

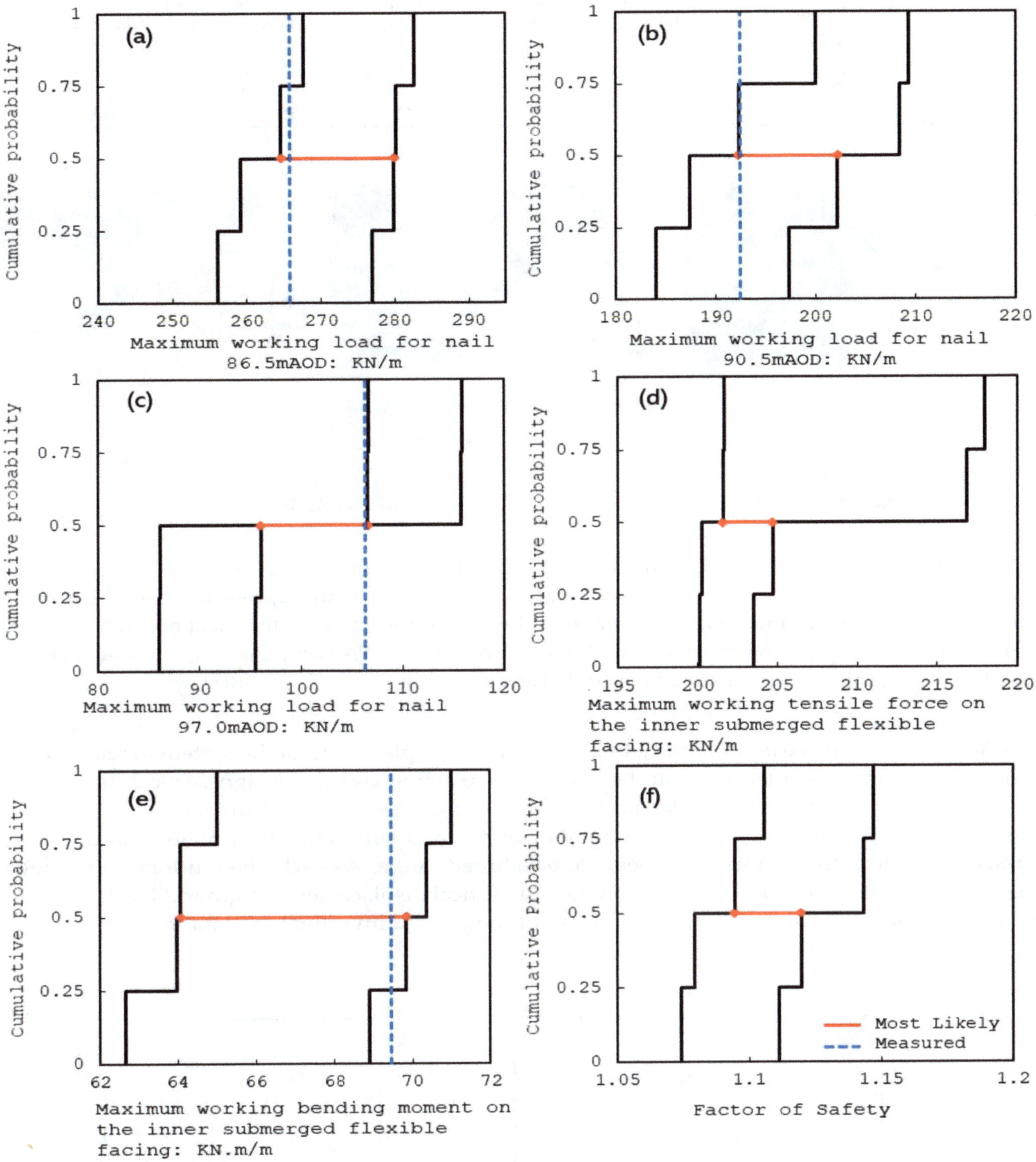

Figure 3. FHWA compliant limit states (index parameters) on RS *p*-Boxes and in relevance to the lower bound, upper bound, and most likely limit states in the face of uncertainty; (**a**) maximum mobilised working axial load in the eighth nail from ground level on the submerged inner wall, (**b**) maximum mobilised working axial load in the sixth nail from ground level on the middle wall, (**c**) maximum mobilised working axial load in the third nail from ground level on the upper wall, (**d**) maximum mobilised axial load along the inner submerged shotcrete facing, (**e**) maximum mobilised bending moment along the inner submerged shotcrete facing, and (**f**) global factor of safety.

Table 4. Maximum normal force in the soil nail results.

	Diameter: mm	Distance from Crest: m	Horizontal Spacing: m	Vertical Spacing: m	Maximum Normal Force: kN/m	Failure Load: kN/m
	D	H	S_h	S_v	T_{max}	T_f
Nail 8	28	21.5	2.5	2.25	265.8	255.4
Nail 6	32	16.2	2.5	2	192.46	333.6
Nail 3	32	9.68	2.5	2.5	106.37	333.6

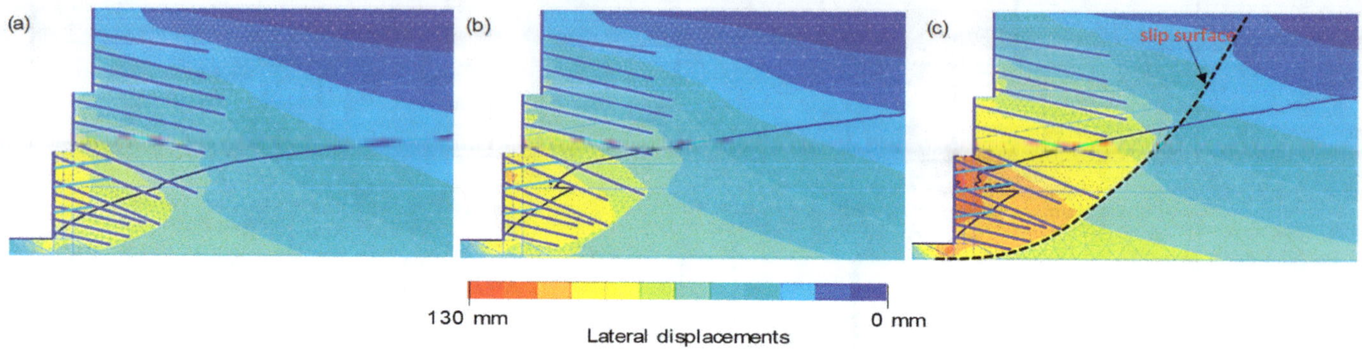

Figure 4. Re-distributed phreatic line following steady state flow and with relevance to drain points; input pre-excavation hydraulic gradients into the FE models: (**a**) i = 0.1, (**b**) i = 0.15, (**c**) i = 0.20.

Using Equations (10) and (11), the mobilised bending moment on the critical submerged inner facing is 69.5 kNm·m^{-1}, an order which is within the bounds of the upper and lower limit of the cumulative distribution function. It appears that the upper bound offers the most accurate estimation of mobilised moments, being consistent with the equivalent FHWA compliant index value (Figure 3e). Most likely the global factor of safety ranges between 1.094 to 1.120 (Figure 3f).

Figure 5 presents the *p*-box of the lateral and vertical displacement of the system in relevance with the measured lateral displacement at the end of construction and the design vertical displacement, taking the mean material properties and actions. The design's vertical displacement and in-situ measured lateral displacements were within the bounds and within the most-likely values, which, in the case of vertical displacements, appears to be a broad range. As such, the random set framework does not appear to be as useful in determination of vertical displacements at ground level. This can be due to the poor choice of variables or small size of the ground investigation data sets [18].

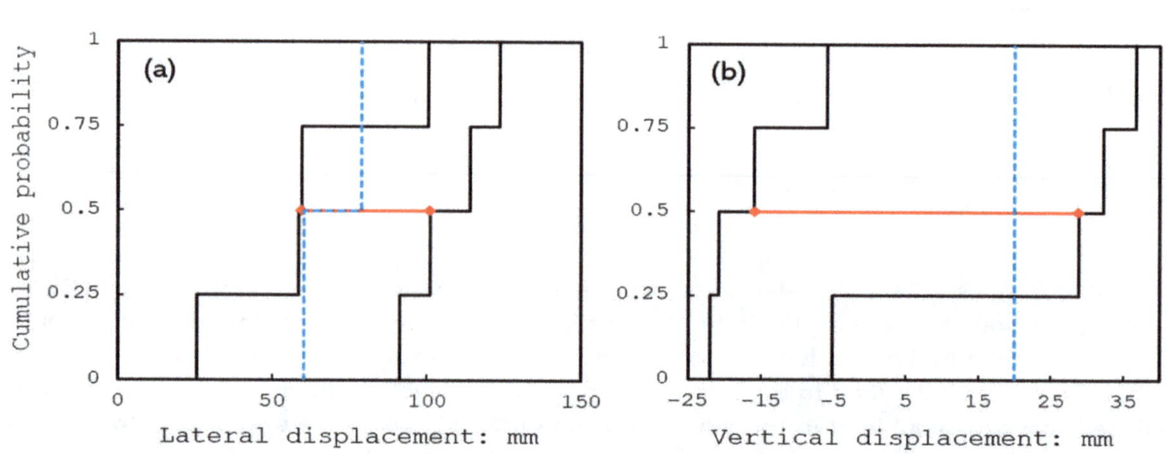

Figure 5. Serviceability limit state; post-construction (**a**) lateral displacements, (**b**) vertical displacement.

4.2. Design Scenario 2: Stochastically Independent Variables

Implications of Seepage

Following seepage analysis (i.e., steady state flow analysis for end-of-construction stage) and for a low 0.1 hydraulic gradient and water level varying from 79.75 to 88 mAOD (scenario 2, set 1, lower-bound flow rate), it appears that the phreatic line passes from the toe of the excavation with a discharge rate of 0.018 to 0.037 m^3/day/m, with no outflow through the drains. With an increase in the hydraulic gradient from 0.1 to 0.15 and 0.2, a re-distributed phreatic line appears to raise towards the drains, forming an outflow zone between the middle drain and the bed of the excavation (Figure 4). Discharge rates gained at drain points range between 0.018 to 0.037 m^3/day/m. For a hydraulic gradient of 0.2, in particular, the re-distributed water table rose further to reach the upper drain, which is consistent with the imposed greater flow rates towards the drains. This is just above the design's mean water level (under no seepage). The combined impact of flow through the inclined drains and change in geometry due to excavation led to alterations of the water level, causing drag and possible excessive hydrodynamic stresses. The equivalent enhanced hydraulic gradient appeared to be in the range of 0.16 to 0.19 (for pre-construction, original gradient of 0.15) and 0.22 to 0.24 (for pre-construction, original gradient of 0.2). This can have significant impacts on the lateral displacement of the retained wall. Figure 6 shows the impact on the lateral displacement at the reflect point (i.e., crest) and maximum lateral deflection of the water level and hydraulic gradient combined. The more pronounced impact of the hydraulic gradient as compared with the water level is obvious. Generally, the re-distributed water table is observed to strike the drains instead of the lining; this naturally yields lower pore water pressures immediately behind the facing, which is welcomed.

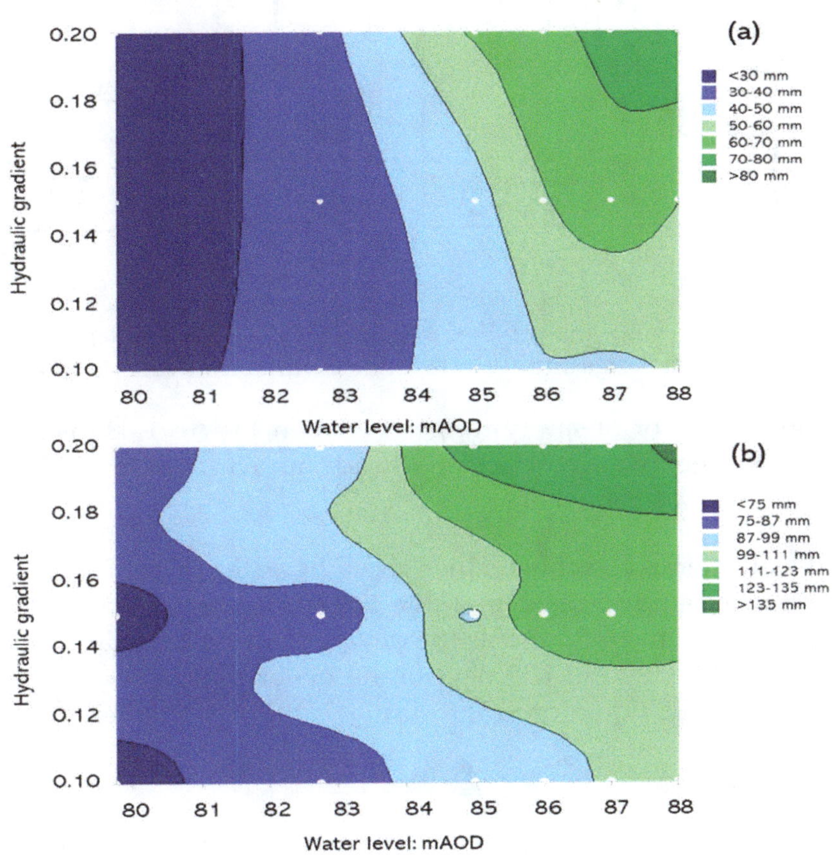

Figure 6. Impact on the lateral displacement of the water level and hydraulic gradient; (**a**) displacement at the reflector level (97 mAOD, near crest), (**b**) maximum displacement.

Limit states values pertinent to four realizations for each combination were sorted to derive the focal point extremes (i.e., the minimum and maximum values) and are presented in Figure 7 in the form of a series of probability boxes in the form of one step in a cumulative distribution function. Every step on the probability boxes represents extreme values for four random set combinations. Extreme values were sorted to form the upper and lower bound stepped line for each limit state.

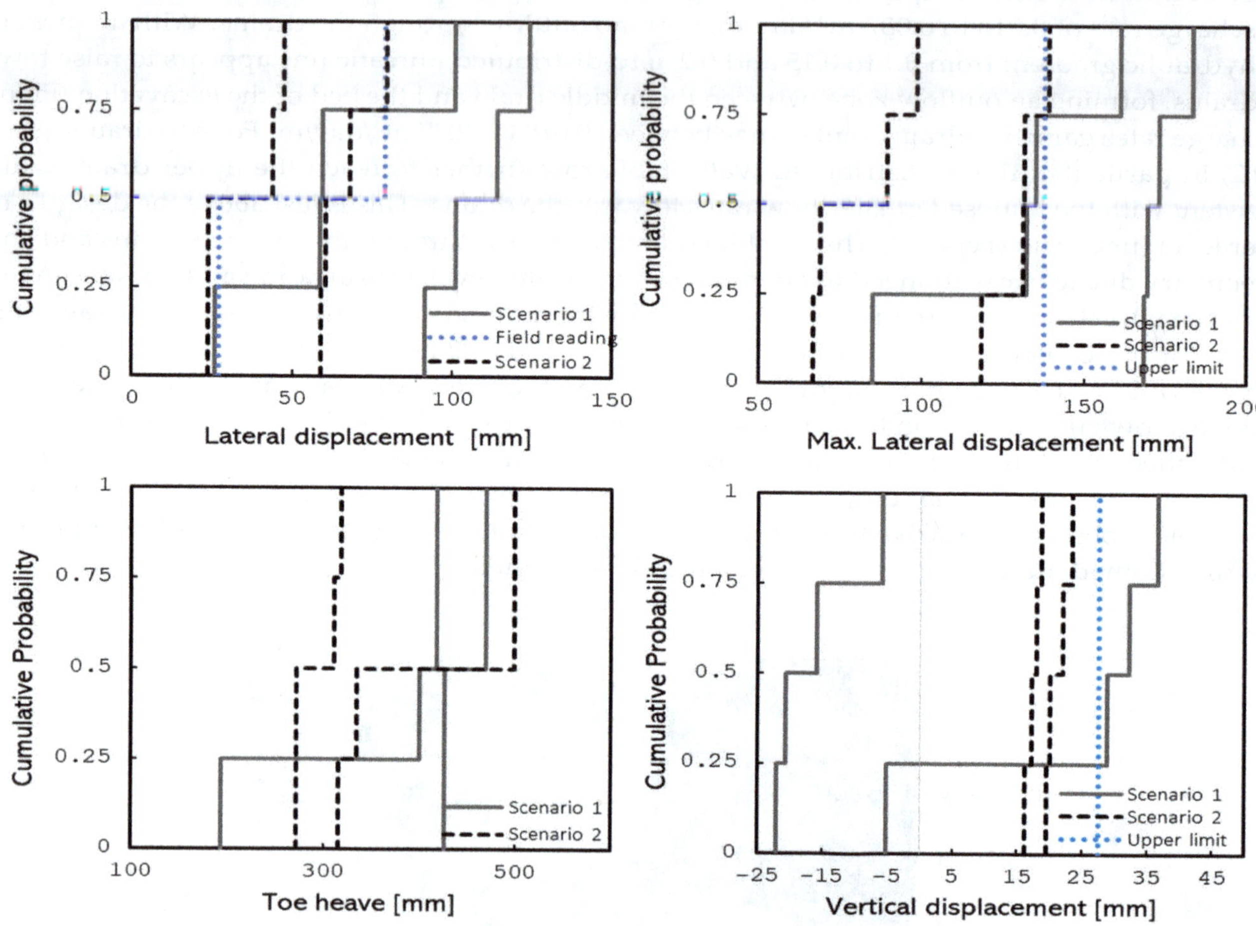

Figure 7. RSFEM modelling results on the *p*-Box; scenario 1: Stochastically dependent variables: water level and cohesion, scenario 2: Stochastically independent variables: water level and flow rate.

The most likely global factor of safety ranges between 1.01 to 1.08 (as compared with design scenario 1, 1.094 to 1.120). The lower and upper bounds appear to offer a relatively wider range of most likely values, which is a welcomed event.

The FE predicted wall lateral deflection values for eight realisations (u_{xR}) at the crest of excavation (i.e., the reflector point) are summarized in Table 5 for seepage conditions. In Table 5, z is the standardised normal variable, which represents the deviation of wall lateral deflections from the mean value, and subscripts LB, LQ, UB, and UQ denote the lower-bound, lower-quartile, upper-bound, and upper-quartile, respectively. Equation (15) formulates the z parameter:

$$z = \frac{u_{xR} - \overline{u_x}}{\delta} \tag{15}$$

where $\overline{u_x}$ represents the mean deflection value and δ the standard variation, based on the FE-predicted and field actual values combined ($\overline{u_x} = 61.1\ mm$ and $\sigma = 16.5\ mm$).

Table 5. Lateral deflections from FE analysis: maximum values at reflector point.

	G_{LB},i_{LB}	G_{LQ},i_{LB}	G_{UQ},i_{LB}	G_{UB},i_{LB}	G_{LB},i_{UB}	G_{LQ},i_{UB}	G_{UQ},i_{UB}	G_{UB},i_{UB}
u_{xR}: mm	41.1	45.9	49.0	55.2	58.2	64.8	69.2	76.5
z	−1.2	−0.9	−0.7	−0.3	−0.1	0.2	0.5	0.9

G = Water level, I = hydraulic gradient, LB = lower-bound, UB = upper-bound, LQ = lower quartile, UQ = upper quartile.

According to Table 5, realisations with the lowest z-values (arguably more realistic) represent combinations of the upper-bound head gradient with the lower-bound and lower-quartile water level, respectively. This suggests that [1]: When maximum hydraulic gradient occurs at a low water level (i.e., the dry season), the predicted serviceability is closest to the actual systems' serviceability at the ground surface level, where lateral displacements should crucially fall within admissible limits to restrict any damage to neighbouring structures (G_{LB}, i_{UB} in Table 5). Changing climatic conditions are generally believed to appear in the form of periods of intense rainfall over prolonged dry seasons; Ref [2] FE contour outputs suggest that drains are ineffective for the flow of groundwater under a lower-bound hydraulic gradient. This suggests that for piping to be mitigated, toe drains are required to be designed for lower-bound hydraulic gradient values. Drains, on the other hand, appeared to be fully functional for flow under upper-bound values of the hydraulic gradient. Drains are advisable to be designed for upper-bound hydraulic gradient values.

Figure 7 shows the probability distribution of post-construction serviceability of the system as a function of both stochastically dependent and independent variables.

Figure 8 illustrates the variation of the extreme values of the working normal force, together with the extreme values of the bending moment for the two modelling scenarios. The FE-predicted values can be contrasted with the computed values using the common FHWA code of practice.

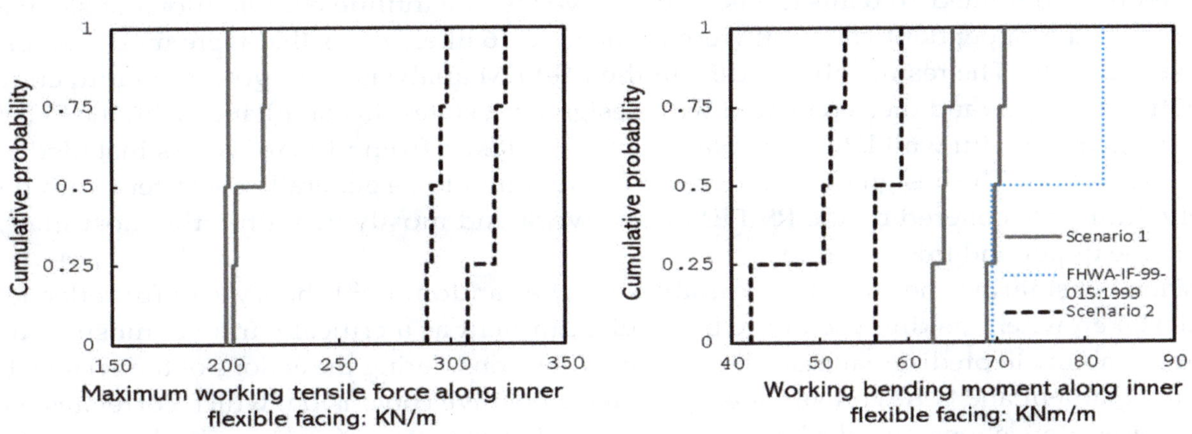

Figure 8. Post-construction deterministic FE predicted mobilised actions on the inner lining: Scenario 1: Stochastically dependent variables: water level and cohesion, scenario 2: Stochastically independent variables: water level and flow rate.

Following the variation of the water level and possible mineral dissolution, the simplifications adopted in the conventional design practice appear to be admissible, as the calculated bending moment values match the upper-bound predicted values. In this, although the conventional design tends to neglect the effect of water level variation and possible impacts on soil strength, the resulting moment values are reasonably conservative.

For scenario 1, the below observations were made:

- Lower-bound water level and upper-bound cohesion (representing dry season with maximum cementation) yielded the maximum factor of safety among the deterministic models.

- Figure 8 shows the variation of the maximum bending moment on the inner wall (i.e., submerged) for the eight modelling scenarios. Solid lines represent the FE-predicted values, whereas the dashed line infers the designed mean bending moment according to the FHWA-IF-99-015 (1999) recommendations. Bending moment values derived from the FHWA formulation match those of the upper-bound RS values and as such are of reasonably good credibility.

For soil nails above the water level (at all times), the designed mobilised tensile forces' (FHWA compliant) formulation closely matched those of the upper-quartile FE-predicted RS values. Where soil nails are located within the zone of water level variation, the design's maximum tensile forces match the lower-bound values from the random set FE analysis. This shows the relatively small significance of the water level and cohesion variation in the determination of design actions of reinforcement nails.

5. Conclusions

A real-world deep excavation problem incorporating soil nailing was analysed and revisited, employing the random set finite element method to test the viability of the probabilistic frameworks in solving complex geotechnical problems, where spatial variability is prevalent. Two input data sets were established from a range of available ground investigation reports, and analysed together with field wall lateral displacement data that was captured over several months during and post-construction. The drained cohesion, hydraulic gradient, and water level were adopted as uncertain parameters for a critical cemented granular layer of soil stratum in the light of its susceptibility to mineral dissolution.

To study the impact of water level variation and subsequent alterations to the apparent cohesion of cemented soils (i.e., design scenario 1: stochastically dependent variables), random set realizations were defined using two datasets of equal probability. A further two input data sets were defined using stochastically independent hydraulic gradient and water level (i.e., design scenario 2) to revisit the stability and performance of the support system. For each scenario, four combinations were generated that contained 16 realisations; in other words, the infinite combinations of stochastically dependent and independent variables were reduced to 16 likely cases that were modelled using the FE Plaxis 2D code. The results obtained from the RS-FEM analysis were processed into cumulative distribution functions and then compared with design limit states (in compliance with the FHWA code of practice), and in-situ wall lateral displacement data taken from nine reflectors installed near the crest of excavation. The design limit state values were found to be generally consistent with the range of likely limit states offered by the RS-FEM framework and mostly lie within the most-likely range between the upper and lower bounds.

When considering the stochastic variables for the random field theory, careful reflection must be undertaken when making use of the hydraulic gradient as a critical variable, mostly due to the variables' inability to predict material uncertainty. When considering the actions of the system, both the hydraulic gradient and cohesion variables produce a positive conclusion, which correlates well with the measured deflections and design limit states. A key positive of using cohesion as a stochastic variable is the insight that both the material and action responses match positively with the measured and observed results.

Retrieving good quality undisturbed samples to determine the shear strength of cemented granular soils continues to be a challenge. In a probabilistic framework, the tendency is to limit the uncertainty in the measurement of soil properties and to emphasise the variability of accurately measured soil properties; this is not plausible for the drained cohesion of cemented granular alluviums. A scope for replacing cohesion with a more reliably measurable alternative parameter is welcomed. This work outlines the benefits and limitations in using seepage and water level parameters as stochastically independent variables.

Author Contributions: Conceptualization, A.A.L and S.D; methodology A.A.L, S.G; software, J.W, K.T., M.M.,; validation, J.H; formal analysis J.W, K.T, M.M., A.L.L; investigation, S.G; resources, S.G and S.D; data curation, A.A.L; writing—original draft preparation, J.H; writing—review and editing, J.H and A.A.L.; supervision, A.A.L; project administration, A.A.L and S.D.

Acknowledgments: Authors would like to thank POR Consulting Engineers and undergraduate research students in Newcastle University and University of East London for their technical inputs.

References

1. UN. *World Population Prospects: The 2012 Revision*; UN Technical Report; UN: New York, NY, USA, 2013.
2. The London Borough of Camden. *Camden Planning Guidance CPG 4: Basements and Lightwells*; The London Borough of Camden: London, UK, 2011.
3. Assadi-Langroudi, A.; Theron, E. Gaps in particulate maters: Formation, mechanisms, implications. In Proceedings of the 17th African Regional Conference on Soil Mechanics and Geotechnical Engineering, Cape Town, South Africa, 7–9 October 2019; Geotechnical Division, South African Institute of Civil Engineers: Cape Town, South Africa, 2019.
4. Assadi-Langroudi, A.; Jefferson, I. Constraints in using site-won calcareous clayey silt (loam) as fill materials. In *XVI ECSMGE Geotechnical Engineering for Infrastructure and Development*; ICE Institution of Civil Engineers Publishing: London, UK, 2015; pp. 1947–1952.
5. *Soil Nail Walls Reference Manual*; FHWA-NHI-14-007; US Department of Transportation, Federal Highway Administration: Washington, DC, USA, 2015.
6. Duncan, J.M. Factors of safety and reliability in geotechnical engineering. *J. Geotech. Geoenviron. Eng.* **2000**, *126*, 307–316. [CrossRef]
7. Wong, F.S. First Order Second Moment Method. *Comput. Struct.* **1985**, *20*, 779–791. [CrossRef]
8. Kunstmann, H.; Kinzelbach, W.; Siegfried, T. Conditional first order second moment method and its application to the quantification of uncertainty in groundwater modelling. *Water Resour. Res.* **2002**, *38*, 1–14. [CrossRef]
9. Du, H.; He, W.; Sun, D.; Fang, Y.; Liu, H.; Zhang, X.; Cheng, Z. Monte Carlo simulation of magnetic properties of irregular Fe islands on Pb/Si(111) substrate based on the scanning tunneling microscopy image. *Appl. Phys. Lett.* **2010**, *96*, 1–3. [CrossRef]
10. Griffiths, D.V.; Fenton, G.A. Three dimensional seepage through spatially random soil. *J. Geotech. Geoenviron. Eng.* **1997**, *123*, 153–160. [CrossRef]
11. Goldsworthy, J.S.; Jaksa, M.B.; Fenton, G.A.; Griffiths, D.V.; Kaggwa, W.S.; Poulos, H.S. Measuring the risk of geotechnical site investigations. In *Probabilistic Applications in Geotechnical Engineering*; ASCE: Denver, CO, USA, 2007; pp. 1–12.
12. Jiang, S.H.; Li, D.Q.; Cao, Z.J.; Zhou, C.B.; Phoon, K.K. Efficient system reliability analysis of slope stability in spatially variable soils using Monte Carlo simulation. *J. Geotech. Geoenviron. Eng.* **2014**, *141*, 1–13. [CrossRef]
13. Tonon, F.; Mammino, A. A random set approach to the uncertainties in rock engineering and tunnel lining design. In Proceedings of the ISRM International Symposium on Prediction and Performance in Rock Mechanics and Rock Engineering (EUROCK '96), Torino, Italy, 2–5 September 1996; Volume 2, pp. 861–868.
14. Tonon, F.; Bernardini, A.; Mammino, A. Determination of parameters range in rock engineering by means of Random Set Theory. *Reliab. Eng. Syst. Saf.* **2000**, *70*, 241–261. [CrossRef]
15. Tonon, F.; Bernardini, A.; Mammino, A. Reliability analysis of rock mass response by means of Random Set Theory. *Reliab. Eng. Syst. Saf.* **2000**, *70*, 263–282. [CrossRef]
16. Schweiger, H.F.; Peschl, G.M. Basic Concepts and Applications of Random Sets in Geotechnical Engineering. In *CISM International Centre for Mechanical Sciences*; Griffiths, D.V., Fenton, G.A., Eds.; Springer: Berlin/Heidelberg, Germany, 1975; Volume 491, pp. 113–126.
17. Assadi, A.; Yasrobi, S.; Nasrollahi, N. A comparison between the lateral deformation monitoring data and the output of the FE predicting models of soil nail walls based on case study. In Proceedings of the 62th Canadian Geotechnical Conference and 10th Joint CGS/IAH-CNC Groundwater Conference, Halifax, NS, Canada, 20–24 September 2009.
18. Nasekhian, A.; Schweiger, H.F. Random Set Finite Element method application to tunnelling. In Proceedings of the 4th International Workshop on Reliable Engineering Computing, Singapore, 3–5 March 2010.
19. *BS 8006-2:2011+A1:2017. Code of Practice for Strengthened/Reinforced Soils, Soil Nail Design*; British Standards Institution: London, UK, 2011.
20. Schikora, K.; Ostermeier, B. Two-dimensional calculation model in tunnelling verification by measurement results and by spatial calculation. In Proceedings of the 6th International Conference on Numerical Methods in Geomechanics, Innsbruck, Austria, 11–15 April 1988; Volume 3, pp. 1499–1530.

21. US Department of Transportation, Federal Highway Administration. *Geotechnical Engineering Circular No. 4: Ground Anchors and Anchored Systems*; FHWA-IF-99-015; US Department of Transportation: Washington, DC, USA, 1999.
22. Phear, A.; Dew, C.; Ozsoy, N.I.; Wharmby, N.J.; Judge, J.; Barley, A.D. *Soil Nailing—Best Practice Guidance C637D*; Construction Industry Research and Information Association CIRIA Publications: London, UK, 2005.
23. *BS 8002:1994. Code of Practice for Earth Retaining Structures*; BSI: London, UK, 1994.

Permissions

 Every chapter published in this book has been scrutinized by our experts. Their significance has been extensively debated. The topics covered herein carry significant findings which will fuel the growth of the discipline. They may even be implemented as practical applications or may be referred to as a beginning point for another development.

The contributors of this book come from diverse backgrounds, making this book a truly international effort. This book will bring forth new frontiers with its revolutionizing research information and detailed analysis of the nascent developments around the world.

We would like to thank all the contributing authors for lending their expertise to make the book truly unique. They have played a crucial role in the development of this book. Without their invaluable contributions this book wouldn't have been possible. They have made vital efforts to compile up to date information on the varied aspects of this subject to make this book a valuable addition to the collection of many professionals and students.

This book was conceptualized with the vision of imparting up-to-date information and advanced data in this field. To ensure the same, a matchless editorial board was set up. Every individual on the board went through rigorous rounds of assessment to prove their worth. After which they invested a large part of their time researching and compiling the most relevant data for our readers.

The editorial board has been involved in producing this book since its inception. They have spent rigorous hours researching and exploring the diverse topics which have resulted in the successful publishing of this book. They have passed on their knowledge of decades through this book. To expedite this challenging task, the publisher supported the team at every step. A small team of assistant editors was also appointed to further simplify the editing procedure and attain best results for the readers.

Apart from the editorial board, the designing team has also invested a significant amount of their time in understanding the subject and creating the most relevant covers. They scrutinized every image to scout for the most suitable representation of the subject and create an appropriate cover for the book.

The publishing team has been an ardent support to the editorial, designing and production team. Their endless efforts to recruit the best for this project, has resulted in the accomplishment of this book. They are a veteran in the field of academics and their pool of knowledge is as vast as their experience in printing. Their expertise and guidance has proved useful at every step. Their uncompromising quality standards have made this book an exceptional effort. Their encouragement from time to time has been an inspiration for everyone.

The publisher and the editorial board hope that this book will prove to be a valuable piece of knowledge for researchers, students, practitioners and scholars across the globe.

List of Contributors

Chao Zhang, Quanzhong Zhan and Ting He
Information Center of the Ministry of water resources of the P.R.C, Beijing 100053, China

Qi Wang and Haichao Wu
Beijing Tritalent Intelligence Technology Co., Ltd., Beijing 100078, China

Yi An
School of Control Science and Engineering, Dalian University of Technology, Dalian 116024, China

Remis Balaniuk
Graduate Program in Governance, Technology and Innovation, Universidade Católica de Brasília, Brasília 71966-700, Brazil

Olga Isupova
Department Computer Science, University of Bath, Bath BA2 7PB, UK

Steven Reece
Department Engineering Science, Oxford University, Oxford OX1 3PJ, UK

Hao Chen
College of Water Conservancy and Hydropower Engineering, Hohai University, Nanjing 211100, China
Centre of Research and Development, Huaneng Lancang River Hydropower Co., Ltd., Kunming 650214, China

Yingchi Mao, Longbao Wang and Hai Qi
School of Computer and Information, Hohai University, Nanjing 211100, China

L. Furgani
Mott MacDonald, Croydon CR0 2EE, UK

M. A. Hariri-Ardebili
Department of Civil Engineering, University of Colorado, Boulder, CO 80309, USA
X-Elastica, LLC, Boulder, CO 80303, USA
Department of Civil Environmental and Architectural Engineering, University of Colorado, Boulder, CO 80302, USA

S. M. Seyed-Kolbadi
X-Elastica, LLC, Boulder, CO 80303, USA

M. Meghella
Ricerca Sistema Energetico (RSE), 20134 Milan, Italy

Jonas Enzell
KTH Royal Institute of Technology, SE 100 44 Stockholm, Sweden

S.M. Seyed-Kolbadi and M. Mirtaheri
K. N. Toosi University of Technology, Tehran 1996715433, Iran

M.A. Hariri-Ardebili
University of Colorado Boulder, CO 80818, USA
University of Maryland, College Park, MD 20743, USA

F. Pourkamali-Anaraki
University of Massachusetts, Lowell, MA 02142, USA

P. Boodagh
Department of Civil Environmental and Architectural Engineering, University of Colorado, Boulder, CO 80302, USA

M. Heshmati
X-Elastica LLC, Boulder, CO 80303, USA

J.W. Salamon
U.S. Bureau of Reclamation, Denver, CO 80225, USA

Stella Pytharouli
Department of Civil and Environmental Engineering, University of Strathclyde, Glasgow G1 1XJ, UK

Panagiotis Michalis
Department of Civil and Environmental Engineering, University of Strathclyde, Glasgow G1 1XJ, UK
School of Engineering, University of Glasgow, Glasgow G12 8LT, UK

Spyridon Raftopoulos
Power Public Corporation, Athens 104 36, Greece

Martina Colombo and Claudia Comi
Department of Civil and Environmental Engineering, Politecnico di Milano, Piazza Leonardo da Vinci 32, 20133 Milan, Italy

Giacomo Sevieri
Department of Civil, Environmental & Geomatic Engineering, University College London, London WC1E 6BT, UK

Anna De Falco
Department of Energy, Systems, Territory and Construction Engineering, University of Pisa, 56126 Pisa, Italy

Giovanni Marmo
Directorate of Dams, Italian Ministry of Infrastructure and Transport, 00161 Rome, Italy

Rocio L. Segura and Patrick Paultre
Department of Civil Engineering and Building Engineering, University of Sherbrooke, Sherbrooke, QC J1K 2R1, Canada

Carl Bernier
Department of Civil and Environmental Engineering, Rice University, Houston, TX 77005, USA

Capucine Durand
Institut des Sciences de la Terre, Université Grenoble Alpes, 38058 Grenoble, France

Rikard Hellgren, Richard Malm and Anders Ansell
Division of Concrete Structures, Department of Civil and Architectural Engineering, KTH the Royal Institute of Technology, 100 44 Stockholm, Sweden

Jacob B. Herridge, Arya Assadi-Langroudi and Sohrab Donyavi
Department of Computing and Engineering, University of East London, London E16 2RD, UK

Konstantinos Tsiminis
Fairhurst, Newcastle Upon Tyne NE4 6DB, UK

Jonas Winzen
bei AECOM, 45127 Essen, Germany

Michael McHugh
MESLEC Ltd., Elstree WD6 3DP, UK

Soheil Ghadr
Faculty of Engineering, Civil Engineering Department, Urmia University, Urmia 5756151818, Iran

Index

Printed in the USA
CPSIA information can be obtained
at www.ICGtesting.com
LVHW080050211023
761596LV00010BA/1991